Brain Inflammation
in Chronic Pain, Migraine and Fibromyalgia

The Paradigm-Shifting Guide for Doctors and Patients Dealing with Chronic Pain

A Three-Part Learning System of Text, Images, and Video

Alex Vasquez D.C. N.D. D.O. F.A.C.N.

- Doctor of Osteopathic Medicine, graduate of University of North Texas Health Science Center, Texas College of Osteopathic Medicine (2010)
- Doctor of Naturopathic Medicine, graduate of Bastyr University (1999)
- Doctor of Chiropractic, graduate of University of Western States (1996)
- Fellow of the American College of Nutrition (2013-present)
- Former Overseas Fellow of the Royal Society of Medicine
- Editor, *International Journal of Human Nutrition and Functional Medicine* IntJHumNutrFunctMed.org. Former Editor, *Naturopathy Digest*; Former/Recent Reviewer for *Journal of Naturopathic Medicine, Alternative Therapies in Health and Medicine, Autoimmune Diseases, International Journal of Clinical Medicine,* and *PLOS One*
- Private practice of integrative and functional medicine in Seattle, Washington (2000-2001), Houston, Texas (2001-2006), Portland, Oregon (2011-2013), consulting practice (present)
- Consultant Researcher and Lecturer (2004-present), Biotics Research Corporation
- Teaching and Academics:
  - Director of Programs, International College/Conference on Human Nutrition and Functional Medicine ICHNFM.org
  - Founder and Former Program Director of the world's first accredited university-affiliated graduate-level program in Functional Medicine
  - Professor, Integrative and Functional Nutrition in Immune Health, Doctor of Clinical Nutrition program
  - Former Adjunct Professor (2009-2013) of Laboratory Medicine, Master of Science in Advanced Clinical Practice
  - Former Faculty (2004-2005, 2010-2013) and Forum Consultant (2003-2007), The Institute for Functional Medicine
  - Former Adjunct Professor (2011-2013) of Pharmacology, Evidence-Based Nutrition, Immune and Inflammatory Imbalances, Principles of Functional Medicine, Psychology of Wellness
  - Former Adjunct Professor of Orthopedics (2000), Radiographic Interpretation (2000), and Rheumatology (2001), Naturopathic Medicine Program, Bastyr University
- Author of more than 100 articles and letters published in *JAMA—Journal of the American Medical Association, BMJ—British Medical Journal,* TheLancet.com, *JAOA—Journal of the American Osteopathic Association, Annals of Pharmacotherapy, Journal of Clinical Endocrinology and Metabolism, Alternative Therapies in Health and Medicine, Nutritional Perspectives, Journal of Manipulative and Physiological Therapeutics, Integrative Medicine, Current Allergy and Asthma Reports, Nutritional Wellness, Evidence-based Complementary and Alternative Medicine, Nature Reviews Rheumatology* and *Arthritis & Rheumatism*: Official Journal of the American College of Rheumatology

International College of Human Nutrition & Functional Medicine
ICHNFM.ORG

Copyrights: © 2004-present by Dr Alex Vasquez. All rights reserved by the author and enforced to the full extent of legal and financial consequences internationally. No part of this book may be reproduced, stored in a retrieval system, used for the creation of derivative works, or transmitted by any means (electronic, mechanical, photocopying, recording, or otherwise) without written permission from the author.

Trademarks: ® 2013-present by Dr Alex Vasquez and International College of Human Nutrition and Functional Medicine. The functional immunology/inflammology protocol discussed in this series of videos/notes/books/audios is recalled by the F.I.N.D.S.E.X. acronym trademarked™ in association with Dr Vasquez's books and videos including but not limited to *Functional Immunology and Nutritional Immunomodulation* (2012), *F.I.N.D. S.E.X. The Easily Remembered Acronym for the Functional Inflammology Protocol* (2013), *Integrative Rheumatology and Inflammation Mastery, 3rd Edition* (2014). Portland, Oregon; Integrative and Biological Medicine Research and Consulting, LLC. All rights reserved and enforced. For additional information and resources, see InflammationMastery.com and/or FunctionalInflammology.com. Additional trademarks referenced/cited in this work include International College of Human Nutrition and Functional Medicine®, International Conference on Human Nutrition and Functional Medicine®, and *International Journal of Human Nutrition and Functional Medicine*®.

Intellectual property: This book contains the creative work and intellectual property of Dr Alex Vasquez, owned and protected internationally by Dr Alex Vasquez, Integrative and Biological Medicine Research and Consulting ("IBMRC") LLC, and International College of Human Nutrition and Functional Medicine ("ICHNFM" and "ICHNFM.ORG"), based in North America and Europe. Except for quotes and excerpts from other sources, all of the information and images are protected by copyright ©; phrases and terms such as the FINDSEX ® acronym are additionally protected by registered trademark. The book is the means of licensed transmittal of this intellectual property; ownership of the book as an instance of licensed private transmittal and access does not equate to ownership of the property. The book also provides individual, private access to proprietary video archives. This work is supported and made possible by revenue from book sales, and readers/purchasers are asked and expected to respect the author's ownership of the work—specifically to not inappropriately copy or distribute—so that this work can continue. Violations of intellectual property rights, copyrights, and trademarks will be pursued to the highest extent possible internationally. For use permissions and to report violations, please contact admin@ichnfm.org.

Notices: The intended audiences for this book are health science students and doctorate-level licensed medical clinicians. This book has been written with every intention to make it as accurate as possible, and each section has undergone peer-review by an interdisciplinary group of clinicians. In view of the possibility of human error and as well as ongoing discoveries in the biomedical sciences, neither the author nor any party associated in any way with this text warrants that this text is perfect, accurate, or complete in every way, and all disclaim responsibility for harm or loss associated with the application of the material herein. Information and treatments applicable to a specific *condition* may not be appropriate for or applicable to a specific patient; this is especially true for patients with multiple comorbidities and those taking pharmaceutical medications, which are generally associated with multiple adverse effects and drug/nutrient/herb interactions. Given that this book is available on an open market, lay persons who read this material should discuss the information with a licensed medical provider before implementing any treatments and interventions described herein.

Sections	Page(s)
Preamble	i
1. Migraine, Cluster and Other Headaches	1 – 38
2. Fibromyalgia (FM, FMS, FMD) & Complex Regional Pain Syndrome (CRPS)	39-118
3. Article reprint: Intracellular Hypercalcinosis	119-121
4. Index	122

Acknowledgments for Peer and Editorial Review of Earlier Versions of This Work: Most of the sections that comprise the current work have been previously reviewed/published/presented; peer/editorial reviews are acknowledged below. Acknowledgement here does not imply that the reviewer fully agrees with or endorses the material in this text but rather that they were willing to review specific sections of the book for clinical applicability and clarity and to make suggestions to their own level of satisfaction.

- 2016 Edition of *Inflammation Mastery* and the excerpt *Pain Revolution for Migraine and Fibromyalgia*: Sabrina Piper BSc (2016 ND candidate), John Bartemus DC BCIM CFMP DACBN, Elizabeth Busetto DC ND, Kenneth Cintron MD
- 2015 Edition of *Human Microbiome and Dysbiosis in Clinical Disease*: Julie Jean BS BSN RN, Joseph Iaccino DC MSc
- 2014 Edition of *Antiviral Strategies and Immune Nutrition*: Annette D'Armata ND, Elizabeth Busetto DC ND
- 2014 Edition of *Naturopathic Rheumatology*: Annette D'Armata ND
- 2012 Edition of *Fibromyalgia in a Nutshell*: Lisa Scholl BA, Annette D'Armata ND
- 2012 Edition of *Migraine Headaches, Hypothyroidism, and Fibromyalgia*: Holly Furlong DC
- 2011 Edition of *Integrative Chiropractic Management of High Blood Pressure and Chronic Hypertension*: Barry Morgan MD, Holly Furlong DC, Kris Young DC, Erika Mennerick DC, and J William Beakey DOM
- 2011 Edition of *Integrative Medicine and Functional Medicine for Chronic Hypertension*: Erika Mennerick DC, JoAnn Fawcett DC, Ileana Bourland MSOM LAc, James Bogash DC, J William Beakey DOM
- 2010 Edition of *Chiropractic Management of Chronic Hypertension*: Joseph Paun MS DC, David Candelario OMS4 (TCOM c/o 2010), James Bogash DC, Bill Beakey DOM, Robert Richard DO
- 2009 Edition of *Chiropractic and Naturopathic Mastery of Common Clinical Disorders*: Heather Kahn MD, Robert Richard DO, James Leiber DO, David Candelario (UNT-HSC TCOM OMS4)
- 2007 Edition of *Integrative Orthopedics*: Barry Morgan MD, Dennis Harris DC, Richard Brown DC (DACBI candidate), Ron Mariotti ND, Patrick Makarewich MBA, Reena Singh (SCNM ND4), Zachary Watkins DC, Charles Novak MS DC, Marnie Loomis ND, James Bogash DC, Sara Croteau DC, Kris Young DC, Joshua Levitt ND, Jack Powell III MD, Chad Kessler MD, Amy Neuzil ND
- 2006 Edition of *Integrative Rheumatology*: Amy Neuzil ND, Cathryn Harbor MD, Julian Vickers DC, Tamara Sachs MD, Bob Sager BSc MD DABFM (Clinical Instructor in the Department of Family Medicine, University of Kansas), Ron Mariotti ND, Titus Chiu (DC4), Zachary Watkins (DC4), Gilbert Manso MD, Bruce Milliman ND, William Groskopp DC, Robert Silverman DC, Matthew Breske (DC4), Dean Neary ND, Thomas Walton DC, Fraser Smith ND, Ladd Carlston DC, David Jones MD, Joshua Levitt ND
- 2004 Edition of *Integrative Orthopedics*: Peter Knight ND, Kent Littleton ND MS, Barry Morgan MD, Ron Hobbs ND, Joshua Levitt ND, John Neustadt (Bastyr ND4), Allison Gandre BS (Bastyr ND4), Peter Kimble ND, Jack Powell III MD, Chad Kessler MD, Mike Gruber MD, Deirdre O'Neill ND, Mary Webb ND, Leslie Charles ND, Amy Neuzil ND

Format and Layout: The format/layout of this book is designed to efficiently take the reader through the clinically relevant spectrum of considerations for each condition that is detailed. Important topics are given their own section within each chapter, while other less important or less common conditions are only described briefly in terms of the four "clinical essentials" of 1) definition/pathophysiology, 2) clinical presentation, 3) assessment/diagnosis, and 4) treatment/management. Each of the expanded sections that details the more important/common conditions maintains a consistent format, taking the reader through the spectrum of primary clinical considerations: definition/pathophysiology, clinical presentations, differential diagnoses, assessments (physical examination, laboratory, imaging), complications, management, and treatment. As my books have progressed, I am increasingly using an article-by-article review format (especially in the sections on management and treatment) so that readers have more direct access to the information so as to understand and *incorporate* more deeply what the research actually states; the goal and general approach here is to use a *representative sampling* of the research literature.

References and Citations: Citations to articles, abstracts, texts, and personal communications are footnoted throughout the text to provide supporting information and to provide interested readers the resources to find additional information. Many of the cited articles are available on-line for free, and often I have included the website addresses so that readers can easily access the complete article.

Peer-review and Quality Control: Peer-review is essential to help ensure accuracy and clinical applicability of health-related information. Consistent with the importance of these goals, I have employed several "checks and balances" to increase the accuracy and applicability of the information within my textbooks:

- Reliance upon authoritative references: Nearly all important statements are referenced to peer-reviewed biomedical journals or authoritative texts, examples of the latter include *The Merck Manual*, *Current Medical Diagnosis and Treatment*, and *5-Minute Clinical Consult*. Each citation is provided by a footnote at the bottom of each page so that readers will know quickly and easily exactly where the information was obtained.
- Extensive cross-referencing: Readers will notice the supranormal number of references and citations. Many important statements have several references. Many references (especially textbooks) are referenced several times even on the same page; the purpose of this extensive referencing is three-fold: 1) to guide you—the reader—to additional information, 2) to help me (as writer) stay organized, and 3) to help you and me (the practicing physicians) employ this information with confidence. In more recent updates/revisions, I have started shortening the number of listed authors by frequent use of *et al* with an interest in keeping each citation to one line of text on the page, likewise reducing mental and eye strain; quite obviously I respect each of the authors—even those whose names are not listed in the citation—and am implementing this solely for the sake of efficient book formatting (aiming for one citation per line) and information density (fewer lines dedicated to citations allows more space for text and images). Given hundreds of pages and thousands of citations, formatting considerations such as these are summatively significant.
- Periodic revision: Any significant errors that are discovered will be posted at InflammationMastery.com/volume1 (…volume2, etc); please check these folders periodically to ensure that you are working with the most accurate information of which I am aware.
- Peer-review: The peer-review process for my books takes several forms. First, colleagues and students are invited to review new and revised sections of the text before publication; every section of the book that you are holding has been independently reviewed by health science students and/or practicing clinicians from various backgrounds: allopathic, chiropractic, osteopathic, naturopathic. Second, you - the reader - are invited to provide feedback about the information in the book, typographical errors, syntax, case reports, new research, etc. If your ideas truly change the nature of the material, I will be glad to acknowledge you in the text (with your permission, of course). If your contribution is hugely significant, such as reviewing three or more chapters or helping in some important way, I will be glad to not only acknowledge you, but to also send you the next edition at a discount or courtesy when your ideas take effect. Third, I keep abreast of new literature by constantly perusing new research and advancements in the health sciences. Having been successful in three separate doctoral programs in the health sciences, I have learned not only to master large amounts of material but to also separate and integrate different viewpoints as appropriate. I also "field test" my protocols with patients in the various clinical arenas in which I work and also with professionals and academicians via presentations and critical dialogue. By implementing these quality control steps, I hope to create a useful text and advance our professions and practices by improving the quality of care that we deliver to our patients.

How to Use This Book Most Effectively: Ideally, these books should be read cover-to-cover within a context of coursework that is supervised by a clinically experienced professor. For post-graduate professionals, they might consider forming a local or virtual "book club" and meeting for weekly or monthly discussions to check their understandings and share their clinical experiences to refine the application of clinical knowledge, perceptions, and skills. Virtual groups and internet forums—such as those hosted by International College of Human Nutrition and Functional Medicine at ICHNFM.ORG—can provide access to an assembly of international professional peers wherein sharing of clinical questions and experiences are synergistic. This book is not intended to extensively cover all aspects of clinical medicine, such as clinical pharmacology and prescribing (for which I recommend Epocrates.com and its associated app) and medical management (for which I recommend *5-Minute Clinical Consult* via book, website, and app).

Video access: Video access is provided via notices and footnotes appropriately placed and indicated throughout the book. Readers actually have to read the book to access the information and gain knowledge.
- Sample: vimeo.com/ichnfm/drv-functional-inflammology-intro2013
- Password: DrVprotocol

Notices: The intention and scope of this text are to provide health science students and doctorate-level clinicians with useful information and a familiarity with available research and resources pertinent to the management of patients in integrative primary care and specialty care settings. Specifically, the information in this book is intended to be used by licensed healthcare professionals who have received hands-on/residential clinical training and supervision at accredited health science colleges. Additionally, information in this book should be used in conjunction with other resources, texts, and in combination with the clinician's best judgment and intention to "*first, do no harm*" and second to provide effective healthcare. Information and treatments applicable to a specific *condition* may not be appropriate for or applicable to a specific *patient* in your office; this is especially true for patients with multiple comorbidities and those taking pharmaceutical medications with potential for multiple adverse effects and drug/nutrient/herb interactions. In my books and articles, I describe treatments—manual, dietary, nutritional, botanical, pharmacologic, and occasionally surgical—and their research support for the clinical condition being discussed; each practitioner must determine appropriateness of these treatments for his/her individual patient and with consideration of the doctor's scope of practice, education, training, skill, and—occasionally—the appropriateness of "off label" use of medications and treatments. This book has been carefully written and checked for accuracy by the author and professional colleagues. However, in view of the possibility of human error and new discoveries in the biomedical sciences, neither the author nor any party associated in any way with this text warrants that this text is perfect, accurate, or complete in every way, and we disclaim responsibility for harm or loss associated with the application of the material herein. With all conditions/treatments described herein, each physician must be sure to consider the balance between what is best for the patient and the physician's own level of ability, expertise, and experience. When in doubt, or if the physician is not a specialist in the treatment of a given severe condition, referral is appropriate. These notes are written with the routine "outpatient" in mind and are not tailored to severely injured patients or "playing field" or "emergency response" situations; consult your First Aid and Emergency Response texts and course materials for appropriate information. These notes represent the author's perspective based on academic education, experience, and post-graduate continuing education and are not inclusive of every fact that a clinician may need to know. This is not an "entry level" book except when used in an academic setting with a knowledgeable professor who can explain the concepts, tests, physical exam procedures, and treatments; this book requires a certain level of knowledge from the reader and familiarity with clinical concepts, laboratory assessments, and physical examination procedures. Suggested doses—if any—are for adults (not infants and children) unless otherwise specified in context; the responsibility for appropriate dosing is of course that of the prescribing clinician in view of the patient's age, weight, overall state, hepatic and renal function, comorbidities, polypharmacy, etc.

Purpose, scope, recommended companion resources

The purpose of this book is not to serve as a stand-alone "recipe book" for the complete management of all reviewed conditions; rather the focus of this book is the delivery of clinically important concepts and facts to enhance the management of various clinical disorders, in particular by documenting and explicating this author's naturopathic, allopathic, integrative and functional medicine approach. Readers and instructors using this book are encouraged to use whichever additional resources they choose, including but not limited to the supporting videos at Vimeo.com/DrVasquez and Vimeo.com/ICHNFM; in particular, *5-Minute Clinical Consult* and *Epocrates* are excellent and strongly advised companion guides for overall medical diagnosis/management and clinical pharmacology/prescribing, respectively. Clinicians need to have a good understanding of clinical medicine before applying many of the approaches described in this book; cross-referencing and double-checking management strategies and drug doses are essential components of quality care. Both *5-Minute Clinical Consult* and *Epocrates* are available as point-of-care references, and their use is advised.

This work is best used with the relevant videos from DrV available online, some of which are linked and made password-accessible via this book; additional videos by Dr Vasquez are available online (occasionally with accompanying printed presentation slides); please see the following examples and locations:
- vimeo.com/ichnfm
- vimeo.com/drvasquez

Updates, Corrections, and Newsletter: When and if omissions, errata, and the need for important updates become clear, I will post these at the website InflammationMastery.com. A reader might access this page periodically to ensure staying informed of any corrections that might have clinical relevance. This book consists not only of the text in the printed pages you are holding, but also the footnotes and any updates at the website. If any clinically important corrections are made, they will be distributed by newsletter InflammationMastery.com/join_email.html and/or placed in the folder FunctionalInflammology.com/volume1/ (with analogous folders for subsequent volumes, e.g., volume2, etc) for constant availability. Be alerted to new integrative clinical research, updates to this textbook and other news/publications/conferences/videos by registering for the free newsletter at ICHNFM.ORG.

Language, Semantics, and Perspective: As a diligent student who previously aspired to be an English professor, I have written this text with great (though inevitably imperfect) attention to detail. Individual words were chosen with care. I confess to knowing, pushing, and creatively breaking several rules of grammar and punctuation. With regard to the he/she and him/her debacle of the English language, I've occasionally mixed singular and plural pronouns for the sake of being efficient and so that the images remain gender-neutral to the extent reasonable. In several previous publications, the subtitle *The art of creating wellness while effectively managing acute and chronic musculoskeletal/health disorders* was chosen to emphasize the intentional creation of wellness rather than a limited focus on disease treatment and symptom suppression; for the 2009 printing of *Chiropractic and Naturopathic Mastery of Common Clinical Disorders*, this subtitle was slightly modified from "creating" to "co-creating" to emphasize the team effort required between physician and patient. *Managing* was chosen to emphasize the importance of treating-monitoring-referring-reassessing, rather than merely *treating*. *Disorders* was chosen to reflect the fact that a distinguishing characteristic of *life* is the ability to regularly create *organized structure* and *higher order* from chaos and *disorder*. For example, plants organize the randomly moving molecules of air and water into the organized structure of biomolecules which eventually take shape as plant structure—fiber, leaves, flowers, petals. Similarly, the human body creates organized structure of increased complexity from consumed plants and other foods; molecules ingested and inhaled from the environment are organized into specific biochemicals and tissue structures with distinct characteristics and definite functions. Injury and disease *result in* or *result from* a lack of order, hence my use of the word "disorders" to characterize human illness and disease. For example, a motor vehicle accident that results in bodily injury, for example, is an example of an external chaotic force, which, when imparted upon human body tissues, results in a disruption (disorder) of the normal structure and organization that previously defined and characterized the now-damaged tissues of the body; likewise, an autoimmune disease process that results in tissue destruction is an *anti-evolutionary* process that takes molecules of higher complexity and reverts them to simpler, fragmented, and non-functional forms. From the perspective of "health" as *organized structure and meaningful function* and "disease" as *the reversion to chaos, destruction of structure, and the loss of function*, the task of healthcare providers is essentially to restore order, and to acutely reduce and proactively prevent/eliminate clinical-biochemical-biomechanical-emotional chaos insofar as it adversely affects the patient's life experience as an individual and our collective experience as an interdependent society. What is required of clinicians then is the ability *first* to create conceptual order from what appears to be chaotic phenomena, and then *second* to materialize—make real and practically applied for patients/people seeking improved health—that conceptual order into our physical world; this is our task, and no small task it is. Also under this heading of Semantics and Language, I will make readers aware of the following additional facts. First, I tend to write very long sentences, both in general and at times when I want to connect two or more complex ideas; rather than be dismayed or discouraged by this occurrence, readers are encouraged to read these longer sentences more than just once and to engage actively, perhaps by asking, "Why is DrV making an effort to connect these ideas?" "**What is the conceptual advantage to the binding of these ideas together**?" I am aware of most of the rules of grammar, and I am generally—but not always—compliant. Second, I create new words and phrases as needed; an index of some of these is provided toward the back of the book, whereas some of these new terms are self-explanatory, e.g., *hypoinsulinreception*—underreception or lack of receptor responsiveness to insulin. When possible, I strongly prefer to use single words when discussing concepts, rather than multiple disparate words for singular concepts. I have started to prefer using *italics* rather than "quotation marks" when introducing new terms or when using terms/phrases/words with emphasis; the main purpose of this is to reduce the number of punctuation marks and character spaces, both of which over the course of a multi-volume work of 2,000 pages and hundreds of thousands of words are numerically significant. Last for this section, the *colorization* process that I began in April 2014 for my (larger) books is intended to 1) bring out more detail in my increasingly complex diagrams, 2) bring emphasis and highlighting to areas of particular interest, 3) make the work more visually stimulating/pleasing over the previous black/white/grayscale versions, and—relatedly—4) to keep the work interesting as readers tread through a remarkable amount of complex and detailed information; I realize that some readers may at times find the colorization to be a small distraction, but I think this is better than the alternative of monotony induced by several hundred dense pages of grayscale.

Integrity and Creativity: I have endeavored to accurately represent the facts as they have been presented in texts and research, and to specifically resist any temptation to embellish or misrepresent data as others have done.[1,2] Conversely, I have not endeavored to make this book appeal to the "average" student or reader; my goal is to write and teach to the students at the top of the class, thereby affirming them and pulling the other students forward and upward. While I offer *explanations*, I intentionally resist *simplifications*, except when one simplification might facilitate the comprehension of a more complex phenomenon, or when such a simplification might facilitate the conveyance of information from clinician to patient. I have allowed this text to be unique in format, content, and style, so that the personality of this text can be

---

[1] Vasquez A. Zinc treatment for reduction of hyperplasia of prostate. *Townsend Letter for Doctors and Patients* 1996; January: 100
[2] Broad W, Wade N. <u>Betrayers of the Truth: Fraud and Deceit in the Halls of Science</u>. New York: Simon and Schuster; 1982

contrasted with that of the instructor and reader, thus enabling the learner to at least benefit from an intentionally different – and intentionally honest – perspective and approach. Students using this text with the guidance of a qualified professor will benefit from the experience of "two teachers" rather than just one.

Linearity, Nonlinearity, Redundancy, Asynchronicity: Although the overall flow of the text is highly linear and sequential, occasionally I place a conclusion before its introduction for the sake of foreshadowing and therefore for preparing the reader for what is to come. The purpose of this is not simply one of preparation for the sake of allowing the reader to know what is already lying ahead on the path, but more to begin creating new "shelf space" in the reader's intellectual-neuronal "library" so that when the new—particularly if *neoparadigmatic*—information is encountered, a space will already exist for it; in other words: the intent is to make learning easier. Likewise, for the sake of *information retention*—or what is physiologically understood as synaptogenesis—important points are presented more than once, either identically or variantly. Given that "*No one ever reads the same book twice*"[3] (because the "person who starts" the reading of a meaningful book is changed into the "person who finishes" the reading of that book (assuming proper intentionality and application of one's "self"), the person reading these words might consider a second glace after the first. For the sake of efficient use of space I have tried to minimize redundancy; however, in a few locations, redundancy of text and images proved necessary as—for example—viewing the same diagram within two different conversations allows the reader to gain a more profound understanding of the concepts by viewing them from two different contexts.

Bon Voyage: All artists and scientists—regardless of genre—grapple with the divergent goals of *perfecting* their work and *presenting* their work; the former is impossible in the ultimate sense, while the latter is the only means by which the effort can create the desired effect in the world, whether that is pleasure, progress, or both. At some point, we must all agree that it is "good enough" and that it contains the essence of what needs to be communicated. While neither this nor any future edition of this book is likely to be "perfect", I am content with the literature reviewed, presented, and the new conclusions and implications which are described—many for the first time ever—in this text. Firstly in and progressively from my *Integrative Rheumatology* (2006), each chapter achieved/achieves a paradigm shift which distanced/distances us farther from the simplistic pathocentric and pharmacocentric model and toward one which authentically empowers both practitioners and patients. With time, I will make future editions more complete, consistently passionate, and either more or less polemical. I hope you are able to implement these conclusions and research findings *into your own life* and into the *treatment plans for your patients*. Hopefully this work's value and veracity will promote patients' vitality via the vigilant and virtuous clinicians viewing this volume; to the more attentive and thoroughgoing reader, more is revealed (for example, the last sentence is a reference to the descriptive and prophetic movie *V for Vendetta* (2006).

Thank you for engaging with this work, and I wish you and your patients the best of success and health.

Alex Vasquez, D.C., N.D., D.O., F.A.C.N.
Work updated: March 9, 2016

Work as love made tangible

"You work that you may keep pace with the earth and the soul of the earth.
For to be idle is to become a stranger unto the seasons, and to step out of life's procession. ...
Work is love made visible."

Kahlil Gibran (1883-1930). *The Prophet*, 1973

---

[3] Davies R. *Reading and Writing*. Salt Lake City: University of Utah Press; 1992, page 23

Reviews of previous and recent works:
- "I love this course and your approach to the material. I am learning so much. Each article you assigned was strategically chosen and offered support and insight. I was pleasantly surprised by the exam and thought it was very fair. ... Thank you for sharing your knowledge and experience with us!" *Doctorate Student under Dr Vasquez, 2016*
- "I appreciate the lecture yesterday and I am truly fascinated by your topic and your vast knowledge. ... I for one feel having people like you on our faculty can only strengthen the credibility of our school. ... I appreciate your education, knowledge and clearly you are the authority in your field. I have listened to all your lectures on YouTube - fantastic!" *University Faculty and Doctorate Student under Dr Vasquez, 2016*
- "Thank you most kindly for your incredible dedication and kindness in sharing your knowledge with us. I am due to start med school next semester and thanks to you and all those who have taught you, I'll be way ahead of the curve." *Premedical/Medical student 2015*
- "Dr Vasquez, I have followed your work extensively and admire your intellect and passion. Thank you for your passion for teaching with integrity!" *Chiropractic doctor 2015*
- "I just wanted to tell you how much I appreciate the information I have received from you. I am still digesting most of it. I feel I have learned quite a bit already yet also feel I have barely scratched the surface." *Doctor and Graduate student under Dr Vasquez, 2013*
- "Dr. Vasquez, Thank you for all you do. **Your conference was simply amazing**. No one wanted to leave the room. I met medical professionals and very interesting lay people who were stimulated and invigorated to change their lives and the lives of others. **I am in awe at your intellectual integrity and veracity.** Best of luck to you in all of your future endeavors." *Medical physician and ICHNFM 2013 Conference Attendee*
- **2014 review of Functional Inflammology, Volume 1:** **"A truly comprehensive text on the vast subject of inflammation. I consider this book to be an essential addition to any health care practitioner who wishes to operate within the realm of Function Medicine. Please be aware that this book is dense in its content, and its 700 plus pages are full of deeply insightful information. I think Dr. Vasquez is one of the most prolific functional medicine contributors and books such as this should cement his reputation as such."**
- "I attended the last ICHNFM conference in Portland (and am still basking in the amazing information received)." *Email from Clinical Oncology Dietitian, in late February 2014*
- "Thanks for a fantastic conference!" *ICHNFM 2013 Conference Attendee*
- "Your discourse today reflected not only your passion and commitment to the wellness of our planet but most importantly the clarity and sincerity of your spirit/ heart/ mind. Always good to be with you and look forward to seeing you soon. Hope we can spend more time then." *Medical physician attendee 2014*
- "I was so refreshed by the 'unfiltered excellence.' What humanness. Breaths of fresh air." *ICHNFM 2013 Attendee*
- "Keep in mind Alex, that humanity is a better place because of you. I know you can't undo it all, but think about how many people would be worse off if it wasn't for your wonderful knowledge being shared with all us docs. Things that I have learned from you have changed peoples' lives for the better." *Naturopathic physician, 2014*
- "Just got back to Guam. Great experience at the International Conference on Human Nutrition and Functional Medicine. Exciting concepts on functional medicine. Thanks Dr Alex Vasquez and team!" *ICHNFM 2013 Conference Attendee*
- "Already waiting in line to buy next year's ticket! **Dr. Vasquez you crushed it!** The future is looking fun already ☺" *ICHNFM 2013 Conference Attendee*
- "Had an incredible time at the 2013 International Conference on Human Nutrition and Functional Medicine. Got to meet some amazing people and hear from some of the top researchers/health professionals about human nutrition and functional medicine approaches. It was definitely worth every penny and can't wait to go back next year!" *ICHNFM 2013 Conference Attendee*
- "I miss you! Your confidence in a program you believed in. I miss your live classes where we would get off topic on a clinical pearl. I miss your way of teaching in a laid back atmosphere that made me feel comfortable, not intimidated. I just needed to let you know, this program is not the same, I am almost done, otherwise, I would have bailed out! I am grateful for the last 18 months I did have with you at the helm. ... You ignited in me my passion for learning again. You sparked the minds of all of us with your enthusiasm. Don't ever let anyone take that away. It has given birth to your new endeavor, and we will follow where you lead. Enjoy your new surroundings and celebrate your new beginnings. I know I look forward to what is ahead." *Doctor and Graduate student under Dr Vasquez, 2013*
- "Wonderful conference! Thanks so much." *ICHNFM 2013 Conference Attendee*
- "Really wonderful conference! Lots of material ready to implement Monday morning! **Congrats to Alex Vasquez on a herculean job very well done!**" *ICHNFM 2013 Conference Attendee*
- "Thanks for a great conference. I really enjoyed all of the speakers, but your lectures were by far the most useful for implementing ideas into my clinical practice. And the most entertaining." *ICHNFM 2013 Conference Attendee*
- "Thank you for your life-changing work." *Physician, 2011*

- "I want Dr. Vasquez to know that I have just received his book, *Chiropractic and Naturopathic Mastery of Common Clinical Disorders*. **It is a treasure. The best book in my library.** Thank you for the contribution that you are giving to the world of health care." *Clinician, 2010*
- "I appreciate the resources you offer the profession. I use your books and articles regularly." *Doctor, 2011*
- "Dr. Vasquez, I greatly appreciate your efforts. I am a student at ___, 8th trimester, and would like to express my gratitude for your research and works. After coming across your texts in the library, **I quickly found your insight and explanations of the current health care crisis, and in depth coverage and algorithms for inflammatory diseases as a profound inspiration and call to action. I appreciate your attention to detail, and have been taken back several times by the potency and meaning of your sentences. Thank you for your hard work, I will enjoy these books and will surely share with those that have the same drive for true and competent patient care.**" *Health Sciences Student, 2008*
- "I never told you this, but whenever I need to research a particular disease, **besides going on Pubmed and checking some classic Pathophysiology and Clinical Nutrition books, I use your books and I find them extremely well organized, concise, and up-to-date and with the functional/integrative medicine thinking I enjoy and believe it is the future of Health Care.**" *Nutrition Research Consultant and University Faculty in Europe, 2009*
- "Thanks so much. You are a great asset to our profession." *Doctor, 2010*
- "As a 7th trimester student quickly approaching 8th trimester and student clinic, I know I will be utilizing your books often. **Your "Chiropractic and Naturopathic Mastery of Common Clinical Disorders" book is referenced very frequently by many clinicians and faculty members at [our university]. Your work is highly regarded**, and I look forward to clinically utilizing the information I will obtain from your writings." *Health Sciences Student, 2011*
- "I am a chiropractic student at ___ Chiropractic College. I just wanted to drop a quick line thanking you for your thorough and accessible textbook Integrative Orthopedics. We are using it in our Differential Diagnosis class, and **it is the best book I've come across in Chiropractic College bar none. The writing is concise, informative and refreshingly eloquent. The material is super practical.** I hope you continue putting out great resources." *Health Sciences Student, 2011*
- "I appreciate the resources you offer the profession. **I use your books and articles regularly.**" *Doctor, 2011*
- "**Your Integrated Orthopedics book is magnificent**. I wish all textbooks were structured and as thoughtful as that one." *Health Sciences Student, 2008*
- "By reading the introduction I realize that calling it an orthopedics book; does not do it justice. **It is far more than that. It looks to me that you have created, or are creating, the bible of Integrative Orthopedics and physical medicine**. *Physician, 2007*
- "First of all let me say how honored I am that you have allowed me to review this work. You have done an amazing job! In my opinion **every healthcare provider SHOULD have this on their bookshelf**." *Physician, 2007*
- "Your work on Chapter 12: Hip and Thigh is very good. The chapter is inclusive of the typical pathologies seen in private practice and I particularly liked the separation of juvenile from adult pathologies. Your choice of tests to assess hip and thigh pathology on page 320 is very nice and inclusive. I appreciate your use of algorithms and find them very useful in teaching and in practice. In general, **I thought this chapter represents a quality, state of the art presentation**!" *Clinician and Professor in Clinical Sciences, 2007*
- "I saw your books in a colleague's office and was really impressed. Really appreciate the thoroughness you've put into them." *Doctor, 2010*
- "**It is with great interest and fascination that I have been reading your material both in your two books (Integrative Orthopedics and Integrative Rheumatology) and online. I consider myself very fortunate to have come across your work**, as many of the basic elements of health which you discuss I never learnt or even heard about while in chiropractic college." *Doctor, 2010*
- "I appreciate the resources you offer the profession. I use your books and articles regularly." *Doctor, 2011*
- "**I'm so pleased with your books and was inspired to let you know they have already been incredibly useful! Good index; well organized algorithms. Sometimes I buy educational material and it just sort of sits there... Your books now live on my main desk. Thanks.**" *Physician and Journal Editor, 2009*
- "I just wanted to let you know how much I am enjoying reading **your book Integrative Rheumatology. It is having an extremely positive impact in the way I view health and am having a tough time putting it down. It is very inspirational.** I have long felt that it is very important to set a good example for your patients and now try my best to be one for my future patients. I like how you stress this in your book. In order to be the best example for my patients I am going to need to address some problems with my own health. I look healthy from the outside but I have been suffering from fatigue for about 4 years. It has a very negative impact on my health. People say that doing the same thing and expecting different results is the definition of insanity so I think it is time that I attempt to make some

- changes. ... **Thanks again for writing such a great book. I feel it is a must have for anyone in a musculoskeletal practice**." *Health Sciences Student, 2010*
- "My name is [recent graduate], and I've been a fan of your books since I was in chiropractic college at [university] campus. Dr. [Author, Presenter] made your book, Integrative Rheumatology, required reading for his 9th quarter nutrition class. I never looked back, and have since purchased Chiropractic & Naturopathic Mastery of Common Clinical Disorders as well as Chiropractic Management of Chronic Hypertension." *Doctor, 2010*
- "I saw your books in a colleague's office and was really impressed. Really appreciate the thoroughness you've put into them." *Doctor, 2010*
- "Reading the new integrative management of high blood pressure book and I am thoroughly enjoying it; excellent job. **I am feeling so empowered I'm opening another office focusing on 'restoring the foundations of health' for the community** that I open it in. I am looking for a location and networking to find an internist and cardiologist that are forward thinking; I'm very excited!" *Doctor, 2011*
- "Thank you for the presentation at [the university] this past weekend. **My horizons about what can be done to help people were greatly expanded. I am now still studying the notes from the seminar and am looking forward to more study and learning on how to** *correctly* **manage diabetes and hypertension**." *Doctor, 2011*
- "Thank you for exposing so many people to the results of our research on the treatment of hypertension. I hope you can pay us a visit during your next trip to our area so we can give you the tour of our new 50+ bed inpatient facility." *Dr Alan Goldhamer, Chief of Health Promoting Clinic, 2010*
- "**I always enjoy reading your work**. I personally gain a lot of knowledge through being a peer-reviewer for you and am better because of it!" *Doctor, Faculty Member, and Postgraduate Instructor, 2011*
- "**I attended your seminar at [University] in June and have been utilizing your hypertension protocols. In that short time, I have seen some marked progress with various patients**." *Doctor, 2010*
- "I want to personally thank you for your expertise and books on...everything. I'm in my last year at SCNM (taking rheumatology right now) and I truly admire your research and ability to compile valuable information. Thank you." *Naturopathic Medical Student, 2014*
- "Doc, I really want to thank you for sharing some of the most important-relevant Facebook posts. **If we had more doctors, leaders and informed human beings (like yourself) our world would be a better place. Thank you for your commitment to truth and doing the right thing.**" *Doctorate Clinician, 2016*
- "I love your No BS approach to everything you do. I loved it in 2013 when you hosted the most informative conference I have ever had the opportunity to attend (because I could afford it at the time thank you). I wish there were more scientists/authors/academics/doctors like you! You are a breath of fresh air among the smell of BS and one can almost "smell" your intolerance to corruption. Please don't ever stop speaking your mind, disseminating information, and rebutting the "experts" because sadly, you're a rare breed." *Doctorate Clinician, 2016*

Begin at the beginning

"He who wishes one day to *fly*, must first learn *standing*
   and *walking*
      and *running*
         and *climbing*
            and *dancing*.
One does not *fly* into *flying*."

Friedrich Nietzsche (1845-1900). *Thus Spoke Zarathustra—A Book for All and None*, 1883-1885

Read, pause, consider, struggle, think, grow

"If this book is incomprehensible to anyone and jars on his ears, the fault, it seems to me, is not necessarily mine. It is clear enough, assuming, as I do assume, that one has first read my earlier writings and has not spared some trouble in doing so: for they are, indeed, not easy to penetrate. Regarding my *Zarathustra*, for example, I do not allow that anyone knows the book who has not at some time been profoundly wounded and at some time profoundly delighted by every word in it; for only then may he enjoy the privilege of reverentially sharing in the halcyon element out of which that book was born and in its sunlight clarity, remoteness, breadth, and certainty. In other cases, people find difficulty with the aphoristic form: this arises from the fact that today this form is not taken seriously enough. An aphorism, properly stamped and molded, has not been "deciphered" when it has simply been read; rather, one has then to begin its exegesis—its explanation, its extraction, for which is required an art of deciphering.

To be sure, one thing is necessary above all if one is to practice reading as an art in this way, something that has been unlearned most thoroughly nowadays—something for which one has almost to be a cow and in any case not a "modern man": *rumination*—taking time to pause, to reflect, to consider...

Friedrich Nietzsche, *On the Genealogy of Morals*, Preface, Section #8, Sils-Maria, Upper Engadine, July 1887

Examples of commonly used abbreviations:

- **25-OH-D** = serum 25-hydroxy-vitamin D(3)
- **ACEi** = angiotensin-2 converting enzyme inhibitor
- **Alpha-blocker** = alpha-adrenergic antagonist
- **ANA** = antinuclear antibodies
- **ARB** = angiotensin-2 receptor blocker/antagonist
- **ARF** = acute renal failure
- **BB** = beta blocker or beta-adrenergic antagonist
- **bHB, BHB** = beta-hydroxy-butyrate
- **BMP** = basic metabolic panel, includes serum Na, K, Cl, CO2, BUN, creatinine, and glucose
- **BP** = blood pressure, relatedly HBP = high blood pressure
- **BUN** = blood urea nitrogen
- **C and S** = culture and sensitivity
- **CAD** = coronary artery disease
- **CBC** = complete blood count
- **CCB** = calcium channel blocker/antagonist
- **CE** = cardiac enzymes, including creatine kinase (CK), creatine kinase myocardial band (CKMB), and troponin-1, with the latter being the most specific serologic marker for acute myocardial injury; for the evaluation of acute MI, these are generally tested 2-3 times at 6-hour intervals with ECG performed at least as often.
- **CHF** = congestive heart failure
- **CHO, carb** = carbohydrate
- **CK** = creatine kinase, historically named creatine phosphokinase (CPK)
- **CKD** = chronic kidney disease, generally stratified into five stages based on GFR of roughly <90, 90-60, 60-30, 30-15, and >15, respectively
- **CMP** = comprehensive metabolic panel, also called a chemistry panel, includes the BMP along with markers of hepatic status albumin, protein, ALT, AST, may also include alkaline phosphatase and rarely GGT; panels vary per laboratory and hospital.
- **CNS** = central nervous system
- **COPD** = chronic obstructive pulmonary disease
- **CRF, CRI** = chronic renal failure/insufficiency
- **CRP** = c-reactive protein, hsCRP = high-sensitivity c-reactive protein

- **CT** = computed tomography
- **CVD** = cardiovascular disease
- **CXR** = chest X-ray
- **DM** = diabetes mellitus
- **DMARD** = disease-modifying antirheumatic drugs
- **ECG** or **EKG** = electrocardiograph
- **Echo** = echocardiography
- **ERS** = endoplasmic reticulum stress
- **GFR** = glomerular filtration rate
- **HDL** = high density lipoprotein cholesterol
- **HTN** = hypertension
- **Ig** = immune globulin = antibodies of the G, A, M, E, or D classes.
- **IHD** = ischemic heart disease
- **I+D** = incision and drainage
- **IM, IV** = intramuscular, intravenous
- **LPS** = bacterial lipopolysaccharide, endotoxin
- **MCV** = mean cell volume
- **MI** = myocardial infarction
- **Mito** = mitochondria(l)
- **MRI** = magnetic resonance imaging, MRI = magnetic resonance angiography
- **mTOR** = mechanistic or mammalian receptor of rapamycin; **TOR** is also reasonable
- **NFkB** = nuclear transcription factor kappa beta
- **PNS** = peripheral nervous system
- **PRN** = from the Latin "pro re nata" meaning "on occasion" or "when necessary"
- **PTH** = parathyroid hormone, iPTH = intact parathyroid hormone
- **PVD** = peripheral vascular disease
- **RA** = rheumatoid arthritis
- **RAD** = reactive airway disease, asthma
- **SIBO** = small intestine bacterial overgrowth
- **SLE** = systemic lupus erythematosus
- **TLR** = Toll-like receptor
- **TRIG(s)** = serum triglycerides
- **UA** = urinalysis
- **UPR** = unfolded protein response
- **US** = ultrasound

Dosing shorthand:

- **bid** = twice daily
- **cc** = with meals
- **hs** = at bedtime
- **ic** = between meals
- **po** = per os = by mouth
- **prn** = as needed

- **q** = each
- **qd** = each day, also /d or /day
- **qid** = four times per day
- **tid** = thrice daily
- **yo** = years old

Orientation to Excerpts from *Inflammation Mastery*:
Introduction to DrV's Functional Inflammology Protocol:
The Seven Major Modifiable Factors in
Systemic Inflammation, Allergy, and Autoimmunity

Major Modifiable Influences on Immune and Inflammatory Balance

Chapter 4 of the larger textbooks—*Inflammation Mastery* (630 pages printed in 2014; 1,200 pages printed in 2016) and *Functional Inflammology* (700 pages printed in 2014)—details and organizes a massive amount of information, organized in my "functional inflammology protocol." I have developed this clinical protocol over many years of working with patients clinical practice, teaching at the graduate, doctorate, and post-graduate levels since 2000, publishing more than 110 articles and letters, and writing and re-writing more than a dozen books, the largest of which—*Inflammation Mastery, 4th Edition*—reached for the publisher's limit of 1,200 pages. I anticipate that my books are a bit of a challenge to read although I make no effort to make them unduly complicated; I simply write the information as it occurs to me, trying to add what I consider to be necessary explanations while not dumbing-down the information nor defining every term. The purpose of reading is, or at least traditionally has been, the quest for new views and information; occasionally we all have to reach for the dictionary or do some background work to enhance our understanding while exploring a new subject. Not everyone needs to or wants to read a textbook of 900 pages; hence, I occasionally excerpt sections that can stand alone as separate books.

However and obviously, these excerpted sections do not and by definition cannot contain all of the previous materials (ie, clinical overview in Chapter 1, wellness promotion and lifestyle medicine in Chapter 2, nonpharmacologic pain management in Chapter 3, the entire functional inflammology protocol in Chapter 4) that leads to the conclusions and clinical applications in Chapter 5, which details the assessment and treatment of a variety of inflammatory disorders, which I categorize as ❶ metabolic inflammation—hypertension, diabetes, migraine, fibromyalgia, ❷ allergic inflammation—allergies in general and asthma in particular, and ❸ autoimmune inflammation—all of the rheumatic conditions ranging from rheumatoid arthritis to lupus/SLE to spondylitis, psoriasis, and vasculitis. The only way to understand the foundational information in Chapters 1,2,3 and 4 is to read those chapters; the most efficient way to grasp an introductory understanding to the overall clinical approach is to see my presentation videos, two of which from the 2013 International Conference on Human Nutrition and Functional Medicine are available per these links and passwords:

- Protocol introduction, part 1: https://vimeo.com/100089988 Password: "DrVprotocol_volume1"
- Protocol introduction, part 2: https://vimeo.com/99857164 Password: "DrVprotocol_volume1"

Following my review and perusal of thousands of research articles in addition to the attentive application of my interest in these conditions throughout three doctoral programs, I have come to appreciate seven major modifiable factors that are chiefly relevant for the initial and long-term management of patients with inflammatory conditions and rheumatic diseases. These seven factors are:

1. <u>Food intake and nutritional status</u>: The pro/anti-inflammatory effects of diet, including food allergies and intolerances, nutrient deficiencies and dependencies,
2. <u>Infections and dysbiosis</u>: Chronic exposure to microbial effectors/effects,
3. <u>Nutritional modulation of the immune system</u>: Nutrigenomic modification of immunocyte phenotype,
4. <u>Dysmetabolism and Dysfunctional organelles, most notably mitochondria</u>: Especially the pro-inflammatory, pro-oxidant, and anti-apoptotic consequences of dysfunctional mitochondria (DysMito or MitoDys); more recently the conversation has extended beyond mitochondrial dysfunction to include endoplasmic reticulum stress/dysfunction (ERS) and resultant unfolded protein response (UPR),
5. <u>Stress, sleep deprivation vs sleep sufficiency, spinal health, social and psychological considerations</u>: Included in this section is a collection of important considerations which—in the first draft of this acronym—started with stress management, sleep hygiene, and pSychological and social factors. Later versions have included spinal health (chiropractic model), somatic dysfunction (osteopathic model), surgery, specialized supplementation, and "stamp your passport"—sometimes we all just need to vacate for a while and implement some *geographic cure* for the sake of inspiration, life enhancement, exposure to new ideas and lifestyles, and the breaking of (dysfunctional) thought patterns and routines,
6. <u>Endocrine imbalances</u>: Hormones can promote or retard the genesis and perpetuation of inflammation/allergy/autoimmunity; therapeutic correction with prescription or nonprescription interventions can have a profound anti-inflammatory benefit.
7. <u>Xenobiotic immunotoxicity</u>: Exposure to and accumulation of toxic chemicals and/or toxic metals can alter immune responses toward allergy and autoimmunity and away from immunosurveillance against infections and cancer.

The above-listed seven modifiable factors—Food, Infections, Nutri-immunology, Dysmetabolism, Society, Endocrine, Xenobiotics—can be recalled by my FINDSEX® acronym which outlines and organizes my Functional Inflammology Protocol. The overall model is represented graphically in the image below.

With regard to the model and my books as a whole, readers should appreciate that the information in various sections likely applies either conceptually or specifically to conditions described in other sections and that therefore the best way to understand inflammatory/allergic/autoimmune disorders in their totality is to appreciate the nuances of each and the common themes among all.

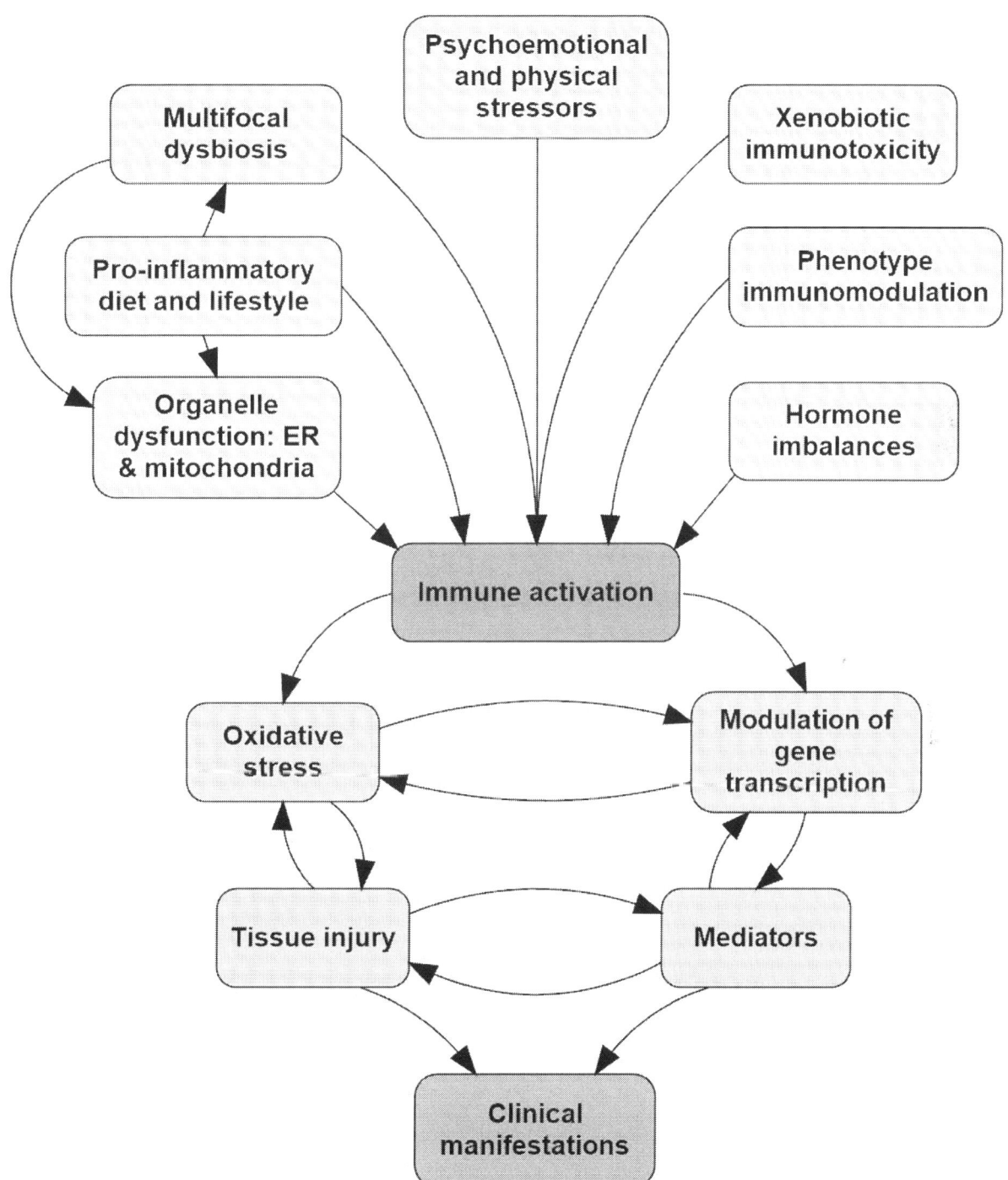

Inflammation in a simple cause-and-effect diagram: The major causative factors amenable to clinical implementation are represented, along with the pathophysiologic consequences and clinical effects. Molecular details, clinical assessments, and therapeutic interventions are introduced/reviewed in this chapter; in later volumes of this work, clinical protocols detail the drugs and doses, etc.

Affirmation and consistency of common themes in an interconnected reality; the importance of transitioning from reception to comprehension to conception to behavior

"The fact that today I still stand by these ideas, **that in the intervening time they themselves have constantly become more strongly associated with one another, even to the point of growing into each other, intertwining, and becoming *one***, that has reinforced in me the joyful confidence that they may not have originally developed in me as single, random, or sporadic ideas, but up out of common roots, from some fundamental *will for knowledge* ruling from deep within, always speaking with greater clarity, always demanding greater clarity.

In fact, this is the only thing appropriate and proper for a philosopher. **We have no right to be isolated in any way: we are not permitted to make isolated mistakes or to run into isolated truths**. Our ideas, our values, our affirmations and denials, our *if*s and *but*s—these rather grow out of us from the same necessity which makes a tree bear its fruit—totally related and interlinked amongst each other: witnesses of one will, one health, one soil, one sun."

Nietzsche FW. <u>On the Genealogy of Morals</u>, 1887, Preface essay #2

"In order for a particular species to maintain itself and increase its power, **its conception of reality must comprehend enough of the calculable and constant** for it to **base a scheme of behavior on it**."

Nietzsche FW. <u>Will to Power</u>, 1901, #480

Functional Inflammology
F.I.N.D.S.E.X.® acronym

"enough of the calculable and constant to base a scheme of behavior"

- Infection, dysbiosis
- Nutritional immunomodulation
- Food, Nutrition
- Dysmetabolism, dysfunctional mitochondria
- Style of living (lifestyle): psychology, sociology, politics, sweat/exercise, stress, sleep, special considerations such as surgery
- Xenobiotic load
- Endocrine

Image copyright © 2015 by Dr Alex Vasquez per the "Inflammation Mastery" series of books and videos InflammationMastery.com and vimeo.com/DrVasquez and vimeo.com/ICHNFM. Trademark: The "Functional Inflammology Protocol" and FINDSEX™ acronym discussed in this series of videos and notes is Trademarked ® in association with Dr Vasquez's books including Functional Inflammology and Functional Medicine Rheumatology and other books and videos.

This work is a stand-alone monograph and yet at the same time is an updated excerpt from Dr Vasquez's larger works <u>Functional Inflammology</u> (700 pages) and <u>Inflammation Mastery, 4th Edition</u> (1,180 pages). While this monograph is complete in itself, reference is made to other sections and chapters for those who have or might be interested in the complete model and "functional inflammology protocol."

Preface & Introduction

Nutrition & FxMed for chronic immune-inflammatory disorders

Causes of Inflammation-Immune-Metabolic Imbalance:

1. Food, Lifestyle
2. Infection, Dysbiosis
3. Nutritional Immunomodulation
4. Dysfunctional mitochondria
5. Stress, Emotions, Psychology, Sociology, Lifestyle
6. Endocrine, Hormones
7. Xenobiotics, Toxins

Notice that these 7 factors can be remembered by the acronym: **F.I.N.D. S.E.X.**

♥ First presented in Paris in 2012 ♥

The "Functional Inflammology Protocol" and FINDSEX® acronym: As the clinical protocol expanded from five components (diet, dysbiosis, xenobiotics, hormones, and stress) published in 2006 and 2007 to seven components (adding nutritional immunomodulation and mitochondrial dysfunction) in 2012, I realized that the time had come to attempt an acronym in order to facilitate student memorization and clinician application. I applied some priority to the sequence of the categories, and then experimented with a few acronyms. The rest, as is said, is history. This occurred just before a series of presentations in France (starting in Paris), Holland, and Belgium in March of 2012. The FINDSEX acronym is a registered trademark (e.g., ® and ™) in association with *Functional Immunology and Nutritional Immunomodulation*[4], *F.I.N.D.S.E.X. The Easily Remembered Acronym for the Functional Inflammology Protocol*[5], *Integrative Rheumatology and Inflammation Mastery, Third Edition*[6], and other books, videos[7], audios[8], and presentations by Dr Vasquez since 2012. One of the more recent introductions to this protocol was delivered at the International Conference on Human Nutrition and Functional Medicine in Portland Oregon in September 2013 and is posted here vimeo.com/ichnfm/drv-functional-inflammology-intro2013 and accessed with the password "DrVprotocol"; access to new videos, book updates, and articles are periodically distributed by email newsletter from ICHNFM.ORG.

---

[4] Published Jun 2012, ISBN-10: 1477603859, ISBN-13: 978-1477603857
[5] Published Apr 2013, ISBN-10: 1484046765, ISBN-13: 978-1484046760
[6] Published Jan 2014, ISBN-10: 1495272621, ISBN-13: 978-1495272622
[7] vimeo.com/drvasquez and vimeo.com/ichnfm
[8] itunes.apple.com/us/artist/dr-alex-vasquez/id475526413 and cdbaby.com/Artist/DrAlexVasquez

CORRESPONDENCE

Neuroinflammation in fibromyalgia and CRPS is multifactorial

Alex Vasquez

In his Review article (Neurogenic neuroinflammation in fibromyalgia and complex regional pain syndrome. *Nat. Rev. Rheumatol.* 11, 639–648; 2015)[1], Geoffrey Littlejohn ascribes neuroinflammation to a "neurogenic" origin, presumably triggered by pain and stress. However, attribution of neuroinflammation and central sensitization to a primary neurogenic origin is premature without integrating the well-documented coexistence of small intestine bacterial overgrowth (SIBO, one type of gastrointestinal dysbiosis), vitamin D deficiency, and mitochondrial dysfunction.

Littlejohn[1] notes that chronic pain has been associated with lipopolysaccharide (LPS)-stimulated proinflammatory cytokines (particularly IFN-γ and TNF); however, he does not pursue this line of thought to connect it to relevant literature showing clear evidence of gastrointestinal dysbiosis and increased intestinal permeability in patients with fibromyalgia and complex regional pain syndrome (CRPS). The gastrointestinal tract is the most abundant source of LPS, systemic absorption of which is increased by SIBO and increased intestinal permeability. In 1999, Pimentel et al.[2] showed that oral administration of antibiotics led to alleviation of pain and other clinical measures of fibromyalgia. In 2004, Pimentel et al.[3] showed that among 42 fibromyalgia patients, all (100%) showed laboratory evidence of SIBO, severity of which correlated positively with severity of fibromalgia. In that same year, Wallace and Hallegua[4] showed that eradication of SIBO with antimicrobial therapy led to clinical improvements in fibromyalgia patients in direct proportion to antimicrobial efficacy. In 2008, Goebel et al.[5] documented that patients with fibromyalgia and CRPS have intestinal hyperpermeability; mucosal "leakiness" was highest in patients with CRPS, indicating a strong gastrointestinal component to the illness. In 2013, Reichenberger et al.[6] showed that CRPS patients have a distinct alteration in their gastrointestinal microbiome characterized by reduced diversity and significantly increased levels of Proteobacteria. LPS from Gram-negative bacteria is powerfully proinflammatory and is known to trigger microglial activation via Toll-like receptor 4; experimental studies have shown that LPS promotes muscle mitochondrial impairment, peripheral hyperalgesia, and central sensitization[7].

Vitamin D deficiency is prevalent in chronic pain and fibromyalgia patients and promotes pain sensitization, myalgia and bone pain (osteomalacia)[8]. Human clinical trials have shown that vitamin D supplementation can alleviate inflammation[9], intestinal hyperpermeability[10], fibromyalgia pain[11] and other neuromusculoskeletal pain. Vitamin D reduces experimental microglial activation[12], a component of neuroinflammation and central sensitization.

Mitochondrial dysfunction, noted in fibromyalgia[13] and CRPS[14], may be triggered by gastrointestinal dysbiosis via LPS, D-lactate, hydrogen sulfide, and inflammation; mitochondrial dysfunction exacerbates and perpetuates microglial activation and glutaminergic neurotransmission[15], thereby promoting pain sensitization centrally while also contributing to muscle pain peripherally[7]. Treatment of mitochondrial dysfunction with ubiquinone alleviates many biochemical and clinical manifestations of fibromyalgia[13].

Thus, neuroinflammation in fibromyalgia and CRPS has biological contributions including gastrointestinal dysbiosis, vitamin D deficiency, and mitochondrial dysfunction. These independent contributions commonly co-exist, and each of these is additive/synergistic with the others in the promotion of peripheral and central hyperalgesia. The consistent pain-alleviating benefits of treatments for intestinal dysbiosis (antibiotics), vitamin D deficiency (supplementation) and mitochondrial dysfunction (ubiquinone) establish that these painful conditions are multifactorial and maintained by ongoing physiologic insults, each of which is treatable.

Alex Vasquez is at the International College of Human Nutrition and Functional Medicine, Calle Balmes 184, 3° 3°, Barcelona, Spain 08006.
avasquez@ichnfm.org

doi:10.1038/nrrheum.2016.25
Published online 3 Mar 2016

1. Littlejohn, G. Neurogenic neuroinflammation in fibromyalgia and complex regional pain syndrome. *Nat. Rev. Rheumatol.* 11, 639–648 (2015).
2. Pimentel, M. et al. Improvement of symptoms by eradication of small intestinal overgrowth in FMS: a double-blind study [abstract]. *Arthritis Rheum.* 42, S343 (1999).
3. Pimentel, M. et al. A link between irritable bowel syndrome and fibromyalgia may be related to findings on lactulose breath testing. *Ann. Rheum. Dis.* 63, 450–452 (2004).
4. Wallace, D. J. & Hallegua, D. S. Fibromyalgia: the gastrointestinal link. *Curr. Pain Headache Rep.* 8, 364–368 (2004).
5. Goebel, A., Buhner, S., Schedel, R., Lochs, H. & Sprotte, G. Altered intestinal permeability in patients with primary fibromyalgia and in patients with complex regional pain syndrome. *Rheumatology* 47, 1223–1227 (2008).
6. Reichenberger, E. R. et al. Establishing a relationship between bacteria in the human gut and complex regional pain syndrome. *Brain Behav. Immun.* 29, 62–69 (2013).
7. Vasquez, A. Human Microbiome and Dysbiosis in Clinical Disease 2015 (International College of Human Nutrition and Functional Medicine, 2015).
8. von Känel, R., Müller-Hartmannsgruber, V., Kokinogenis, G. & Egloff, N. Vitamin D and central hypersensitivity in patients with chronic pain. *Pain Med.* 15, 1609–1618 (2014).
9. Timms, P. M. et al. Circulating MMP9, vitamin D and variation in the TIMP-1 response with VDR genotype: mechanisms for inflammatory damage in chronic disorders? *QJM* 95, 787–796 (2002).
10. Raftery, T. et al. Effects of vitamin D supplementation on intestinal permeability, cathelicidin and disease markers in Crohn's disease: results from a randomised double-blind placebo-controlled study. *United European Gastroenterol. J.* 3, 294–302 (2015).
11. Wepner, F. et al. Effects of vitamin D on patients with fibromyalgia syndrome: a randomized placebo-controlled trial. *Pain* 155, 261–268 (2014).
12. Hur, J., Lee, P., Kim, M. J. & Cho, Y. W. Regulatory effect of 25-hydroxyvitamin D₃ on nitric oxide production in activated microglia. *Korean J. Physiol. Pharmacol.* 18, 397–402 (2014).
13. Cordero, M. D. et al. Oxidative stress correlates with headache symptoms in fibromyalgia: coenzyme Q₁₀ effect on clinical improvement. *PLoS One* 7, e35677 (2012).
14. Tan, E. C. et al. Mitochondrial dysfunction in muscle tissue of complex regional pain syndrome type I patients. *Eur. J. Pain* 15, 708–715 (2011).
15. Nguyen, D. et al. A new vicious cycle involving glutamate excitotoxicity, oxidative stress and mitochondrial dynamics. *Cell Death Dis.* 8, e240 (2011).

Competing interests statement
The author declares that he has worked as a consultant for Biotics Research Corporation (a nutraceutical company based in the USA), and that he has lectured and written for this company on various topics, including fibromyalgia.

2016 publication in *Nature Reviews Rheumatology* substantiating the model (at least partly, per the space limitations) described in this monograph: Provided here in printed format in accord with publisher's copyright agreement; citation details: Vasquez A. Neuroinflammation in fibromyalgia and CRPS is multifactorial. *Nat Rev Rheumatol.* 2016 Mar 3. doi: 10.1038/nrrheum.2016.25. [Epub ahead of print]
- Publisher site: http://www.nature.com/nrrheum/journal/vaop/ncurrent/full/nrrheum.2016.25.html
- Pubmed citation: http://www.ncbi.nlm.nih.gov/pubmed/26935282

Chapter 5.1—Functional Inflammology Protocol for Metabolic Inflammation: Migraine and Fibromyalgia

Migraine, Cluster and Other Headaches

> **Introduction:**
> This section focuses on migraine headaches in their classic and prototypic manifestations, originating primarily from the additive/synergistic combination of glial activation (and the related neurogenic inflammation and neuroinflammation) and mitochondrial dysfunction; understood as such, migraine and its closely related variant cluster headache can reasonably and very accurately be categorized within my model of metabolic inflammation—the manifestation of inflammation strongly associated with or caused by metabolic impairment, most generally noted in causal/contributory association with mitochondrial dysfunction. Per the origination of this work in my clinical textbooks (*Integrative Orthopedics*, 2004, 2007, 2012) and clinical monographs (*Musculoskeletal Pain*, 2008), this section includes clinical evaluation and differential diagnosis.

Description/pathophysiology:
- Introduction: Headaches are a common symptom-based diagnosis with a wide variety of underlying causes ranging from commonplace and benign (e.g., muscle tension headache) to catastrophic (e.g., meningitis or stroke). This section deals only with the pathophysiology and amelioration of routine benign headaches (migraine, cluster, allergic, tension, and cervicogenic); emphasis is placed on migraine headaches as the prototype for these disorders. Once serious pathological causes of headache have been excluded, the headache can be treated with symptom-suppressing drugs (which have associated risks and expenses without collateral benefits) or by biological/nutritional/natural interventions that address the underlying causative mechanisms and thereby improve overall health.
- Social and medical significance: Headaches and migraine—while seemingly insignificant compared to life-threatening diseases such as cancer and autoimmune diseases—account for huge losses in quality of life and productivity. Headache is a diagnosis based on the patient's subjective report of pain *in* (deep) or *on* (superficial) the head. The potential causes are numerous, ranging from benign muscle tension to life-threatening intracranial hemorrhage or meningitis. Of important note: migraine headache patients have increased risk for neurologic and cardiovascular diseases[1,2]; simply treating the *pain* of migraine does not address the underlying biochemical, physiologic, and inflammatory disturbances whereas nutritional and anti-inflammatory interventions hold great potential for both *alleviation of pain* and *improvement of overall health* via correction of the underlying pathophysiology.
- Mechanism of pain sensation in headache: The final common pathway for "primary" headaches (e.g., migraine and cluster headaches) is currently reported to be neurogenic/brain inflammation: inflammatory mediators from the brain generally and nerves specifically activate trigeminal (cranial nerve V) neurons to produce both vasoconstriction and the sensation of pain.[3]
- A contemporary integrated model of migraine: The task of intellectuals is the creation of cohesion, integration, and understanding; as such, one of the first tasks in the conversation on migraine is to define and characterize the disorder. Effective treatment, excepting blind luck, must be based on a comprehensive and cohesive understanding of the disorder in its *essential* totality. The major

> **Dysfunction precedes disease; understanding precedes efficacy**
>
> The best model of any disease is one that incorporates all of the major known facts into a cohesive and sequential understanding, predicting and being supported by efficacious treatments that are known to address the abnormal physiology—the dysfunction that precedes and causes the disease. The intellectual error of previous models of migraine is that they had no specific starting point, other than to attribute the genesis to "genetic traits and environmental triggers." As such, these earlier descriptions "started from the middle" and simply explained ongoing pathophysiology. By failing to start at the beginning, these earlier models likewise based their treatment on the downstream effects rather than on the treatment of the original cause of the disease. As such, the medical treatments based on this faulty model were necessarily ineffective. Here, I present a complete model, facilitating both understanding and treatment of various headache types.

---

[1] "Depression [adjusted OR = 2.12] and migraine [adjusted OR = 3.65] were more commonly recorded before the diagnosis of dementia in the DLB group." Fereshtehnejad et al. Comorbidity profile in dementia with Lewy bodies versus Alzheimer's disease. *Alzheimers Res Ther*. 2014 Oct 6;6(5-8):65
[2] "The migraine cohort had a higher prevalence of diabetes, hypertension, coronary artery disease, head injury and depression at baseline (p < 0.0001). After adjusting the covariates, migraine patients had a 1.33-fold higher risk of developing dementia [hazard ratio (HR) 1.33]. The sex-specific incidence rate of dementia was higher in men than in women in both cohorts, with an HR of 1.09 for men compared to women. Kaplan-Meier analysis shows that the cumulative incidence of dementia was 1.48% greater in the migraine cohort than in the nonmigraine cohort. This study shows that migraines are associated with a future higher risk of dementia after adjusting for comorbidities. Specifically, the association between migraine and dementia is greater in young adults than in older adults." Chuang et al. Migraine and risk of dementia. *Neuroepidemiology*. 2013;41:139-45
[3] Tierney ML. McPhee SJ, Papadakis MA (eds). *Current Medical Diagnosis and Treatment 2006, 45th Edition*. New York: Lange Medical Books; 2006, pages 31-33

themes from experimental studies and clinical trials have to be integrated and reconciled so that the best model of the disorder emerges triumphantly above the trivia of anecdote and the dogma of pharmaceutical profiteering. Beyond, in addition to, and in support of clinical efficacy, we need a grand unified theory (GUT) that helps us perceive the disease and prioritize the treatments; otherwise, a disarticulated understanding will perpetuate the disarrayed medical management and dependency that we currently observe in migraine and headache management. Each of the following components are sequentially ordered, starting with the first most important primary cause: mitochondrial dysfunction. Importantly, given that—as Thoreau noted in *Civil Disobedience* (1849, p. 26)—"We love eloquence for its own sake, and not for any truth which it may utter", we cannot be satisfied with a clear explanation; the explanation has to have high merit in the real world, being proven by the safety and efficacy of the treatments that it advocates. This standard reveals the falsity of the medical model of both migraine and fibromyalgia, since both the models are selectively incomplete in order to justify drug treatment, and the interventions are unnecessarily hazardous and inadequately efficacious.

1. <u>Mitochondrial impairment is the origin of migraine and cluster headache</u>: Patients with migraine (and cluster headache) have very clear and consistent defects in mitochondrial performance, leading to cellular energy/ATP deficiency, excess production of free radicals (reactive oxygen species—ROS, which promote cellular damage and inflammation). Patients with migraine are often deficient in coenzyme Q-10 (CoQ10), and this causes mitochondrial dysfunction and reduced antioxidant protection against the harmful and pro-inflammatory effects of ROS. Nutritional treatments, such as riboflavin and CoQ10, which support mitochondrial function are consistently the safest and most effective anti-migraine treatments available, thereby proving the mitochondrial origin of migraine. Defects in cellular energy/ATP production cause neurons to be more unstable, resulting in excessive activation, resulting in pain sensation and sensitization. Mitochondrial dysfunction always promotes inflammation, at the very least by increasing formation of free radicals and the liberation of free ATP via leaky mitochondrial membranes; these molecules are perceived by cellular receptors as danger signals, thereby triggering the nonspecific alarm response of inflammation. The metabolic impairment likely contributes to vasodilation, as arteries dilate to bring more oxygen to support metabolic demand (in physiology, this is termed "reactive hyperemia").

    Increasingly over the past several years, the model of microglial activation along with mitochondrial dysfunction is gaining strength and due popularity; this model helps explain many divergent aspects of migraine and provides unification of previously fragmented models and disconnected facts. One of the strongest primary drivers of migraine is mitochondrial dysfunction, which shows a severity-response relationship and is maternally inheritable. Mitochondrial dysfunction is sufficient to promote (micro)glial activation, and the two then form a vicious cycle, ultimately promoting neocortical excitation and the resultant pain sensitization, thereby again promoting continuance and reinforcement of this vicious cycle. By analogy, the brains of these patients are "physiologically fragile" with a constant smoldering sterile inflammation; the brain is either constantly smoldering (e.g., chronic neuroinflammation) or actively "on fire" (e.g., migraine attack).

2. <u>Sustained glial activation results from mitochondrial dysfunction and causes brain inflammation and hyperexcitation</u>: Glial cells are the "glue"—the interconnecting cells—of the brain comprised chiefly of microglia and astrocytes. Microglia (the immune cells of the brain) are sensitive to ROS and inflammatory signals, and become "activated" (microglial activation) in response to peripheral inflammation (including obesity, trauma, infection and vaccination) and central "within the nervous system" events such as trauma and stress. When microglia become activated, they signal the astrocytes (cells that physically and chemically support neurons) to change behavior by providing *less protective support* to brain neurons and *causing more stimulation* of these same neurons; more stimulation with less protection causes the neurons to become "sensitized" and hyperresponsive and eventually promotes the "burn-out" of these neurons. The combinations of more excitation, more inflammation, less energy/ATP and less protection is called "excitotoxicity" (neuronal injury by overstimulation) and eventually leads to neurodegeneration, damage to neurons, brain structures, and the brain as a whole. Stated again and differently: microglia cells in the brain receive inflammatory stimuli and then trigger astrocytes to increase stimulation of neurons via the excitatory neurotransmitter glutamate, which activates a receptor called the NMDA receptor (NMDAr, detailed later); in this manner, inflammatory signals are converted into altered levels of neurotransmitters

Chapter 5.1—Functional Inflammology Protocol for Metabolic Inflammation: Migraine and Fibromyalgia

(especially glutamate, also quinolinic acid [QUIN], a metabolite of tryptophan produced during conditions of inflammation, discussed later) which stimulate neurons to perceive more pain. Excessive stimulation of neurons feeds-back into causing more microglial activation, resulting in a vicious cycle.

3. <u>Brain inflammation, neuroinflammation, neurogenic inflammation all result from glial activation and promote additional brain inflammation</u>: Nerve cells become inflamed in response to any insult; this is called **neuroinflammation**, and it promotes various neurologic and psychiatric disorders, such as pain and depression (e.g., the components of **sickness behavior**), respectively. The neurons themselves can also release inflammatory mediators; this is called **neurogenic inflammation** because the inflammation is coming from the nerve cells while also affecting those same nerve cells. When **brain inflammation** is triggered, it affects all of the major cells types of the brain and becomes a self-reinforcing cycle, sometimes called "**brain on fire**."[4] Inflammation in the brain has many consequences; for example, ❶ inflamed neurons release neuropeptides and inflammatory mediators that activate endothelial cells (thereby causing vasoconstriction) and promote additional inflammation, and ❷ activation of mast cells and platelets causes these cells to secrete inflammatory/vasoactive amines, arachidonate metabolites (such as prostaglandins, leukotrienes, isoprostanes); these substances promote additional inflammation and also promote constriction of blood vessels. Remarkably, the brain inflammation and metabolic impairment seen in migraine known as **cortical spreading depression** triggers release of the inflammation-associated and tissue-destructive enzyme matrix metalloproteinase (MMP), which causes leakiness of the blood-brain barrier (BBB), leading to brain edema and enhanced uptake of inflammatory molecules from the blood.[5]

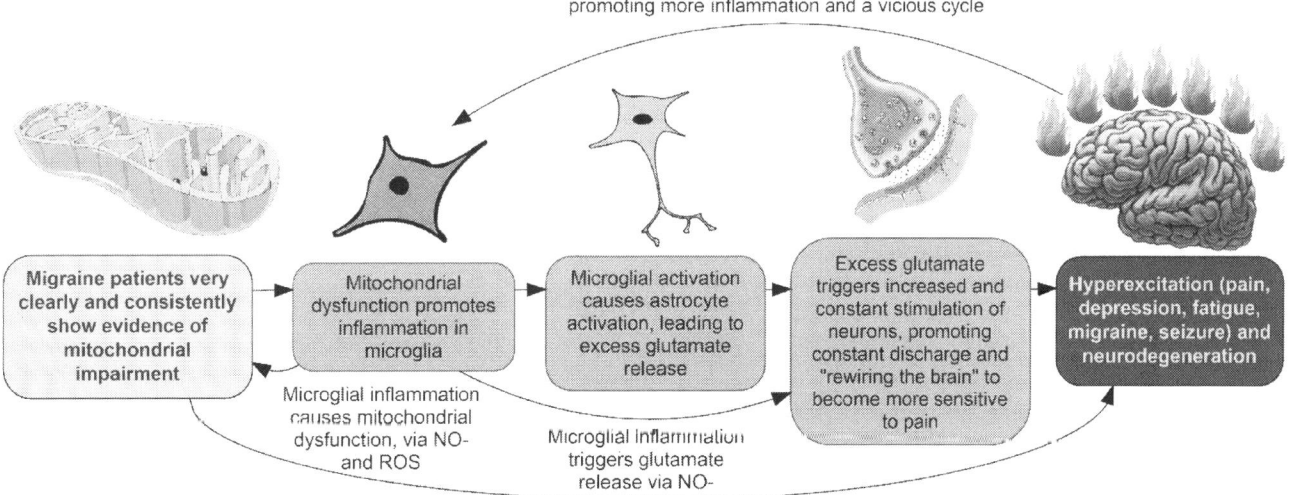

Foundational model of migraine: ❶ Migraine patients very clearly and consistently show evidence of mitochondrial impairment: This genotropic mitochondrial dysfunction can be due to different factors, including 1) defects in CoQ10 synthesis, 2) defects in the citric acid cycle, and 3) defects in the function of the electron transport chain (ETC). The majority of these problems can be partly/largely circumvented by use of nutritional interventions. ❷ Mitochondrial dysfunction promotes inflammation in microglia: Sterile inflammation promoted by excess free radicals produced by dysfunctional mitochondria promote microglial activation. Microglial inflammation causes mitochondrial dysfunction via NO- (nitric oxide, causes impairment of Complex #4 in the electron transport chain[ETC], leading to reduced cellular energy/ATP production), ROS (reactive oxygen species, free radicals), and perhaps directly via inflammatory cytokines thereby creating a vicious cycle. ❸ Microglial activation causes astrocyte activation, leading to excess glutamate release. ❹ Excess glutamate triggers increased and constant stimulation of neurons, promoting constant discharge and "rewiring the brain" to become more sensitive to pain. ❺ Hyperexcitation promotes pain, depression, fatigue, seizure, migraine and neurodegeneration. The combination of mitochondrial dysfunction with excess excitation is particularly devastating to neurons, leading to neuron death: neurodegeneration. Image of brain by IsaacMao per Flickr.com via creativecommons.org/licenses/by/2.0. See educational videos and updates at www.inflammationmastery.com/pain

---

[4] Cohen G. The brain on fire? *Ann Neurol*. 1994 Sep;36(3):333-4
[5] Moskowitz MA. Genes, proteases, cortical spreading depression and migraine: impact on pathophysiology and treatment. *Funct Neurol*. 2007 Jul-Sep;22(3):133-6

Brain neuron excitation

- Glutamate/NMDA receptors are activated by glutamate, QUIN and homocysteine
- Reduced mitochondrial performance impairs homeodynamics in response to excessive NMDAr stimulation
- Free radicals from mitochondria, neurons, and glia promote molecular and cellular damage, thereby triggering inflammation and metabolic collapse of neurons
- Excess intracellular calcium triggers inflammation
- Hyperexcitation promotes pain, pain sensitization, and neurodegeneration

Mitochondrial dysfunction

- Increases in oxidant production promote progressive inflammation and metabolic collapse
- Promotes inflammation in all affected cells
- Depletes antioxidant nutrients such as CoQ10, thereby leading to additional vulnerability, inflammation, and mitochondrial dysfunction
- Makes neurons vulnerable/fragile to excessive activation (eg, lowered depolarization threshold) due to oxidant damage and increased intracellular calcium
- Mitochondrial ROS cause mitochondrial damage, which increases ROS production in a vicious cycle

Central sensitization, cortical spreading depression, metabolic collapse
Image © by Dr Alex Vasquez, ICHNFM.ORG

Microglial activation, astrocyte activation

- Increased oxidant production promotes progressive inflammation and metabolic collapse
- NO- causes mitochondrial dysfunction at ETC #4 and triggers glutamate release to activate NMDAr
- Inflammatory response creates QUIN to activate NMDAr
- Inflammation promotes leaky blood-brain barrier and additional inflammation, edema

The vicious cycles of migraine, free radical production/damage, "metabolic collapse" and cortical spreading depression: Illustration of the interconnecting cycles that promote persistent and additive/synergistic brain inflammation, neuron dysfunction, stemming from primary mitochondrial dysfunction. Integrated models of migraine are now available, providing better understanding, more effective treatment, and less medical dependency. This model of *brain hypersensitivity induced by mitochondrial dysfunction* helps explain why migraine patients 1) are vulnerable to otherwise innocuous stimuli such as hormonal fluctuations and changes in weather, 2) respond poorly to drug treatments, which generally fail to address these components sufficiently, and 3) respond brilliantly to nutritional interventions that support mitochondrial function and reduce inflammation. Independently from my own models of migraine (described and illustrated in this section), Malkov et al[6] proved the merit of the model I have illustrated by showing that "Reactive oxygen species initiate metabolic collapse" in brain neurons, and that this free radical damage that tripartitely damages neurons, glia, and mitochondria is a major cause of cortical spreading depression. See educational videos and updates at www.inflammationmastery.com/pain

4. <u>Mitochondrial dysfunction and glial activation combine to cause altered brain neuron function—brain destabilization, metabolic fragility; in its entirety, this three-part combination of mitochondrial dysfunction, glial inflammation, and neuronal dysfunction is called cortical spreading depression (CSD)</u>: Neurons are simultaneously hyperactive due to (NMDA) neurotransmitter receptor activation and also hyporesponsive due to the mitochondrial impairment; this "physiologic confusion" contributes to the altered brain function seen in migraine, especially migraine with aura. In migraine, the brain is "destabilized" (per Moskowitz[7]), leading to what I call "metabolic fragility" or "brain fragility" that makes migraine patients more sensitive to changes in diet, climate, hormones, stress and sleep. These combinations of ❶ metabolic/mitochondrial impairment with ❷ increased/altered brain activity (e.g., specifically mediated by glutamate at the NMDA

> **Mitochondrial dysfunction is a key component of migraine**
>
> "In migraine, the degree of the mitochondrial impairment ... is related to the severity of the clinical phenotype."
>
> Lodi et al. *J Neurol Sci.* 1997 Feb

---

[6] Malkov et al. Reactive oxygen species initiate a metabolic collapse in hippocampal slices: potential trigger of cortical spreading depression. *J Cereb Blood Flow Metab*. 2014 Sep;34(9):1540-9
[7] Moskowitz MA. Genes, proteases, cortical spreading depression and migraine: impact on pathophysiology and treatment. *Funct Neurol*. 2007 Jul-Sep;22(3):133-6

receptor) and ❸ glial/neuronal/brain inflammation is what creates the wave of abnormal brain function—cortical spreading depression—that typifies migraine and which promotes its exacerbation; cortical spreading depression (CSD) leads to elaboration of the inflammatory and destructive enzyme MMP9 which causes leakiness of the BBB and subsequent brain edema (secondary to protein and water entry into the brain) and increased brain entry of substances from the blood, such as peripherally derived proinflammatory cytokines.[8,9] Brain edema in migraine is associated with and likely contributes to reduced brain perfusion (ie, reduced blood flow).[10] Very importantly, enhanced glutaminergic neurotransmission is itself sufficient to induce cortical spreading depression in experimental models. In an insightful article published in 2014 that supports the model that I have proposed, Malkov et al[11] showed that cortical spreading depression is caused by elaboration of reactive oxygen species (ROS, free radicals) and that these initiate "metabolic collapse" in brain cells.

> **All of these components are interconnected and thus the terms and components become (ultimately/practically) conceptually synonymous**
> Since microglial activation causes astrocyte activation, these terms can be summarized as glial activation. Microglial activation triggers formation of QUIN, which along with glutamate from astrocyte activation, causes stimulation of the NMDA receptor, promoting excitation of neurons. Since glial activation causes neuronal excitation, we can generally state that glial activation is synonymous with hyperexcitation of neurons, which segues into excitotoxic death of neurons and neurodegeneration. Persistent and prolonged hyperexcitation of neurons causes these neurons to strengthen their connections with each other, leading to facilitated pain perception, called central sensitization, as noted in **migraine**, **fibromyalgia**, and **complex regional pain syndrome**. Microglial activation—via release of nitric oxide (NO-) —causes mitochondrial dysfunction and additional glutamate release, causing the combination of metabolic impairment (e.g., reduced ATP formation) and increased metabolic demand, because activation of the NMDA receptor by glutamate and QUIN imposes increased metabolic demand on the neuron cells as they must control the resulting influx of calcium, which if not controlled will promote additional inflammation, impairment, and neuronal cell death.

5. Pain route—the covering of the brain is sensitive to metabolic and inflammatory changes within the brain, and interprets the inflammatory substances as pain signals: The trigeminal nerve (cranial nerve V, #5) receives transmissions from nerve endings surrounding the blood vessels of the membrane surrounding the brain (pia mater) and inside of the skull (dura mater). Recall as previously discussed and cited that the blood-brain barrier becomes more permeable when the brain is inflamed, thereby promoting passage/diffusion of inflammatory mediators from the brain to nearby neurons that receive noxious stimuli and convert the reception of those substances into nerve impulses received and interpreted as pain signals (nociception). While sensory innervation of the supratentorial dura mater membrane is via small meningeal branches of the trigeminal nerve, the innervation for the infratentorial dura mater is via upper cervical nerves, thereby establishing a bidirectional relationship between neck pain (and other subconscious neurologic inputs) and intracranial stimuli and structures.

6. Pain sensitization: As more pain signals are received, the brain facilitates the reception of these messages and thereby becomes more sensitive to the reception of pain; this is called central sensitization, and is greatly facilitated by brain inflammation and mitochondrial impairment.

7. Blood vessel dilation and constriction, and the role of serotonin-1D receptors: Metabolic impairment can trigger vasodilation, while inflammatory mediators promote vasoconstriction; both vasodilation and vasoconstriction have been noted in migraine.

8. Nuances in the contribution of various factors leads to different clinical presentations (e.g., migraine headaches vs cluster headaches); however, in the main primary headache conditions—migraine and cluster—the main themes of mitochondrial dysfunction and brain inflammation dominate the causal pathophysiology and therefore guide treatment: The model presented and used here is that cluster headache is simply a variant of migraine headache, with secondary rather than primary causes of the

---

[8] Moskowitz MA. Genes, proteases, cortical spreading depression and migraine: impact on pathophysiology and treatment. *Funct Neurol*. 2007 Jul-Sep;22(3):133-6
[9] Wilson CJ, Finch CE, Cohen HJ. Cytokines and cognition--the case for a head-to-toe inflammatory paradigm. *J Am Geriatr Soc*. 2002 Dec;50(12):2041-56
[10] For evidence of brain edema (with associated hypoperfusion) in patients with migraine: Kim et al. Recurrent steroid-responsive cerebral vasogenic edema in status migrainosus and persistent aura. *Cephalalgia*. 2015 Jul;35:728-34. See also: Bereczki et al. Cortical spreading edema in persistent visual migraine aura. *Headache*. 2008 Sep;48:1226-9
[11] Malkov et al. Reactive oxygen species initiate a metabolic collapse in hippocampal slices: cortical spreading depression. *J Cereb Blood Flow Metab*. 2014 Sep;34(9):1540-9

underlying mitochondrial dysfunction, and with a greater contribution by psychoemotional stress, muscle tension, and nutritional deficiencies. Mitochondrial dysfunction is seen in both migraine and cluster headache.[12] Nutritional deficiencies (e.g., folic acid) and excesses of systemic inflammation and serum homocysteine are noted in various types of headache, in both adult and pediatric populations.[13]

- <u>Pathophysiology—from past to current models</u>: The sensation of headache pain results from activation and sensitization of sensory trigeminal pain neurons that service intracranial blood vessels and meninges. For many years, the debate focused on whether *vasculogenic* or *neurogenic* influences predominated; most if not all headaches appear to involve *both* of these main components, thus allowing for the consensus that headaches have a *neurovascular* component. That said, the weight of evidence increasingly shifted to support the *neurological* origin—from within the brain and neurons (rather than the blood vessels)—of headaches in general and migraines in particular. "Brain-initiated events" such as cortical spreading depression—a wave of electrical and metabolic disturbance that sweeps across the brain surface, making the brain tissue physiologically unstable, and thus more fragile and vulnerable to various insults—culminate in the release of pain-inducing nociceptive substances including hydrogen ions and arachidonate metabolites, which irritate trigeminovascular sensory neurons surrounding pial vessels.[14,15]

Both dilation and constriction of arteries has been noted in migraine. Dilation of arteries may be an early compensatory response to impaired cellular energy/ATP production as mitochondrial dysfunction progresses from mild to more severe as vicious cycles exacerbate an ever-present primary defect; vasodilation in response to impaired energy/ATP production is well known in physiology as "reactive hyperemia." As mitochondrial function deteriorates before and during a migraine attack, it segues from a metabolic problem to an inflammatory problem, and the consequences of mitochondrial dysfunction (e.g., ROS, inflammation, failure of calcium homeostasis) plus brain neuron dysfunction due to excessive excitation (e.g., ROS, inflammation, failure of calcium homeostasis) lead to vasoconstriction specifically via increased intracellular calcium in

---

Brain sensitization to pain: 4 main components

1. <u>Pain signals</u>: Defects in cellular energy/ATP production cause neurons to be more unstable, resulting in excessive activation, resulting in pain sensation and sensitization. Mitochondrial dysfunction always promotes inflammation, at the very least by increasing formation of free radicals, which are perceived by cellular receptors as danger signals. The metabolic impairment likely contributes to vasodilation, as arteries dilate to bring more oxygen to support metabolic demand (in physiology, this is termed "reactive hyperemia").
2. <u>Brain inflammation</u>: Microglia and astrocytes in the brain transform inflammatory signals into altered levels of neurotransmitters which further activate neurons.
3. <u>Mitochondrial dysfunction</u>: Nerve cells become inflamed in response to any insult; this is called **neuroinflammation**, and it promotes various neurologic and psychiatric disorders. The neurons themselves can also release inflammatory mediators; this is called **neurogenic inflammation** because the inflammation is coming from the nerve cells. When **brain inflammation** is triggered, it affects all of the major cell types of the brain and becomes a self-reinforcing cycle that has been called **brain on fire**. Released neuropeptides activate endothelial cells, mast cells, and platelets to then increase extracellular levels of amines, arachidonate metabolites, peptides, and ions; these substances promote additional inflammation and also promote constriction of blood vessels.
4. <u>Free radicals, reactive oxygen species (ROS)</u>: "We show that ROS accumulation...is capable of triggering an abrupt metabolic collapse (MC) that reproduces most features of cortical spreading depression (CSD). This suggests that oxidative stress may be the primary cause of CSD and not just its consequence. In pathological conditions, the failure to neutralize ROS during the excessive ROS surge and/or deficiency of the neuronal antioxidant system may result in the MC and subsequent ignition of CSD. Indeed, our in vivo results show that when the oxidative stress-induced ROS accumulation is suppressed by an exogenous antioxidant, CSD occurrence is strongly reduced."*

*Malkov et al. Reactive oxygen species initiate a metabolic collapse: potential trigger of cortical spreading depression. *J Cereb Blood Flow Metab.* 2014 Sep

---

[12] "The maximum rate of mitochondrial ATP production (Qmax), calculated from the rate of post-exercise PCr recovery and the end-exercise [ADP], was low in cluster headache patients as well as in migraine patients except MwoA. In migraine the degree of the mitochondrial impairment, that apparently is associated with a reduced glycolytic flux, is related to the severity of the clinical phenotype." Lodi et al. Quantitative analysis of skeletal muscle bioenergetics and proton efflux in migraine and cluster headache. *J Neurol Sci.* 1997 Feb 27;146(1):73-80
[13] "Mean values for body mass index, C-reactive protein, and homocysteine were higher in children with than without headaches, and more children with headaches were in the highest quintile of risk for these factors. Serum and red blood cell folate levels were lower in children with headache. More children with headache were in the highest quintile of risk for 3 or more of these factors. Several important risk factors for long-term vascular morbidity cluster in children and adolescents with severe or recurrent headache or migraine. Further study and screening of children with headaches may permit improved preventive management." Nelson et al. Headache and biomarkers predictive of vascular disease in a representative sample of US children. *Arch Pediatr Adolesc Med.* 2010 Apr;164(4):358-62
[14] Moskowitz MA. Pathophysiology of headache—past and present. *Headache.* 2007 Apr;47 Suppl 1:S58-63
[15] Moskowitz MA. Genes, proteases, cortical spreading depression and migraine: impact on pathophysiology and treatment. *Funct Neurol.* 2007 Jul-Sep;22(3):133-6

astrocytes and inflammation-triggered phospholipase-A2-catalyzed formation of vasoconstrictive prostaglandins, specifically prostaglandin E2 (PGE2) and F2-alpha (PGF2), which are also elaborated from endometrial tissue, thereby supporting the biochemical basis of menstrual migraine.[16]

Neurogenic inflammation (in this conversation, the release of neuropeptides from trigeminal nerve [cranial nerve V] neurons to local blood vessels and meninges) is also important and contributes to a vicious cycle of pain and inflammation.[17] Elevated intracellular calcium levels that trigger inflammatory pathways can be promoted by arachidonate, secondary hyperparathyroidism due to vitamin D deficiency, a relative insufficiency of magnesium, and also by mitochondrial impairment. Mast cell degranulation releases inflammatory mediators such as serotonin, prostaglandin I-2, and histamine, which induce local inflammation and activation of meningeal nociceptors[18,19] and might serve as a pathophysiological link between emotional stress or allergen exposure and headache (i.e., the link between environmental stressors and headache pain). Mast cells can also be activated by neuropeptides that originate from neurons in the brain parenchyma/tissue. Further substantiating the role of local inflammation in migraine is the finding of increased activity of nuclear transcription factor-kappa B (NFkB) in jugular blood of migraine patients during migraine episode[20]; NFkB is an important mediator of inflammation through its ability to enhance transcription of genes that encode for inflammatory mediators.[21] This model provides for the often observed continuum between external and biopsychosocial factors such as exposure to bright lights, hypoglycemia, stress, anxiety, allergen exposure, and hormonal fluctuations with the triggering of new or recurrent headaches. An appreciation for the intraneuronal genesis of headaches such as migraines sharpens our focus on events occurring *within the neuronal cell*, particularly impaired mitochondrial bioenergetics, increased intraneuronal calcium, and the elaboration of inflammatory mediators derived from omega-6 (n6) polyunsaturated fatty acids. With the realization of mitochondrial and eicosanoid contributions to headache, clinicians can intervene with nutritional intervention and fatty acid supplementation to enhance mitochondrial function and modulate eicosanoid production, respectively. Failure to appreciate these underlying pathophysiological mechanisms forces clinicians and patients to rely on pharmacological symptom suppression while the underlying processes remain unaddressed.

The historically documented failure of migraine treatments has arisen largely from the incomplete model of the disease upon which those treatments are founded. Without raw luck, a treatment based on an erroneous or incomplete model has no chance of providing *major*—let alone *optimal*—benefit. Any listing of medical treatments for migraine reveals a catalog of chaos: bits and pieces of incomplete and inconsistent models and the resulting therapy—ie, drugs—which address a small fraction of the problem and therefore have to be overpowered in effect to compensate for their minor significance; hence the low efficacy and high risk of adverse effects.

Important for the perpetuation of any ongoing disease are the vicious cycles that are initiated and maintained; skilled clinicians focus on breaking these vicious cycles because failure to do so allows the disease condition to re-initiate and perpetuate, even after limited therapeutic efficacy of incomplete treatment. I might introduce the concept of "double-stranded" or "triple-stranded" (etc) therapies that simultaneously break multiple vicious cycles, in contrast to treatments such as with drugs which focus only on a single molecule or a single pathway, ie, single-stranded therapy. In this metaphor, the "strands" are biochemical and physiologic pathways; the more that we can optimize the maximum number of pathways, the greater our opportunities for restoring and enjoying optimal health.

A 2015 review discussing fibromyalgia (FM) and complex regional pain syndrome (CRPS) focused on "neurogenic neuroinflammation", the essential definition/concept of which is that that neuronal activity in general and its inflammatory effects in particular can become autonomous and self-perpetuating; neuroinflammation could be initiated externally so-to-speak by stress or trauma and then become a vicious

---

[16] Shaik MM, Gan SH. Vitamin supplementation as possible prophylactic treatment against migraine with aura and menstrual migraine. *Biomed Res Int*. 2015;2015:469529
[17] Tierney ML. McPhee SJ, Papadakis MA (eds). *Current Medical Diagnosis and Treatment 2006, 45th Edition*. New York: Lange Medical Books; 2006, pages 31-33
[18] Levy D, Burstein R, Kainz V, Jakubowski M, Strassman AM. Mast cell degranulation activates a pain pathway underlying migraine headache. *Pain*. 2007 Jul;130(1-2):166-76
[19] Zhang XC, et al. Sensitization and activation of intracranial meningeal nociceptors by mast cell mediators. *J Pharmacol Exp Ther*. 2007 Aug;322(2):806-12
[20] Sarchielli et al. NF-kappaB activity and iNOS expression in monocytes from internal jugular blood of migraine without aura during attacks. *Cephalalgia*. 2006 Sep; 1071-9
[21] Tak PP, Firestein GS. NF-kappaB: a key role in inflammatory diseases. *J Clin Invest*. 2001 Jan;107(1):7-11

cycle within the nervous system promoting chronic pain and neurodegeneration.[22] The existence of neurogenic neuroinflammation is physiologically *likely* and becomes *probable* within dysfunctional and predisposed (i.e., "primed") metabolic and physiologic systems; such "priming factors" clearly include a pro-inflammatory diet, nutrient deficiencies (especially of vitamin B6, magnesium, vitamin D, and CoQ10), mitochondrial dysfunction, and dysbiosis. Hence, the treatment of persistently painful and inflammatory disorders—including but not limited to migraine, recurrent headaches, fibromyalgia and CRPS—needs to focus on the treatment of factors which continue to sustain these disease processes.

- <u>The importance of glutamate and the NMDA receptor in headache, migraine, and chronic pain syndromes</u>: The excitatory neurotransmitter glutamate stimulates neurons by binding to the NMDA (N-methyl-D-aspartate) receptor (NMDAr). As shown in the diagram, excitatory glutamate (which promotes pain, seizure, migraine, anxiety and depression) can be converted into inhibitory GABA (gamma-amino butyric acid, which has an inhibitory, relaxing effect on neurons and the brain as a whole) via the enzyme glutamic acid decarboxylase, which is dependent upon and also dose-dependently stimulated by the vitamin pyridoxine (vitamin B6). Stimulation of the NMDA receptor by glutamate and other receptor activators such as QUIN (quinolinic acid, a "dysfunctional metabolite" of the amino acid L-tryptophan which is formed in response to inflammation and which causes additional inflammation, oxidative damage, and neurotoxicity) causes calcium to enter into the stimulated neuron cells to trigger activation or "firing" of the neuron. A moderate amount of NMDAr stimulation is a normal part of learning and the formation of memories—normal and healthy neurologic function; however, too much NMDAr stimulation causes overstimulation (excitotoxicity) of neurons thereby promoting pain, depression, anxiety, migraine, seizure/epilepsy, and neurodegeneration. Magnesium and zinc partly block the NMDAr calcium channel to reduce/modulate calcium entry into neurons; in this way, magnesium and zinc might be thought of as "softening the effect" of NMDAr activation. The safety and efficacy of supplemental pyridoxine (vitamin B6) in reducing glutamate levels—and thus reducing excessive stimulation of the NMDAr by glutamate—necessitates its inclusion in the treatment of any and all chronic pain disorders, especially migraine and fibromyalgia. Pyridoxine does more than simply lower glutamate levels, as pyridoxine also helps to lower homocysteine (HYC) levels. Glutamate and HYC are both amino acids that activate the NMDAr and mGluR[23]—the metabotropic glutamate receptor (detailed shortly). Generally, higher homocysteine levels correlate with fatigue and pain in patients with fibromyalgia and chronic fatigue syndrome, and with headache pain and increased cardiovascular disease risk in patients with migraine.[24]

- <u>In the treatment of pain—including headaches and fibromyalgia—reducing the effects of glutamate-mediated neurotransmission and cellular effects is of very high importance</u>: Glutamate is an amino acid with many functions, including serving as a precursor to the antioxidant glutathione (GSH), serving as a precursor to alpha-keto-glutarate (a substrate for energy production in the Krebs/citrate cycle in mitochondria) and serving as an excitatory neurotransmitter. Our concern in this conversation is with glutamate's role as a stimulator of neurotransmission in the peripheral and central nervous system; while some minimal glutaminergic stimulation is normal and necessary, excess glutaminergic neurotransmission very clearly promotes anxiety, depression, fibromyalgia pain, myofascial pain and myofascial trigger points, migraine and headaches, seizures and epilepsy; in the extreme, excess glutamate in the brain causes over-excitation of neurons leading to cell death—neurodegeneration—and either mild or massive, acute or chronic, brain damage. In the following image and subsequent descriptors, I provide an accurate and yet simplified overview of important concepts, but I will state plainly here what everyone needs to know about this section: Because glutaminergic neurotransmission promotes pain/anxiety/depression/neurodegeneration, our therapeutic goals are to 1) reduce glutamate levels with vitamin B6 and by avoiding/treating microglial activation (ie, "brain inflammation"), 2) reduce glutamate-triggered influx of calcium with zinc and magnesium, also vitamin D, alkalinization (increased consumption of base-forming foods, such as fruits and vegetables which contain citrate which is converted to bicarbonate to promote alkalinization, one effect of which is to promote magnesium retention, thereby alleviating pain[25]), omega-3 fatty acids such as from fish oil, 3) reduce the effects

---

[22] Littlejohn G. Neurogenic neuroinflammation in fibromyalgia and complex regional pain syndrome. *Nat Rev Rheumatol*. 2015 Nov;11(11):639-48
[23] Abushik et al. NMDA and mGluR5 in calcium mobilization and neurotoxicity of homocysteine in trigeminal/cortical neurons and glial cells. *J Neurochem*. 2014 Apr; 264-74
[24] "Mean homocysteine plasma levels - as well as the proportion of subjects with hyperhomocysteinaemia - were significantly higher in patients with MA than in healthy controls." Moschiano et al. Homocysteine plasma levels in patients with migraine with aura. *Neurol Sci*. 2008 May;29 Suppl 1:S173-5
[25] Vormann et al. Supplementation with alkaline minerals reduces symptoms in patients with chronic low back pain. *J Trace Elem Med Biol*. 2001;15(2-3):179-83

Chapter 5.1—Functional Inflammology Protocol for Metabolic Inflammation: Migraine and Fibromyalgia

of glutamate/NMDA receptor activation by counterbalancing with benzodiazepine/GABA receptor activation by promoting conversion of glutamate to GABA and perhaps also by using niacinamide and botanicals that act as ligands for the GABA receptor. Because much of this information is both important and a bit complicated, I will create some teaching videos on this material and make them available per the following internet link/redirect: www.inflammationmastery.com/pain

Glutamate is increased by dietary MSG, deficiencies of B6 and Mg, glial activation, genotropic enzyme defects

- Other agonists (eg, QUIN, homocysteine) and co-agonists
- **Glutamate (neuroexcitatory)**
- Glutamic acid decarboxylase requires P5P, the active form of vitamin B6
- **GABA (neuroinhibitory)**
- Other agonists (eg, niacinamide, ethanol, several botanicals)

NMDA-type glutamate receptor

Zn

NMDA receptors are located in the brain, spinal cord, and peripheral tissues, including nerves, muscles and skin

Mg

Neuron cell membrane

Glutamate-activated calcium channel

GABA receptor

Calcium entry following activation of glutamate receptor: excitatory/stimulatory effect

Chloride entry following activation of GABA receptor: inhibitory effect

Excess dietary arachidonate and insufficient EPA increases intracellular calcium

Excess intracellular calcium: relative to magnesium, secondary to true magnesium deficiency or mild chronic metabolic acidosis with resultant loss of intracellular magnesium

Mild hyperparathyroidism (most commonly due to vitamin D3 insufficiency)

Increased/excess intracellular calcium; nerve depolarization activation of intracellular pathways

GABA receptor activation increases intracellular chloride which has inhibitory, calming, relaxing, analgesic effects

<u>Glutamate receptors in the brain</u>: Neurocortical excitation, promotion of inflammation (neuroinflammation), promotion of mitochondrial dysfunction, apoptosis/neurodegeneration

<u>Glutamate receptors in the periphery</u>: Glutamate receptors (including NMDAr) are located throughout the body; increased glutamate signaling promotes neurogenic inflammation, peripheral pain sensitization, muscle contraction

Central sensitization, increased pain sensitivity (hyperalgesia), depression and anxiety, migraine and seizure

Muscle contraction/hypertonicity, myofascial trigger points (MFTP), muscle cramps, hypertension due to increased peripheral resistance

Illustration © 2016 by Dr Alex Vasquez, *Inflammation Mastery, 4th Ed*. InflammationMastery.com All rights reserved and protected internationally. No duplication, derivation, or reuse without written permission. ICHNFM.ORG

Clinically relevant conceptual illustration of the NMDA-type glutamate receptor (NMDAr), its activation, effects, and nutritional modulation: The image above provides a conceptually accurate and clinically applicable model of glutamate reception and the effects thereof; categorized details are provided below, listed from top to bottom of the image and also prioritized to clinical relevance (top) and additional details and context (bottom). See instructional videos at ICHNFM.ORG.

Image caption—continued from previous page:
- Various types of glutamate receptors in the central nervous system and periphery share the common themes of promoting pain and inflammation: Glutamate receptors are described in two broad categories; **ionotropic glutamate receptors—iGluR—**(divided into three groups: AMPA, NMDA and kainate receptors) transpose ions such as sodium and calcium upon activation and thus can be considered mostly involved with propagation of nerve impulses, while **metabotropic glutamate receptors—mGluR—**(also with several subtypes, such as mGluR5) lead more to activation of intracellular pathways with results dependent on the cell type but generally consistent with some type of cellular activation and/or inflammation.
- Glutamate reception, with the NMDA receptor (NMDAr) as the prototype receptor: Many types and subtypes of glutamate receptors exist, and the specific subtype NMDA is very clearly the most discussed for its relevance in both chronic pain disorders and neurodegenerative diseases. The NMDA receptor is activated by glutamate, QUIN (quinolinic acid, a metabolite of tryptophan made in inflammatory conditions, discussed in more detail in the section on fibromyalgia), aspartate, homocysteine and other substances which act as agonists/activators or required co-activators/co-agonists (e.g., D-serine, glycine). Different forms of the NMDA receptor exist in the central and peripheral nervous systems, each with slightly different characteristics and sensitivity to agonists and requirement for co-agonists; thus, the image presented here is a generalized version that is conceptually accurate (rather than all-inclusive; for more details see reviews[26]) and clinically relevant. Although we have traditionally thought of glutamate/NMDA receptors as existing separately (ie, on different cell types) from the GABA/benzodiazepine receptors, that fact remains true (ie, some cells are clearly dominated by one receptor type over others) while we are also increasingly appreciating that glutamate/NMDA receptors can coexist with GABA/benzodiazepine receptors on the same cell and that these receptors are interactive, not simply oppositional, and occasionally behave/interact in paradoxical and age-specific manners.[27,28]
- Homocysteine (HYC), a toxic intermediate of amino acid metabolism, activates glutamate receptors, thereby promoting pain, headache/migraine/seizure: Glutamate is the prototypic excitatory neurotransmitter, activating a wide range of ionotropic glutamate receptors (including the NMDA receptor) and metabotropic glutamate receptors (mGluR) which are present throughout the central and peripheral nervous systems and all of which are generally involved in (enhanced) pain processing. We have exacting clarity that both NMDA receptors and mGluR5 are activated by homocysteine with resultant calcium influx just as with glutamate-mediated activation of these same receptors; Abushik et al[29] published in 2014, "Thus, elevation of intracellular calcium (Ca2) by HCY in neurons is mediated by NMDA and mGluR5 receptors while SGC are activated through the mGluR5 subtype. Long-term neurotoxic effects in peripheral and central neurons involved both receptor types. Our data suggest glutamatergic mechanisms of HCY-induced sensitization and apoptosis of trigeminal nociceptors." This is of very high clinical importance, because we gain the mechanistic insight that lowering of homocysteine levels (technique detailed later) will reduce the total stimulation of these glutamate receptors in the brain, spinal cord, and periphery to reduce the pain and fatigue of migraine, fibromyalgia, and other pain conditions.
- Glutamate promotes pain and inflammation; therefore, reducing levels of glutamate or reducing the effects of its reception are important therapeutic goals, especially in the treatment of pain, anxiety/depression, migraine, and seizure/epilepsy: Glutamate levels are increased by microglial inflammation and the subsequent astrocyte activation[30]; therefore, reducing inflammation generally and "brain inflammation" specifically is an important therapeutic goal. Reducing inflammation must always focus on the trigger of the inflammation, most commonly microbial (e.g., gastrointestinal dysbiosis[31]) and/or metabolic (e.g., excess sugar and "junk/fast food" in the diet[32], vitamin D deficiency, lack of phytonutrients due to insufficient intake of fruits and vegetables, insufficient omega-3 fatty acids from fish oil, etc). Glutamate is excitatory to neurons, promoting pain, depression, migraine, seizure and neurodegeneration; glutamate is readily converted by the enzyme glutamic acid decarboxylase to GABA—gamma-amino-butyric acid—which has opposing effects to those of glutamate.
- Modulation of calcium entry/accumulation following NMDA receptor activation: Following NMDAr activation, sodium (Na) enters to propagate nerve impulses, and calcium (Ca) enters and promotes intracellular signaling, including the promotion of pain and inflammatory pathways. Intracellular calcium is a famous "second messenger" responsible for physiologic processes such as the pancreatic release of insulin; however, excess intracellular calcium triggers the activation of pathways that can promote pain, migraine and hypertension, hence the well-established use of calcium-channel blocking (CCB) drugs to treat migraine and hypertension. Calcium entry following glutamate stimulation of the NMDA receptor is reduced or "modulated" by both zinc and magnesium, both of which "adhere" to the NMDA receptor to reduce calcium influx; in fact, magnesium is often described as a "cork" or "plug" of the NMDA receptor. Magnesium (Mg) can also be thought of as competing for space with calcium or otherwise blocking some of the effects of intracellular calcium; as such, Mg reduces the effect of glutamate receptor stimulation. Excess intracellular calcium also challenges or stresses the capacity of mitochondria, while magnesium supports mitochondrial function. Intracellular calcium promotes muscle contraction, important in hypertension (due to systemic constriction of arteries/arterioles) and myofascial trigger points (MFTP, an important cause of and contributor to pain in migraine and fibromyalgia), while magnesium promotes muscle relaxation, arterial dilation, and pain relief. Thus, we would expect—and indeed we see clinically—that magnesium supplementation (typically 600 mg per day for adults) provides many of its benefits by offsetting the adverse effects of glutaminergic stimulation and excess intracellular calcium, while also supporting mitochondrial function. Many of the factors that contribute

---

[26] Vyklicky et al. Structure, function, and pharmacology of NMDA receptor channels. *Physiol Res.* 2014;63 Suppl 1:S191-203
[27] Ben-Ari et al. GABAA, NMDA and AMPA receptors: a developmentally regulated 'ménage à trois'. *Trends Neurosci.* 1997 Nov;20(11):523-9
[28] Ben-Ari Y. Excitatory actions of gaba during development: the nature of the nurture. *Nat Rev Neurosci.* 2002 Sep;3(9):728-39
[29] Abushik et al. The role of NMDA and mGluR5 receptors in calcium mobilization and neurotoxicity of homocysteine in trigeminal and cortical neurons and glial cells. *J Neurochem.* 2014 Apr;129(2):264-74
[30] Béchade C, Cantaut-Belarif Y, Bessis A. Microglial control of neuronal activity. *Front Cell Neurosci.* 2013 Mar 28;7:32
[31] Vasquez A. Nutritional and Botanical Treatments against "Silent Infections" and Gastrointestinal Dysbiosis. *Nutr Perspectives* 2006. Translating Microbiome (Microbiota) and Dysbiosis Research into Clinical Practice. *Int J Hum Nutr Funct Med* 2015 https://ichnfm.academia.edu/AlexVasquez
[32] Aljada et al. Increase in intranuclear nuclear factor kappaB and decrease in inhibitor kappaB in mononuclear cells after a mixed meal: proinflammatory effect. *Am J Clin Nutr.* 2004 Apr;79(4):682-90. Mohanty et al. Glucose challenge stimulates reactive oxygen species (ROS) generation by leucocytes. *J Clin Endocrinol Metab.* 2000 Aug;85(8):2970-3

to excess intracellular calcium—vitamin D deficiency, magnesium deficiency, an acidic acid-base balance, excess omega-6 arachidonic acid relative to omega-3 fatty acids—are easily treated with vitamin D supplementation, magnesium supplementation, promotion of systemic alkalinization, and omega-3 fatty acid supplementation, respectively; proof-of-principle is demonstrated by the observation that each of these interventions provides analgesic, antihypertensive, and other clinical benefits.[33]

- Stimulation of the GABA/benzodiazepine receptor: GABA reception at the GABA receptor—a large multicomponent receptor that also receives benzodiazepine and barbiturate drugs—promotes analgesia, euphoria, relaxation and antiseizure benefits. GABA receptors are also activated by the niacinamide form of vitamin B3 as well as by alcohol/ethanol in beer, wine, and liquors. Botanical medicines that have proven clinical benefit via—at least in large part—their activation of the GABA/benzodiazepine receptor include *Matricaria recutita* (Chamomile), *Melissa officinalis* (lemon balm), *Passiflora incarnata* (passionflower), *Piper methysticum* (kava), *Scutellaria lateriflora* (skullcap), *Valeriana species* (valerian), *Withania somnifera* (ashwagandha).[34]

- Conversion of glutamate to GABA, the importance of vitamin B6 in neuroprotection and pain alleviation: Conversion of glutamate to GABA via glutamic acid decarboxylase requires vitamin B6 (pyridoxine), and the speed/efficiency of this conversion is generally proportionate to the provision of B6. Giving more vitamin B6 results in lower glutamate levels and therefore less activation of the glutamate receptor, thereby providing anti-pain, anti-depression, and anti-seizure benefits. As such, what is obvious is that supplementation with vitamin B6 clearly has an essential role in the treatment of pain, depression, migraine, and seizure; all patients affected by such disorders should receive high-potency vitamin B6 supplementation, at least as a therapeutic trial if not as a default component of therapy. Very importantly, vitamin B6 alleviates pain and the excess brain activity seen with migraine and seizure by means other than serving as a cofactor for glutamic acid decarboxylase; vitamin B6 also provides analgesic and antiinflammatory benefits in peripheral tissues/nerves, in the spinal cord, in the deep brain structures of the brain such as the pain-relaying thalamus, as well as in the neurocortex.

- Activation of the glutamate receptor, especially the NMDA receptor, in the nervous system results in stimulation/depolarization of neurons and the promotion of new connections, promoting memory/learning as well as pain: Activation of the NMDA receptor allows sodium (Na) and calcium (Ca) to enter the cell; entry of Na promotes depolarization of the nerve membrane to allow propagation of the nerve impulse, sometimes called nerve "firing." Entry of Ca following glutamate receptor activation triggers intracellular events, some of which are beneficial for processes such as learning and memory, while others—especially if intracellular calcium levels are too high for too long—are harmful and promote inflammatory responses and mitochondrial stress. While we have typically thought of glutamate receptors and the classic NMDA receptor as existing in the brain and neocortex, we now share the clarity that NMDA receptors exist throughout the nervous system including the spinal cord and peripheral nerves. Activation of NMDAr is important for learning and memory and is also important for the generation of excessive neuronal/brain activity that is seen in seizure/epilepsy, chronic pain, and overactivation of neurons that leads to neuron/brain damage—neuroexcitation, neurodegeneration, and neuroexcitatory neuronal death.

- Activation of the glutamate receptor in peripheral tissues (outside of the nervous system) is not well understood, but again mostly correlates with pain and inflammation: Beyond the NMDA receptor and beyond the nervous system, other types of glutamate receptors are active throughout the body. Given that most tissues and cells are innervated by the nervous system, the distinction between glutaminergic effect via the nervous system and the direct effect on the cells is a bit challenging; however, one of the most important themes observed in the research and science literature is that elevated levels of glutamate in the periphery are clearly and causatively associated with increased pain sensation and to a lesser extent with adverse physiologic changes. For example, activation of mGluR5 in skin is seen with inflammatory and irritating/itchy/pruritic skin disorders, and blockade of the receptor is therapeutic; this same subtype mGluR5 also participates in hypersensitivity to pain in inflammatory diseases.[35] Elevated levels of glutamate are seen in malignant diseases (ie, cancer) and appear to suppress immune function; blockade of iGluR reduces cancer invasiveness. Overall, the associations with excessive glutamate signaling are pain and inflammation; when glutamate is injected into muscles, the result is increased intensity of pain and enlargement of the receptive field (ie, spreading of pain, increased area of heightened sensitivity).[36] Therefore, treatments to reduce glutamate levels (especially vitamin B6) and to reduce the effect of glutamate-triggered increases in intracellular calcium (e.g., magnesium and vitamin D) are expected to reduce glutamate-triggered pain and inflammation.

- Conclusion of this image caption with a few more details and bit of redundancy: Conversion of glutamate to GABA requires vitamin B6 and provides analgesic and "calming" benefits to brain and muscles. Neuroexcitatory glutamate is converted to neuroinhibitory GABA by the enzyme glutamic acid decarboxylase, which—as firmly established in the disease pyridoxine-dependent/responsive epilepsy—shows very clear dose-responsiveness in its ability to reduce glutamate levels in response to high-dose vitamin B6 supplementation. Magnesium and zinc (and perhaps copper) retard the passage of calcium through this channel, thereby mitigating some of the effects of NMDAr activation. Quenching NO- (for example with the hydroxocobalamin form of vitamin B12), which would otherwise trigger glutamate release, and dousing glial activation (for example, with anti-inflammatory nutrients such as vitamin D3 and EPA and DHA from fish oil) which otherwise promotes elaboration of glutamate and QUIN are important considerations not included in this illustration. Glycine is generally considered a necessary co-activator of the NMDAr; but given that glycine is ubiquitous and mostly invariable, it is not immediately malleable and therefore not considered of high relevance as a clinical therapeutic target.

---

[33] Vasquez A. Intracellular Hypercalcinosis. *Naturopathy Digest* 2006 naturopathydigest.com/archives/2006/sep/vasquez.php. Included at the end of this section/chapter.
[34] Sarris et al. Plant-based medicines for anxiety disorders, part 2: a review of clinical studies with supporting preclinical evidence. *CNS Drugs*. 2013 Apr;27(4):301-19
[35] Julio-Pieper et al. Exciting times beyond the brain: metabotropic glutamate receptors in peripheral and non-neural tissues. *Pharmacol Rev*. 2011 Mar;63(1):35-58
[36] Wang et al. Spatial pain propagation over time following painful glutamate activation of latent myofascial trigger points in humans. *J Pain*. 2012 Jun;13(6):537-45

Clinical presentations:
- Headache—general considerations: Head pain, with a wide range of differential diagnoses and possible causes and contributions, ranging from simple "stress" and so-called "reactive hypoglycemia" to life-threatening causes such as stroke, aneurysm, tumor, meningitis. Especially for new headaches or acute-onset headaches, concomitant subjective complaints (e.g., lethargy, sleepiness, mood/cognitive changes, changed vision) and/or objective presentations (e.g., fever, skin rash, galactorrhea, or neurologic deficits) indicate the need for additional evaluation to exclude important intracranial lesions such as pituitary adenoma, meningitis, tumor, or subdural hematoma. Since the trigeminal sensory pathway is activated in any condition associated with brain inflammation, relief of headache with analgesic medications does not exclude serious underlying disease such as hemorrhage or meningitis.
- Migraine: Periodic headache characterized by unilateral distribution, commonly with a pulsatile sensation; severity ranges from moderate to severe and disabling; commonly begin in adolescence; commonly with a maternal inheritance, consistent with inheritance of mitochondrial DNA from the mother; twice as common in women (5-25%) as in men (2-10%); typical duration of migraine "attack" is 4-72 hours. 80% of migraine is "common migraine" or "migraine without aura." Photophobia and phonophobia (excessive sensitivity to light and sound, respectively) are common, as is nausea, sometimes leading to vomiting. Patients with migraine show a consistent pattern of different—additive and synergistic—mitochondrial defects affecting various locations in the pathway of substrate conversion to cellular energy/ATP: ❶ enzymatic impairment of citrate synthase—the first enzyme in the Krebs cycle, ❷ impaired function of complexes #1-4 of the electron transport chain, ❸ deficiency of coenzyme Q-10 due to insufficient endogenous production, thereby promoting failure of performance of the electron transport chain as well as reduced antioxidant and antiinflammatory defense, and ❹ magnesium deficiency—noted in all headache types—which leads to impairment of complex #5 (ATP synthase) of the electron transport chain, and also leading to increased intracellular calcium influx following activation of the NMDA receptor, leading to increased metabolic demand in neurons.
- Migraine with aura: Migraine with aura is characterized by focal neurologic symptoms/deficits; the localization of the neurologic involvement has traditionally been attributed to regional brain vasospasm, leading to reduced blood flow and compromised neuron/brain function in the affected areas. Migraine with aura may present with any combination of the following: blurred/altered vision including scotoma (the perception of "flashing lights"), vertigo/dizziness, hallucinations such as hearing nonreal sounds or seeing nonreal images. Some patients experience the aura as hyper/hypo-activity, depression, food cravings, yawning, mood changes. More than 50% of migraine patients report significant impairment in life tasks and personal relationships as a result.
- Cluster headache: Cluster headache (CH) affects predominantly middle-aged men. Although the "pathophysiology is unclear" according to *Current Medical Diagnosis and Treatment 2014*, triggering of trigeminal pain sensation and vasoconstriction are clearly involved, identically to migraine. CH patients generally lack a family history of headache or migraine. CH manifests with episodes of severe unilateral periorbital pain, generally with one or more of the following: ipsilateral nasal congestion, rhinorrhea (runny nose), lacrimation (tearing), redness of the eye, and Horner syndrome (ptosis/drooping of the eyelid, meiosis/constriction of the pupil, and anhidrosis—reduced sweating on the affected side). CH attacks may occur daily (especially nightly) for several weeks, and patients often feel restless and agitated. CH attacks typically last 15 minutes - 3 hours and occur in clusters for weeks or months, then remit. Triggers include alcohol, stress, glare, or specific foods. The prototypic CH patient is a stressed male entrepreneur who smokes, with varying levels of alcohol intake. Mechanistically, stress promotes muscle tension especially in the neck; ethanol/alcohol and tobacco smoke's cyanide are both mitochondrial toxins, and the prototypical stressed male entrepreneur is not eating sufficient fruits and vegetables to maintain urinary alkalinization and sufficient magnesium intake/retention. The facts that CH patients generally ❶ have no family history of the disorder, ❷ have a characteristic lifestyle pattern known to promote mitochondrial impairment, and ❸ respond acutely to oxygen therapy (obviously a form of mitochondrial support, since the primary function of oxygen in the human body is to drain hydrogen protons from the intramembrane space via the formation of ATP and water), support the contention that these patients have a *secondary* lifestyle-generated mitochondrial impairment leading to their headaches, via the aforementioned glial activation and resultant brain inflammation and the remainder of the pain-inducing cascade of events.

Major differential diagnoses: The differential diagnosis of headache by history, examination, and laboratory and imaging assessments should be familiar to clinicians. In particular, the neurological examination should include psychoemotional assessment, as well as cranial nerve and fundoscopic examination, and any new headache symptoms, even in a patient with a history of headaches, must receive due diligence on the part of the clinician.

- Cervical spondylosis: Cervical spine dysfunction and arthropathy can cause and contribute to head pain and headaches; confer with history and examination.
- Cluster headache: Presents with intense unilateral periorbital pain often associated with ipsilateral nasal congestion, rhinorrhea, lacrimation, eye redness, and transient/chronic Horner's syndrome; more common in men, especially in smokers; exacerbated by alcohol; tend to recur at the same time each day/night.
- Cough headache: Severe transient headache triggered by coughing, straining, sneezing, or laughing; patients with recurrent complaints need to be evaluated with a complete neurologic examination and are candidates for CT/MRI since 10% of patients with persistent cough headache have an intracranial lesion.[37]
- Dental or occlusive disorders: Mouth examination, history, oral/dental exam.
- Depression: Check for history consistent with depression: apathy, recent stressful life events.
- Drug side-effect: Check each drug to see if side-effects correlate with clinical complaints.
- Food allergy: Evaluate with elimination/challenge, history; consider blood tests for recalcitrant cases.
- Head injury: Evaluate with history and examination.
- Hyperparathyroidism: Begin by assessing serum calcium.
- Hypertension: Assess blood pressure; although most patients with hypertension do not have headaches, and most patients with headaches do not have hypertension, acute exacerbations of hypertension commonly precipitate headache. Assess for papilledema and hyperreflexia.
- Hyperthyroidism: Weight loss, tremor; assess TSH (generally low) and free T4 (always high).
- Hypothyroidism: Assess TSH (typically elevated), free T4 (generally low), free T3 (may be low or normal), anti-TPO antibodies (seen with autoimmune hypothyroidism: Hashimoto's disease); effective treatment with thyroid hormone alleviates most headaches in hypothyroid-headache patients.[38]
- HIV infection: Patients with HIV are at increased risk for infections, including intracranial infections, particularly toxoplasmosis; intracranial lymphoma is also more common in HIV-positive patients.
- Intracranial aneurysm: May present with throbbing pain; assessed with contrast angiography. In one large international study with 1449 patients[39], the risk of rupture was less than 1% per year, whereas complications from surgery were seen in approximately 14%. A Japanese study[40] found that 95% of patients had a favorable outcome with surgery, implying that 5% had an unfavorable outcome, which is still greater than the risk of rupture, being less than 1% per year for untreated aneurysms reported previously.[41] A more recent study also suggested that the risks of treatment might exceed the risk of spontaneous rupture.[42] Thus the clinical management of intracranial aneurysms must be determined per patient, neuroanatomic location, available techniques/technology, current research, and experience of the neurosurgeon.
- Iron deficiency: Iron is necessary for function of the mitochondrial electron transport chain as well as for formation of the neurotransmitters dopamine and serotonin, both of which can be said to have an analgesic effect. As such, iron deficiency can promote headaches in general and migraine in particular; iron deficiency might also contribute to the clinical presentation of fibromyalgia.[43] Optimal iron status correlates with serum ferritin values of 40-70 ng/ml; rarely, a person with what can be described as a defect in the blood-brain barrier transport of iron into the brain will need to have a serum ferritin value of 120 ng/ml in order to promote entry of iron into the brain.

---

[37] Tierney ML, McPhee SJ, Papadakis MA (eds). *Current Medical Diagnosis and Treatment 2002, 41st Edition*. New York: Lange; 2002. Page 999-1005
[38] "Thirty-one patients with hypothyroidism of 102 (30%) presented with headache 1-2 months after the first symptoms of hypothyroidism. The headache was slight, nonpulsatile, continuous, bilateral, and salicylate responsive and disappeared with thyroid hormone therapy." Moreau et al. Headache in hypothyroidism. *Cephalalgia* 1998 Dec:687-9
[39] International UIA Investigators. Unruptured intracranial aneurysms—risk of rupture and risks of surgical intervention. *N Engl J Med* 1998 Dec 10;339(24):1725-33
[40] Orz et al. Risks of surgery for patients with unruptured intracranial aneurysms. *Surg Neurol* 2000 Jan;53(1):21-7; discussion 27-9
[41] International UIA Investigators. Unruptured intracranial aneurysms—risk of rupture and risks of surgical intervention. *N Engl J Med* 1998 Dec 10;339(24):1725-33
[42] Risks associated with spontaneous rupture "were often equaled or exceeded by the risks associated with surgical or endovascular repair of comparable lesions." Wiebers DO, et al. Unruptured intracranial aneurysms: natural history, clinical outcome, and risks of surgical and endovascular treatment. *Lancet*. 2003 Jul 12; 362(9378): 103-10
[43] "The mean serum ferritin levels in the fibromyalgia and control groups were 27.3 and 43.8 ng/ml, respectively, and the difference was statistically significant. Binary multiple logistic regression analysis with age, body mass index, smoking status and vitamin B12, as well as folic acid and ferritin levels showed that having a serum ferritin level <50 ng/ml caused a 6.5-fold increased risk for FMS." Ortancil et al. Association between serum ferritin level and fibromyalgia syndrome. *Eur J Clin Nutr*. 2010 Mar;64(3):308-12

- <u>Iron overload, with or without genetic hemochromatosis</u>: For reasons reviewed in Chapter 1 and per my previous reviews[44], all patients must be tested for iron overload. Iron overload causes headaches; iron depletion can relieve headaches.[45,46] Optimal iron status correlates with serum ferritin values of 40-70 ng/ml.
- <u>Magnesium deficiency</u>: Magnesium deficiency is common in industrialized nations[47,48,49,50] and can be assessed clinically (e.g., response to supplementation) or with laboratory tests (e.g., intracellular magnesium). Associated findings common with magnesium deficiency are muscle cramps, bruxism, constipation, and cravings of sweets/candies and especially chocolate.
- <u>Meningitis</u>: Evaluate fundoscopic examination, skin rash, fever, CBC, CRP; immediate transport to emergency department if meningitis is suspected.
- <u>Migraine</u>: Classic presentation includes periodicity, unilaterality, with prodrome, photophobia, nausea, vomiting, visual changes, and positive family history and onset in early teens or adulthood; a large percentage of migraine patients do not have the classic presentation. Migraine can be associated with transient neurologic deficits: numbness, aphasia, clumsiness, and weakness.
- <u>Muscle tension and tension headaches</u>: Assessed with palpation/provocation of cervical/cranial musculature; worse with stress and generally worse at the end of the workday; generally responsive to manual therapies, stress reduction, stretching of affected musculature, and magnesium supplementation.
- <u>Myofascial trigger points</u>: Palpation/provocation of cervical/cranial musculature; treat with post-isometric stretching, ergonomic improvements, and the supplemented Paleo-Mediterranean diet[51] with an emphasis on supplementation with vitamin D, calcium, and magnesium.
- <u>Ocular disorders</u>: Assess with history (e.g., recent change in prescription, new glasses or contacts), and neurologic, eye, and fundoscopic examination; consider diabetes mellitus, multiple sclerosis, and glaucoma and test or refer appropriately.
- <u>Preeclampsia</u>: Headache in a pregnant woman may indicate preeclampsia; assess for hypertension, edema, and proteinuria; emergency or urgent obstetrical referral will be indicated in most cases.
- <u>Pheochromocytoma</u>: Common presentation is periodic headache concurrent with exacerbations of hypertension, sweats, and tachycardia/palpitations.
- <u>Sinusitis or sinus infection</u>: History, fever, pain with palpation of sinuses, nasal discharge; test CBC and CRP; consider radiographic or CT imaging.
- <u>Temporal arteritis (TA), giant cell arteritis (GCA), polymyalgia rheumatica (PMR)</u>: History of diffuse head/shoulder pain and jaw claudication generally with systemic complaints of myalgia and fatigue in a patient over 50 years of age; if suspected, must assess CRP/ESR and palpation of artery. **Remember that temporal arteritis can result in blindness; any visual change in a patient with TA/PMR should be considered a medical emergency.** "Loss of vision is the most feared manifestation and occurs quite commonly."[52]
- <u>Temporomandibular joint (TMJ) dysfunction</u>: Assess with examination, history, oral/dental exam, pain worse with chewing (DDX temporal arteritis); notably associated with excess interstitial glutamate.
- <u>Tumor other intracranial lesion</u>: One-third of brain tumor patients present with headache[53], typically worse upon waking and worse with exertion. Assess with neurologic exam/imaging as indicated.

---

[44] See Chapter 1 of either *Inflammation Mastery* / *Functional Inflammology* (2014 or later) for the most complete reviews, including assessment, management, and radiographic presentations. See also: Vasquez A. Musculoskeletal disorders and iron overload disease: comment on the American College of Rheumatology guidelines for the initial evaluation of the adult patient with acute musculoskeletal symptoms. [Letter] *Arthritis & Rheumatism* 1996; 39:1767-8. Vasquez A. High body iron stores. *Nutr Perspect* 1994 October
[45] In a study involving more than 51,000 patients: "Phenotypic hemochromatosis and the C282Y/C282Y genotype were both associated with an 80% increase in headache prevalence evident only among women. The reason for this association is unclear, but one may speculate that iron overload alters the threshold for triggering a headache by disturbing neuronal function." Hagen K, et al. High headache prevalence among women with hemochromatosis. *Ann Neurol* 2002;51(6):786-9
[46] "…the temporary improvement of headache from depletion of iron stores may indicate a causal relation, possibly mediated by iron deposits in pain-modulating centres in the brainstem." Stovner et al. Hereditary haemochromatosis in two cousins with cluster headache. *Cephalalgia* 2002 May;22(4):317-9
[47] "The American diet is low in magnesium, and with modern water systems, very little is ingested in the drinking water." Innerarity S. Hypomagnesemia in acute and chronic illness. *Crit Care Nurs Q*. 2000 Aug;23(2):1-19
[48] "Altogether 43% of 113 trauma patients had low magnesium levels compared to 30% of noninjured cohorts." Frankel et al. Hypomagnesemia in trauma patients. *World J Surg*. 1999 Sep;23(9):966-9
[49] "There was a 20% overall prevalence of hypomagnesemia among this predominantly female, African American population." Fox et al. An investigation of hypomagnesemia among ambulatory urban African Americans. *J Fam Pract*. 1999 Aug;48(8):636-9
[50] "Suboptimal levels were detected in 33.7 per cent of the population under study. These data clearly demonstrate that the Mg supply of the German population needs increased attention." Schimatschek HF, Rempis R. Prevalence of hypomagnesemia in an unselected German population of 16,000 individuals. *Magnes Res*. 2001 Dec;14(4):283-90
[51] Vasquez A. A Five-Part Nutritional Protocol that Produces Consistently Positive Results. *Nutritional Wellness* 2005Sept.
[52] Tierney ML. McPhee SJ, Papadakis MA (eds). *Current Medical Diagnosis and Treatment 2002, 41st Edition*. New York: Lange; 2002, page 999-1005
[53] Tierney ML. McPhee SJ, Papadakis MA (eds). *Current Medical Diagnosis and Treatment 2002, 41st Edition*. New York: Lange; 2002, page 999-1005

Chapter 5.1—Functional Inflammology Protocol for Metabolic Inflammation: Migraine and Fibromyalgia

Clinical assessment:
- History/subjective:
  - Subacute or chronic/periodic head pain: Most likely benign if course is not progressive and if no neurologic deficits and other findings are present.
  - Acute headache: Recent onset of severe headache in a previously healthy patient suggests intracranial lesion or meningitis.[54] Approximately 1% of patients with acute headache who present to emergency departments will have a life-threatening disorder.[55]
- Physical examination/objective:
  - Neurologic examination should be performed on all patients with a recent onset of new headaches or a change from their previous headache. The finding of any mental abnormality or neurologic deficit indicates immediate need for further evaluation: brain CT/MRI and/or emergency department referral.[56]
  - Muscle strength and reflexes
  - Fundoscopic examination for papilledema
  - Cranial nerve examination
  - Blood pressure
  - Spinal and cervical musculature assessment for joint dysfunction and myofascial trigger points[57]
  - Signs for meningeal irritation:
    - Nuchal rigidity (previously referred to as Soto-Hall maneuver): Patient supine on examining table; doctor gently-yet-assertively forces patient's neck into flexion: positive sign for meningeal irritation is undue pain or resistance. This test must not be performed in patients who may have atlantoaxial instability or cervical spine fracture.
    - Kernig sign: Patient supine with hip flexed, slowly extend knee; positive sign: pain in posterior thigh with or without flexion of opposite knee.
    - Brudzinski sign: Bilateral hip flexion following forced cervical flexion when the patient is supine.
- Imaging & laboratory assessments:
  - Imaging: Rarely required except to assess for or exclude intracranial pathology or cervical spondylosis. Importantly, **new onset of headache in an elderly patient or a patient with HIV warrants neuroimaging even if the neurologic examination is normal**.[58]
  - Lumbar puncture for CSF analysis: This procedure assesses for infection and subarachnoid hemorrhage and must not be performed unwittingly in patients with increased intracranial hypertension/papilledema.
  - Laboratory evaluation is generally routine and includes the following:
    - 25-OH-vitamin D (serum): Should be between 50-100 ng/mL. All patients with pain need to be assessed for vitamin D deficiency and/or supplemented with 2,000 IU/d (children) or at least 4,000 IU/d (adults).[59]
    - CBC: Assess for anemia and evidence of infection.
    - Chemistry panel: Screening for diabetes, hypercalcemia/electrolytes, liver and kidney function.
    - CRP: Helps to exclude an infectious or inflammatory etiology.
    - Ferritin: Should be 40-70 and certainly less than 120 mcg/L for most people. Assessment for iron overload is indicated in African Americans[60,61], white men over age 30 years[62], patients with peripheral

> **Spinal and myofascial assessment**
> "Because treating myofascial problems may be the only way to offer complete relief from certain types of headache, clinicians must learn to diagnose and manage trigger points in neck, shoulder, and head muscles."
>
> Davidoff RA. *Cephalalgia.* 1998 Sep

---

[54] "The onset of severe headache in a previously well patient is more likely than chronic headache to relate to an intracranial disorder such as subarachnoid hemorrhage or meningitis." Tierney ML. McPhee SJ, Papadakis MA (eds). *Current Medical Diagnosis and Treatment 2002, 41st Edition.* New York: Lange Medical Books; 2002, page 999
[55] Tierney ML. McPhee SJ, Papadakis MA (eds). *Current Medical Diagnosis and Treatment 2006, 45th Edition.* New York: Lange Medical Books; 2006, pages 31-33
[56] Tierney ML. McPhee SJ, Papadakis MA (eds). *Current Medical Diagnosis and Treatment 2006, 45th Edition.* New York: Lange Medical Books; 2006, pages 31-33
[57] Davidoff RA. Trigger points and myofascial pain: toward understanding how they affect headaches. *Cephalalgia.* 1998 Sep;18(7):436-48
[58] Tierney ML. McPhee SJ, Papadakis MA (eds). *Current Medical Diagnosis and Treatment 2006, 45th Edition.* New York: Lange Medical Books; 2006, pages 31-33
[59] Vasquez et al. Clinical Importance of Vitamin D: Paradigm Shift for All Healthcare Providers. *Altern Ther Health Med* 2004; 10: 28-37 ichnfm.academia.edu/
[60] Barton JC, Edwards CQ, Bertoli LF, Shroyer TW, Hudson SL. Iron overload in African Americans. *Am J Med.* 1995 Dec;99(6):616-23
[61] Wurapa RK, Gordeuk VR, Brittenham GM, Khiyami A, Schechter GP, Edwards CQ. Primary iron overload in African Americans. *Am J Med.* 1996;101(1):9-18
[62] Baer DM, et al. Hemochromatosis screening in asymptomatic ambulatory men 30 years of age and older. *Am J Med.* 1995 May;98(5):464-8

arthropathy[63], diabetics[64] and is advisable in children[65], women[66], young adults[67] and the general population.[68] Iron overload causes headaches.[69,70]

- **Homocysteine (serum)**: Optimal level is below 7 micromoles/liter in blood/serum; all patients with pain disorders—including but not limited to migraine/headaches, fibromyalgia and CRPS—should be tested for elevated homocysteine. Importantly, we need to appreciate that the most "pathologic" increases in homocysteine occur in the fluid around the brain—the cerebrospinal fluid (CSF) which is typically not subject to laboratory assessment due to the pain, risk, technical needs and skill involved in the procedure.[71]
- **Thyroid assessment**: Especially in patients with classic manifestations of hypothyroidism: fatigue, depression, cold hands and feet, dry skin, constipation, and delayed Achilles return.[72] See Chapter 1.
- **Food allergy testing**: May be helpful when elimination-and-challenge procedures are nonproductive and when other therapeutic measures have failed. I personally (DrV) think food allergy testing is overused and that addressing mucosal barrier defects and phenotype immunomodulation (Chapter 4, Section 3) is more important than laboratory testing for food allergies.

- Establishing the diagnosis:
  o Headache is considered a diagnosis based on the patient's subjective report of head pain. However, the headache is always secondary to some other cause of pain, which is the true diagnosis. A clinical or empirical process of elimination must consider common and dangerous causes of head pain, including meningitis, temporal arteritis, sinus infections, cervicogenic pain, intracranial lesions such as brain tumors, hypertension, drug side-effects, and food intolerances.
  o **Serious causes of head pain must be considered with each recurrence, as a patient with a long-term history of benign headaches may contract meningitis or develop hypertension as a new or additive cause of his/her headaches.**

Complications:
- Pain, nausea/vomiting/diarrhea, secondary inability to engage in work, play, and other daily activities.
- Cost and adverse effects of drugs.
- Complications may arise if an underlying cause (e.g., tumor, meningitis, hemorrhage) is undiagnosed.

Clinical management:
- A complete patient history and the above-mentioned lab tests and a physical examination with neurologic assessment will exclude most of the lethal differential diagnoses, allowing the provisional assessment of "benign headache" or "migraine headache" to be established. New onset of headaches or a progressive headache disorder always requires investigation. Refer to neurologist if clinical outcome is unsatisfactory or if complications become evident. For benign headaches including migraine, standard medical treatment is targeted at the alleviation of symptoms. To this end, analgesics and anti-inflammatory drugs such as acetaminophen, aspirin, ibuprofen, naproxen, and ketoprofen are the medical mainstays. Antidepressant drugs ranging from amitriptyline to fluoxetine also might be used for both migraine and tension headaches. Other drugs used for migraine include beta-adrenergic blockers such as propanolol, calcium-channel antagonists such as verapamil, anticonvulsants such as gabapentin and topiramate, and serotonin-modulating drugs such as methysergide and sumatriptan, as well as monoamine oxidase inhibitors and angiotensin-2 receptor blockers. Treatments unique to cluster headaches include inhaled oxygen, lithium carbonate, and prednisone.

---

[63] Olynyk J, Hall P, Ahern M, KwiatekR, MackinnonM. Screening for hemochromatosis in a rheumatology clinic. *Aust NZ J Med* 1994; 24: 22-5
[64] Phelps G, Chapman I, Hall P, Braund W, Mackinnon M. Prevalence of genetic haemochromatosis among diabetic patients. *Lancet* 1989; 2: 233-4
[65] Kaikov Y, et al. Primary hemochromatosis in children: report of three newly diagnosed cases and review of the pediatric literature. *Pediatrics* 1992; 90: 37-42
[66] Edwards CQ, Kushner JP. Screening for hemochromatosis. *N Engl J Med* 1993; 328: 1616-20
[67] Gushusrt TP, Triest WE. Diagnosis and management of precirrhotic hemochromatosis. *W Virginia Med J* 1990; 86: 91-5
[68] Balan V, Baldus W, Fairbanks V, et al. Screening for hemochromatosis: a cost-effectiveness study based on 12, 258 patients. *Gastroenterology* 1994; 107: 453-9
[69] Hagen K, Stovner LJ, Asberg A, et al. High headache prevalence among women with hemochromatosis: the Nord-Trondelag health study. *Ann Neurol* 2002 Jun;51(6):786-9
[70] Stovner LJ, Hagen K, Waage A, Bjerve KS. Hereditary haemochromatosis in two cousins with cluster headache. *Cephalalgia* 2002 May;22(4):317-9
[71] "The concentration of free HC did not differ significantly from normal controls, but the total HC concentration was significantly higher in MOA and MWA patients (41% increase in MOA and 376% increase in MWA). These findings suggest that an increase of total HC concentration in the brain is commonly seen in migraine patient and is particularly pronounced in MWA sufferers." Isobe C, Terayama Y. A remarkable increase in total homocysteine concentrations in the CSF of migraine patients with aura. *Headache*. 2010 Nov;50(10):1561-9
[72] DeQowin RL. *DeQowin and DeQowin's Diagnostic Examination. Sixth Edition*. New York, McGraw-Hill; 1994, page 900

Migraine patients may become dependent on prescription narcotic drugs, which carry inherent risks of dependence and abuse. Topiramate (Topomax®) is one of the most commonly used pharmaceutical drugs for the treatment of migraine, and a brief description of its efficacy and expense is warranted in order to provide clinical perspective. A recent clinical trial in a leading headache journal concluded that topiramate "resulted in statistically significant improvements" and that the drug is "safe and generally well tolerated"[73]; these statements would appear to support clinical use of the drug. However, more than 10% of patients stopped using the drug due to adverse effects, and the statistically significant benefit largely consisted of a reduction in headache days by 1.5 days per 91 days of treatment compared to placebo. The out-of-pocket cost for 3 months of this drug treatment (not including physician fees, recommended laboratory monitoring, and management of adverse effects) is in the range of $400 to $600. Thus, for a yearly cost of approximately $2000, the total reduction in headache days over placebo would be approximately 6 days per year. This study was funded by the company that makes the drug, and 11 of the 13 authors received funding, employment, or direct payment from Ortho-McNeil Neurologics, Inc. Therapeutic trials are implemented to address the underlying problem(s); natural treatments may be superior to drug treatments especially when used in combination.[74]

> **Patients with migraine headaches—noted in 50% of patients with fibromyalgia and some patients with hypertension—often have food allergies/sensitivities/intolerances**
>
> "The commonest foods causing reactions were wheat (78%), orange (65%), eggs (45%), tea and coffee (40% each), chocolate and milk (37% each), beef (35%), and corn, cane sugar, and yeast (33% each). When an average of ten common foods were avoided there was a dramatic fall in the number of headaches per month, 85% of patients becoming headache-free."
>
> Grant EC. Food allergies and migraine. *Lancet*. 1979 May

- Medical standard for migraine—symptomatic relief: "Management of migraine consists of avoidance of any precipitating factors, together with prophylactic or symptomatic pharmacologic treatment if necessary."[75] The goals of medical management are to reduce pain and other manifestations of migraine such as nausea and aura; this symptom-based approach ignores—conceptually and therapeutically—nearly all of the underlying biochemical and nutritional components of the illness, thereby providing minimal/modest benefit while fostering drug-dependency; additional medical goals are to reduce use of high-cost higher-risk emergency "rescue" drugs as well as the utilization of urgent/emergency medical services.
- Treatments (all benign headaches): Standard medical treatment for headaches is expensive and fraught with adverse effects, drug dependence, and suboptimal efficacy. Further, such symptom-suppressive treatment fails to address the causative food intolerances, nutritional deficiencies, and mitochondrial defects that are common in headache patients and migraineurs. Following the exclusion of serious underlying disease, headache patients should be counseled on allergen identification (free and highly efficacious) and should receive nutritional supplementation with combination fatty acids (e.g., ALA, GLA, EPA, DHA) and therapeutic doses of vitamins and minerals, particularly riboflavin, vitamin D3, and magnesium. CoQ10, 5-HTP, melatonin, spinal manipulation, post-isometric stretching, and the other treatments listed above can be used in combination as appropriate per patient to optimize the therapeutic response.

Food & Nutrition The foundational diet is the 5pSPMD as described previously; this "Paleo template"—a diet of fruits, vegetables, nuts, seeds, berries, and lean sources of protein (thereby excluding grains in general and gluten-containing grains in particular)—immediately helps patients increase potassium and magnesium intake specifically and increase nutritional density and systemic alkalinization generally while reducing intake of sodium chloride, common allergens and triggers such as wheat/gluten and milk/dairy, and chemicals such as MSG and aspartame. Patients should—generally speaking—base the diet on "fruits, vegetables, nuts, seeds, and berries with adequate protein intake." Grains that contain the most inflammatory and allergenic form of gluten are rye, barley, and wheat; these foods should be avoided due to their pro-inflammatory properties, their ability to promote gastrointestinal

---

[73] Silberstein SD, et al. Efficacy and safety of topiramate for the treatment of chronic migraine. *Headache*. 2007 Feb;47(2):170-80
[74] Vasquez A. Interventions need to be consistent with osteopathic philosophy. *J Am Osteopath Assoc*. 2006 Sep;106(9):528-9 jaoa.org/cgi/content/full/106/9/528
[75] Tierney ML, McPhee SJ, Papadakis MA (eds). *Current Medical Diagnosis and Treatment 2002, 41st Edition*. New York: Lange; 2002. Page 999-1005

dysbiosis generally and small intestine bacterial overgrowth (SIBO) in particular, and their promotion of gastrointestinal damage, release of zonulin and—in some patients—promotion of systemic inflammation and brain inflammation, both leading to pain while also likely promoting neurodegeneration.[76] The ability of gluten-containing grains to trigger brain inflammation (triggering migraine and other headaches) and autoimmune brain conditions (mimicking multiple sclerosis [MS] and amyotrophic lateral sclerosis [ALS]) has been well proven for decades and is irrefutable; a gluten-free diet is curative.[77,78,79] Beyond custom, convenience, and the government subsidies that make "junk foods"—many of which contain gluten and other inflammatory dietary components—inexpensive and widely available, no legitimate medical or nutritional reason exists for the consumption of gluten-containing foods; the myth that "whole grain foods promote health" is a lie foisted on an ignorant public and an equally uninformed population of nutritionally ignorant medical professionals.[80]

> **Dr Vasquez's Five-part Nutrition Protocol: The "Supplemented Paleo-Mediterranean Diet" (SPMD)**
>
> 1. **Diet: Emphasize fruits, vegetables, nuts, seeds, berries, and lean sources of protein** (fish, grass-fed lamb/beef). Make modifications for patient-specific food allergies and sensitivities; this is especially important for patients with known allergy-related conditions such as migraine headaches. Patients with kidney disease should use caution when consuming a potassium-rich diet. Vasquez A. Revisiting the Five-Part Nutritional Wellness Protocol: The Supplemented Paleo-Mediterranean Diet. *Nutritional Perspectives* 2011 Jan
> 2. **Multivitamin and multimineral supplement**: Nutrient deficiencies are common and are easily treated with nutritional supplementation. Fletcher and Fairfield. Vitamins for chronic disease prevention in adults. *JAMA* 2002 Jun
> 3. **Vitamin D dosed at 2,000-10,000 IU per day**: The adult requirement for vitamin D3 is approximately 4,000 IU per day; some patients may achieve optimal blood levels with lower doses, but generally daily doses of 4,000-10,000 IU are necessary. Vasquez A et al. The Clinical Importance of Vitamin D. *Alternative Therapies in Health and Medicine* 2004 Sep
> 4. **Combination fatty acid supplementation**: A combination of flax oil, borage oil, and fish oil provides the health-promoting fatty acids (ALA, GLA, EPA, DHA). Patients should consume organic virgin olive oil liberally with foods. Vasquez A. New Insights into Fatty Acid Supplementation and Its Effect on Eicosanoid Production and Genetic Expression. *Nutritional Perspectives* 2005; Jan
> 5. **Probiotics**: Health-promoting bacteria can be consumed in the form of powders, pills, and fermented foods such as yogurt and kefir.
>
> For a video review of this foundational diet and introduction to the functional inflammology protocol, see Dr Vasquez "Functional Inflammology Protocol, part 1" from the 2013 International Conference on Human Nutrition and Functional Medicine (ICHNFM.ORG): https://vimeo.com/100089988 Password: "DrVprotocol_volume1"

- Food allergy elimination (*Lancet* 1979 May): Mitochondrial dysfunction promotes inflammation, including allergic inflammation; as a result of mitochondrial dysfunction, elevated glutamate levels, and microglial activation, headache patients in general and migraine patients in particular are more sensitive to triggers (e.g., emotional, environmental, hormonal, nutritional) that might otherwise not cause problems. Food allergy is among the most common causes/triggers of headaches[81,82], particularly migraine headaches, particularly those that do not respond to drug treatments.[83,84] In the important study by Grant[85], the following foods were identified as the most common headache triggers: wheat (78%), orange (65%), eggs (45%), tea and coffee (40% each), chocolate and milk (37% each), beef (35%), corn, cane sugar, and yeast (33% each); when an average of 10 triggering foods were avoided, patients experienced a "dramatic fall in the number of headaches per month, 85% of patients becoming headache-free." Food allergen identification via the *elimination and challenge technique*[86] is accurate and inexpensive, and problem-causing foods are then eliminated from the diet.

---

[76] Daulatzai MA. Non-celiac gluten sensitivity triggers gut dysbiosis, neuroinflammation, gut-brain axis dysfunction, and vulnerability for dementia. *CNS Neurol Disord Drug Targets*. 2015;14(1):110-31
[77] Finsterer J, Leutmezer F. Celiac disease with cerebral and peripheral nerve involvement mimicking multiple sclerosis. *J Med Life*. 2014 Sep 15;7(3):440-4
[78] "The authors describe 10 patients with gluten sensitivity and abnormal MRI. All experienced episodic headache, six had unsteadiness, and four had gait ataxia. MRI abnormalities varied from confluent areas of high signal throughout the white matter to foci of high signal scattered in both hemispheres. Symptomatic response to gluten-free diet was seen in nine patients." Hadjivassiliou et al. Headache and CNS white matter abnormalities associated with gluten sensitivity. *Neurology*. 2001 Feb 13;56(3):385-8
[79] "CD is an autoimmune-mediated disorder of the gastrointestinal tract. Initial symptom presentation is variable and can include neurologic manifestations that may comprise ataxia, neuropathy, dizziness, epilepsy, and cortical calcifications rather than gastrointestinal-hindering diagnosis and management. We present a case of a young man with progressive neurologic symptoms and brain MR imaging findings worrisome for ALS. During the diagnostic work-up, endomysium antibodies were discovered, and CD was confirmed by upper gastrointestinal endoscopy with duodenal biopsies. MR imaging findings suggestive of ALS improved after gluten-free diet institution." Brown et al. White matter lesions suggestive of amyotrophic lateral sclerosis attributed to celiac disease. *AJNR Am J Neuroradiol*. 2010 May;31(5):880-1
[80] Adams et al. Nutrition education in U.S. medical schools: latest update of a national survey. *Acad Med*. 2010 Sep;85(9):1537-42
[81] Egger J, Carter CM, Wilson J, Turner MW, Soothill JF. Is migraine food allergy? A double-blind controlled trial of oligoantigenic diet treatment. *Lancet* 1983 Oct ;2:865-9
[82] Monro J, Brostoff J, Carini C, Zilkha K. Food allergy in migraine. Study of dietary exclusion and RAST. *Lancet* 1980 Jul 5;2(8184):1-4
[83] Monro J, Carini C, Brostoff J. Migraine is a food-allergic disease. *Lancet* 1984 Sep 29;2(8405):719-21
[84] Finn R, Cohen HN. "Food allergy": Fact or Fiction? *Lancet* 1978 Feb 25;1(8061):426-8
[85] Grant EC. Food allergies and migraine. *Lancet* 1979 May 5;1(8123):966-9
[86] "Elimination diets can be both a diagnostic tool and a therapeutic intervention for people with a suspected food sensitivity or allergy." Denton C. The elimination/challenge diet. *Minn Med*. 2012 Dec;95(12):43-4. The classic book on the topic of the elimination and challenge technique is William G. Crook MD's *Detecting Your Hidden Allergies* or *Tracking Down Hidden Food Allergy*. Professional Books; 2 edition (June 1980).

- <u>Gluten-free diet alleviates migraine and fibromyalgia in a significant proportion of affected patients (*Rheumatol Int.* 2014)</u>: "The level of widespread chronic pain improved dramatically for all patients; for 15 patients, chronic widespread pain was no longer present, indicating remission of FM. Fifteen patients returned to work or normal life. In three patients who had been previously treated in pain units with opioids, these drugs were discontinued. Fatigue, gastrointestinal symptoms, migraine, and depression also improved together with pain. ... For some patients, the clinical improvement after starting the gluten-free diet was striking and observed after only a few months; for other patients, improvement was very slow and was gradually observed over many months of follow-up."
- <u>Avoidance of food additives</u>: Red wine, aged cheeses, sardines, sausage, bacon, and monosodium glutamate (MSG)-containing foods are common triggers for headache and migraine in susceptible patients and should therefore be avoided or at least trialed, ie, avoided and reintroduced to observe for any reduction and recurrence, respectively, of headache or other inflammatory manifestations. Most of these foods contain tyramine, nitrites, or other neuroexcitatory or vasoactive substances, in addition to components (allergens) to which migraine patients tend to be immunologically sensitized. MSG consumption can trigger headache, nausea, and increased blood pressure in apparently normal healthy people.[87] Sulfites in red wine are also noted to trigger migraine and headache in some patients; many wines are available on the market now which contain no detectable sulfites. Sulfites trigger migraine by directly triggering the release of inflammatory mediators and by impairing mitochondrial function; sulfite inhibits the enzyme glutamate dehydrogenase in its conversion of glutamate into alpha-keto-glutarate thereby blocking substrate/fuel entry into the Krebs/citrate cycle (leading to a 50% reduction in cellular energy/ATP production in an experimental study using rat brain cells[88]) while perhaps also leaving excess glutamate present for NMDAr activation—note here again, as previously mentioned, that the combination of mitochondrial dysfunction with glutamate-mediated NMDAr activation is particularly lethal for neurons because the mitochondrial dysfunction starves the neurons of energy/ATP and mitochondria-mediated calcium homeostasis at the exact moment when these same neurons are overstimulated.
- <u>Fish oil supplying 3,000 mg of eicosapentaenoic acid (n3 EPA) and docosahexaenoic acid (n3 DHA) per day, with additional 400 IU mixed tocopherols</u>: Fish oil has been shown to reduce the frequency, duration, and intensity of migraine headaches[89] and the effectiveness of fish oil may be mediated via alterations in cytokine production.[90] More specifically per current research, we appreciate that EPA and especially DHA alleviate glial activation, thereby reducing excessive glutamate-driven pain-inducing excitatory neurotransmission.
- <u>Gamma-linolenic acid (n6 GLA, from plants such as borage and hemp) and alpha-linolenic acid (n3 ALA, notably from flaxseed oil)</u>: Supplementation with GLA and ALA—along with the use of a multivitamin and multimineral supplement and avoidance of dietary arachidonic acid—has been shown to significantly reduce the intensity, frequency, and duration of migraine headaches.[91] Exacerbation of temporal lobe epilepsy (TLE) with GLA combined with n6 linoleic acid has been reported; this exacerbation can be problematic or diagnostically useful (e.g., temporary exacerbation of TLE aids in the differential from schizophrenia).[92] Relatedly and importantly, Al-Khamees et al[93] reported a case of previously well 41-year-old female (i.e., 41yoF) who developed temporal lobe status epilepticus following one week of 1.5-3 g/d borage oil; the amount of GLA and LA and any contaminants in the product were not determined but serum fatty acid analysis showed elevations of GLA 345 microg/g of blood (control 191 microg/g), and LA 259 microg/g of blood (control 165 microg/g). Migraine patients could reasonably limit GLA intake to not more than 500 mg/d; other treatments within this protocol such as vitamin D, pyridoxine, and magnesium have established anti-seizure benefits.

---

[87] "A statistically significant increase in systolic and diastolic blood pressures after MSG administration was observed, as well as a significantly higher frequency of reports of nausea and headache in the MSG group. No robust effect of MSG on muscle sensitivity was found." Shimada et al. Differential effects of repetitive oral administration of monosodium glutamate on interstitial glutamate concentration and muscle pain sensitivity. *Nutrition*. 2015 Feb;31(2):315-23
[88] Zhang et al. A mechanism of sulfite neurotoxicity: direct inhibition of glutamate dehydrogenase. *J Biol Chem*. 2004 Oct 8;279(41):43035-45
[89] "In fact, results of this preliminary study suggest that both fish oil and olive oil may be beneficial in the treatment of recurrent migraines in adolescents." Harel et al. Supplementation with omega-3 polyunsaturated fatty acids in the management of recurrent migraines in adolescents. *J Adolesc Health* 2002 Aug;31(2):154-61
[90] Smith RS. The cytokine theory of headache. *Med Hypotheses* 1992 Oct;39(2):168-74
[91] "In 129 patients available for study, 86% experienced reduction in severity, frequency and duration of migraine attacks, 22% became free of migraine and more than 90% had reduced nausea and vomiting." Wagner et al. Prophylactic treatment of migraine with gamma-linolenic and alpha-linolenic acids. *Cephalalgia*. 1997 Apr;17(2):127-30
[92] Vaddadi KS. The use of gamma-linolenic acid and linoleic acid to differentiate between temporal lobe epilepsy and schizophrenia. *Prostaglandins Med*. 1981 Apr;6(4):375-9
[93] Al-Khamees et al. Status epilepticus associated with borage oil ingestion. *J Med Toxicol*. 2011 Jun;7(2):154-7

Generally, the minimal dose of GLA for systemic anti-inflammatory effect is 500 mg/d; doses up to 2-4 grams per day have been used with safety and efficacy in other inflammatory/metabolic disorders such as cancer, asthma, psoriasis, and rheumatoid arthritis.

- Magnesium supplementation to bowel tolerance (generally with additional pyridoxine): Magnesium deficiency is common, affecting approximately 30% of different populations in various industrialized nations.[94,95,96,97] Regardless of headache etiology or classification, magnesium deficiency is more common in headache patients than in headache-free controls. Magnesium deficiency directly contributes to headache by at least 4 mechanisms: (1) facilitating brain cortex hyperexcitability and hypesthesia due to a reduction in the partial blockade of N-methyl-D-aspartate (NMDA) neurotransmitter receptor sites by magnesium[98], (2) impairing cellular energy production, specifically at the level of mitochondrial electron chain complex #5, the ATP synthase enzyme, (3) promoting vasoconstriction, and (4) promoting increased muscle tension, with the latter 2 mechanisms caused in part by impaired energy production, as well as altered intracellular calcium-to-magnesium ratios. Conversely, adequate magnesium nutriture and use of magnesium supplementation help prevent headaches by modulation of NMDA receptor sensitivity and support of energy production, vasorelaxation, and myorelaxation. Not only is magnesium deficiency common in female patients with menstrual migraine[99] and in patients with post-traumatic headaches[100], but magnesium supplementation is justified in headache patients based on the findings of "disturbances in magnesium ion homeostasis" which appear to contribute to brain cortex hyperexcitability.[101] **Except when contraindicated due to renal failure or drug interaction, magnesium supplementation is safe, effective, and reasonable for essentially all patients with headache.**[102,103,104,105,106] Intravenous magnesium (sulfate) is more effective than drug therapy with dexamethasone/metoclopramide for the treatment of acute migraine headaches.[107] A reasonable clinical approach is to 1) evaluate patient with history, physical examination, and screening laboratory tests to exclude contraindications such as renal insufficiency (assess with BUN, creatinine, and urinalysis), 2) assess for possible drug interactions, and then 3) begin the patient with 200 mg elemental magnesium (citrate or malate) with the dose increased by 200 mg every 1-2 days until bowel tolerance is reached. Reduce dose if excessively loose stools or diarrhea occur. When high-dose magnesium supplementation is used in patients with renal insufficiency or drugs that predispose to hypermagnesemia, cautious professional supervision is warranted, with periodic measurement of serum or ionized magnesium. Efficacy of magnesium supplementation is enhanced with concomitant pyridoxine supplementation (e.g., 100-250 mg per day with food) and with an alkalinizing Paleo-Mediterranean Diet as described in Chapter 2. The Paleo-Mediterranean Diet can promote alkalinization[108] which facilitates systemic mineral and magnesium

> **Clinical Pearl**
> The importance of alkalinization for the renal retention and intracellular uptake of magnesium can hardly be overemphasized; so-called failure of magnesium therapy is generally due to failure to attain systemic alkalinization, without which magnesium is both hyperexcreted in the urine and "underabsorbed" into the intracellular space. As discussed in this context, systemic pH can be assessed by measuring urine pH, which should range from 7.5 up to approximately 8.5.

---

[94] Innerarity S. Hypomagnesemia in acute and chronic illness. *Crit Care Nurs Q*. 2000 Aug;23(2):1-19
[95] Frankel H, Haskell R, Lee SY, Miller D, Rotondo M, Schwab CW. Hypomagnesemia in trauma patients. *World J Surg*. 1999 Sep;23(9):966-9
[96] Fox CH, Ramsoomair D, Mahoney MC, et al. An investigation of hypomagnesemia among ambulatory urban African Americans. *J Fam Pract*. 1999 Aug;48(8):636-9
[97] Schimatschek HF, Rempis R. Prevalence of hypomagnesemia in an unselected German population of 16,000 individuals. *Magnes Res*. 2001 Dec;14(4):283-90
[98] Boska et al. Contrasts in cortical magnesium, phospholipid and energy metabolism between migraine syndromes. *Neurology* 2002 Apr 23;58(8):1227-33
[99] "CONCLUSIONS: The high incidence of IMg2+ deficiency and the elevated ICa2+/IMg2+ ratio during menstrual migraine confirm previous suggestions of a possible role for magnesium deficiency in the development of menstrual migraine." Mauskop A, Altura BT, Altura BM. Serum ionized magnesium levels and serum ionized calcium/ionized magnesium ratios in women with menstrual migraine. *Headache* 2002 Apr;42(4):242-8
[100] "Abnormalities in serum IMg(2+) concentrations and ICa(2+)/IMg(2+) ratios were found in children with post-traumatic headaches, but total magnesium levels were normal." Marcus JC, Altura BT, Altura BM. Serum ionized magnesium in post-traumatic headaches. *J Pediatr* 2001 Sep;139(3):459-62
[101] "...disturbances in magnesium ion homeostasis may contribute to brain cortex hyperexcitability and the pathogenesis of migraine syndromes associated with neurologic symptoms." Boska et al. Contrasts in cortical magnesium, phospholipid and energy metabolism between migraine syndromes. *Neurology* 2002 Apr 23;58(8):1227-33
[102] Mazzotta G, Sarchielli P, Alberti A, Gallai V. Intracellular Mg++ concentration and electromyographical ischemic test in juvenile headache. *Cephalalgia* 1999 Nov;19(9):802-9
[103] Mishima K, et al. Platelet ionized magnesium, cyclic AMP, and cyclic GMP levels in migraine and tension-type headache. *Headache* 1997 Oct;37(9):561-4
[104] "After a prospective baseline period of 4 weeks they received oral 600 mg (24 mmol) magnesium (trimagnesium dicitrate) daily for 12 weeks or placebo… High-dose oral magnesium appears to be effective in migraine prophylaxis." Peikert et al. Prophylaxis of migraine with oral magnesium. *Cephalalgia* 1996 Jun;16(4):257-63
[105] Mauskop A, Altura BT, Cracco RQ, Altura BM. Intravenous magnesium sulfate rapidly alleviates headaches of various types. *Headache* 1996 Mar;36(3):154-60
[106] Wang et al. Oral magnesium oxide prophylaxis of frequent migrainous headache in children: a randomized, double-blind, placebo-controlled trial. *Headache*. 2003;43:601-610
[107] "We gave dexamethasone/metoclopramide to one group and magnesium sulfate to the other group, and evaluated pain severity at 20 min and at 1- and 2-h intervals after infusion. ... According to the results, magnesium sulfate was a more effective and fast-acting medication compared to a combination of dexamethasone/metoclopramide for the treatment of acute migraine headaches." Shahrami et al. Comparison of therapeutic effects of magnesium sulfate vs. dexamethasone/metoclopramide on alleviating acute migraine headache. *J Emerg Med*. 2015 Jan;48(1):69-76
[108] Sebastian et al. Estimation of the net acid load of the diet of ancestral preagricultural Homo sapiens and their hominid ancestors. *Am J Clin Nutr* 2002;76:1308-16

retention[109,110] and increases intracellular magnesium levels; alkalinization and increased intracellular magnesium levels are associated with reductions in low-back pain according to a clinical trial.[111] **Magnesium may decrease the absorption or effectiveness of several drugs**, including: Azithromycin (Zithromax), Cimetidine (Tagamet), Ciprofloxacin (Ciloxan, Cipro), Doxycycline (Atridox, Doryx, Doxy, Monodox, Periostat, Vibramycin), Famotidine (Mylanta-AR, Pepcid, Pepcid AC), Hydroxychloroquine (Plaquenil), Levofloxacin (Levaquin), Nitrofurantoin (Furadantin, Macrobid, Macrodantin), Nizatidine (Axid, Axid AR), Ofloxacin (Floxin, Ocuflox), Tetracycline (Achromycin, Sumycin, Helidac), and Warfarin (Coumadin). Misoprostol (Cytotec, Arthrotec) with magnesium may result in diarrhea. **Spironolactone (Aldactone, Aldactazide) or Amiloride (Midamor, Moduretic) may cause hypermagnesemia.**

- Oral magnesium for migraine prophylaxis. (*J Pak Med Assoc.* 2013 Feb[112]): "In this clinical trial study, effects of 500 mg/day oral magnesium oxide for migraine prophylaxis and serum magnesium concentration in 77 migrainous adults (case=33, control=44) aged 34.10±9.61 years, were assessed. Significant reduction in migraines, migraine days, headache severity and migraine index in both the groups compared with baseline, were observed. In magnesium oxide group compared with control group, 50% or greater reduction in migraines (P<0.01) and headache severity (P<0.05) were significant. ... Magnesium supplementation increased significantly (P<0.001) serum magnesium concentration while in control group no difference was seen. Considering that oral oxide magnesium supplementation resulted in positive outcomes in decreasing frequency and severity of migraine seizures without leaving any serious side effects, it seems that magnesium oxide supplementation associated with the routine treatments may be effective especially in patients with low level of serum magnesium."

- Vitamin C, ascorbate: Ascorbate promotes mitochondrial function at cytochrome c, between complexes 3 and 4. My personal hypothesis is that ascorbate provides analgesic benefits via enhancement of central dopaminergic mechanisms and via its ability to lower histamine levels[113] thereby potentially alleviating neurogenic inflammation.[114] Appreciating its safety and efficacy in treating CRPS, migraine, neuropathic and postsurgical pain[115,116], I think all patients with pain should receive ascorbate 2-6 grams daily in divided doses.

- Hydroxocobalamin (hydroxo-vitamin-B12, OH-B12): Hydroxocobalamin is a nitric oxide (NO-) scavenger and appears to benefit the majority of patients with migraine headaches; this study used OH-B12 1 mg/d via aqueous intranasal administration to obviate the need for parenteral administration.[117] If the route of administration is unimportant, then high-dose oral supplementation with 2,000-6,000 mcg/d may prove to be just as effective, according to comparable research using cyanocobalamin.[118] NO- promotes migraine by two mechanisms—induction of mitochondrial dysfunction and promotion of glutamate release—which function synergistically to promote pain amplification, central sensitization as discussed in great detail in the following section on fibromyalgia. Therefore, hydroxocobalamin's effectiveness in migraine is mediated by NO- scavenging is simultaneously mitoprotective (protective of mitochondria) and neuroprotective (protective of neurons).

> **NO- scavenging with hydroxocobalamin**
>
> "Drugs which directly counteract nitric oxide (NO), such as endothelial receptor blockers, NO-synthase inhibitors, and NO-scavengers, may be effective in the acute treatment of migraine, but are also likely to be effective in migraine prophylaxis. ...This is the first prospective, open study indicating that intranasal hydroxocobalamin may have a prophylactic effect in migraine."
>
> van der Kuy et al. Hydroxocobalamin, a nitric oxide scavenger, in the prophylaxis of migraine. *Cephalalgia*. 2002 Sep

---

[109] Sebastian et al. Improved mineral balance and skeletal metabolism in postmenopausal women treated with potassium bicarbonate. *N Engl J Med.* 1994;330(25):1776-81
[110] Tucker et al. Potassium, magnesium, and fruit and vegetable intakes are associated with greater bone mineral density in elderly men and women. *Am J Clin Nutr.* 1999;69(4):727-36
[111] "The results show that a disturbed acid-base balance may contribute to the symptoms of low back pain. The simple and safe addition of an alkaline multimineral preparate was able to reduce the pain symptoms in these patients with chronic low back pain." Vormann J, Worlitschek M, Goedecke T, Silver B. Supplementation with alkaline minerals reduces symptoms in patients with chronic low back pain. *J Trace Elem Med Biol.* 2001;15(2-3):179-83
[112] Talebi M, Goldust M. Oral magnesium; migraine prophylaxis. *J Pak Med Assoc.* 2013 Feb;63(2):286
[113] Johnston CS, Martin LJ, Cai X. Antihistamine effect of supplemental ascorbic acid and neutrophil chemotaxis. *J Am Coll Nutr.* 1992 Apr;11(2):172-6
[114] Rosa AC, Fantozzi R. The role of histamine in neurogenic inflammation. *Br J Pharmacol.* 2013 Sep;170(1):38-45
[115] Hasanzadeh Kiabi et al. Can vitamin C be used as an adjuvant for managing postoperative pain? A short literature review. *Korean J Pain.* 2013 Apr;26(2):209-10
[116] Mohseni M. Use of vitamin C as placebo in anesthesiology. *Anesth Pain Med.* 2013 Winter;2(3):141
[117] van der Kuy PH, et al. Hydroxocobalamin, a nitric oxide scavenger, in the prophylaxis of migraine: an open, pilot study. *Cephalalgia*.2002; 22:513–519
[118] "In cobalamin deficiency, 2 mg of cyanocobalamin administered orally on a daily basis was as effective as 1 mg administered intramuscularly on a monthly basis and may be superior." Kuzminski et al. Effective treatment of cobalamin deficiency with oral cobalamin. *Blood* 1998 Aug 15;92(4):1191-8 bloodjournal.org/cgi/content/full/92/4/1191

- Folic acid in the form of folinic acid or methylfolate ("5-methyltetrahydrofolate" or "5-MTHF"): Most nutrition-knowledgeable doctors do not use "folic acid" in the form of folic acid due to concerns about increased free radical generation and possible increased risk of cellular damage and malignant disease; we still use the term "folic acid" but nowadays this is—in practice—meant to imply the use of either folinic acid or methylfolate, two forms of folic acid that are considered safer, if not also more effective. Strictly speaking, "folic acid" refers to the synthetic form of the vitamin, whereas "folate" refers to derivatives of tetrahydrofolate that are found in food, especially leafy green vegetables, of which most people do not consume a sufficient amount. Some people develop antibodies against the folic acid transporter (cerebral folate receptor autoantibodies) that facilitates entry of folate into the brain, and they must receive either folinic acid or methylfolate to avoid neurologic devastation due to cerebral folate deficiency, wherein blood/serum levels of folate are normal but the brain (on the other side of the "wall" formed by the blood-brain barrier) is starved for this nutrient.[119] Folic acid from diet and/or supplementation serves many roles and thereby provides numerous benefits, largely centered on the provision of single-carbon methyl groups for metabolic processes (e.g., homocysteine metabolism) and DNA methylation, which regulates/suppresses gene transcription and thereby reduces risk of cancer and viral activation (e.g., cervical cancer following exposure to the human papilloma virus [HPV][120]). In this conversation, we are primarily concerned with optimizing folate intake to optimize "neurologic function" (i.e., generally speaking: normalization of homocysteine-mediated NMDAr activation in the brain, spinal cord and periphery) by reducing homocysteine levels because elevated homocysteine levels will cause excessive pain/fatigue/depression due to activation of the NMDA-receptor, mostly in the brain but also in the periphery. The most important nutrients for reducing homocysteine are folate (vitamin B9), pyridoxine (vitamin B6), cobalamin (vitamin B12) and the amino acid N-acetyl-cysteine (NAC); some people have a defect in their ability to convert folate into its active form via the enzyme methylenetetrahydrofolate reductase (MTHFR) and therefore need more nutritional supplementation to push this sluggish pathway to metabolic completion and reduce/normalize homocysteine levels. Diagrams illustrating these pathways tend to be repulsively complex, immemorably curvaceous, and/or incomplete and thereby clinically valueless; the illustration below is perhaps the most simple for efficient understanding of the means by which nutritional supplementation lowers homocysteine levels.

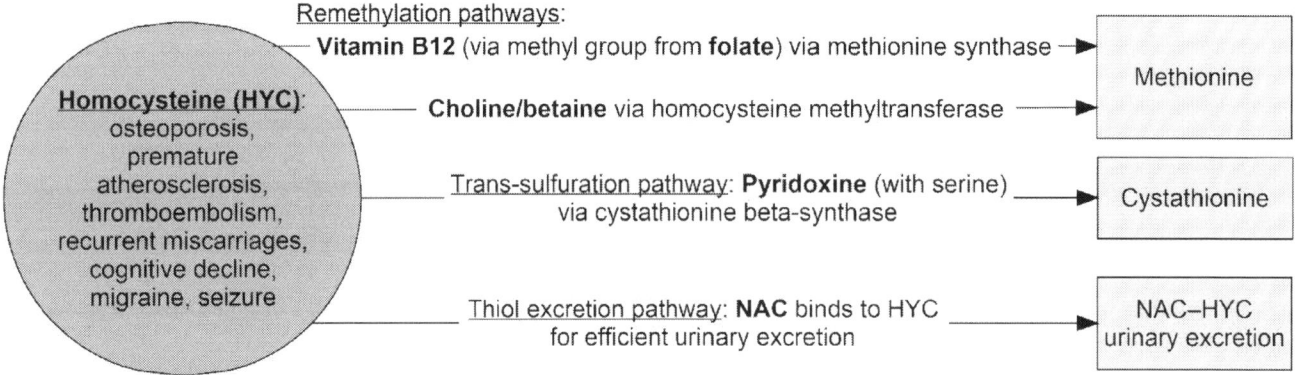

Lowering homocysteine (HYC) via nutritional supplementation: Folate gives methyl group to cobalamin (vitamin B12) to convert HYC via methionine synthase to methionine; choline/betaine can remethylate homocysteine via homocysteine methyltransferase to form methionine. Pyridoxine promotes conversion of HYC via cystathionine beta-synthase to cystathionine. The amino acid N-acetyl-cysteine (NAC) binds to HYC for efficient renal excretion of NAC-HYC.[121]

Increased consumption of folate from diet and/or supplements can alleviate depression, fatigue, and pain and is therefore recommended for all "pain patients", including those with migraine, fibromyalgia, and chronic fatigue syndrome. Adult doses of folate 1-5 mg (1,000-5,000 mcg) per day are reasonable and should be

---

[119] Gordon N. Cerebral folate deficiency. *Dev Med Child Neurol*. 2009 Mar;51(3):180-2
[120] Piyathilake et al. Indian women with higher serum concentrations of folate and vitamin B12 are significantly less likely to be infected with carcinogenic or high-risk (HR) types of human papillomaviruses (HPVs). *Int J Womens Health*. 2010 Aug 9;2:7-12
[121] "NAC intravenous administration induces an efficient and rapid reduction of plasma thiols, particularly of Hcy; our data support the hypothesis that NAC displaces thiols from their binding protein sites and forms, in excess of plasma NAC, mixed disulphides (NAC-Hcy) with a high renal clearance." Ventura et al. N-Acetyl-cysteine reduces homocysteine plasma levels after single intravenous administration by increasing thiols urinary excretion. *Pharmacol Res*. 1999 Oct;40(4):345-50

Chapter 5.1—Functional Inflammology Protocol for Metabolic Inflammation: Migraine and Fibromyalgia

coadministered with a roughly equal amount of vitamin B12. Anti-seizure drugs (especially phenytoin, carbamazepine, barbiturates[122]), some of which are used in the treatment of migraine and chronic pain, are notorious for causing folate deficiency and homocysteine elevation[123]; obviously, the drugs would paradoxically promote pain and seizure if folate deficiency develops—coadministration of folate with anti-seizure drugs should be supervised by the prescribing physician.

- Vitamin supplementation to lower homocysteine levels alleviates migraine. (*Pharmacogenet Genomics.* 2009 Jun[124]): "This was a randomized, double-blind placebo, controlled trial of 6 months of daily vitamin supplementation (i.e. 2 mg of folic acid, 25 mg vitamin B6, and 400 microg of vitamin B12) in 52 patients diagnosed with migraine with aura. Vitamin supplementation reduced homocysteine by 39% (approximately 4 mumol/l) compared with baseline, a reduction that was greater than placebo (P=0.001). Vitamin supplementation also reduced the prevalence of migraine disability from 60% at baseline to 30% after 6 months, whereas no reduction was observed for the placebo group. Headache frequency and pain severity were also reduced, whereas there was no reduction in the placebo group. In this patient group the treatment effect on both homocysteine levels and migraine disability was associated with MTHFRC677T genotype whereby carriers of the C allele experienced a greater response compared with TT genotypes."
- Higher levels of dietary folate intake reduce migraine disability and frequency. (*Headache.* 2015 Feb[125]): "A significant inverse relation was observed between dietary folate equivalent and (folic acid) FA consumption and migraine frequency. It was also observed that in individuals with the CC genotype for the methylenetetrahydrofolate reductase (MTHFR) C677T variant, migraine frequency was significantly linked to FA consumption. The results from this study indicate that folate intake in the form of FA may influence migraine frequency in female MA sufferers."

- Pyridoxine (vitamin B6) 50-250 mg/d taken with food and aided by concomitant supplementation with magnesium and riboflavin: Pyridoxine promotes the conversion, via glutamic acid decarboxylase, of neuroexcitatory glutamate into neuroinhibitory GABA, as previously discussed and diagrammed. Pyridoxine also lowers levels of homocysteine, which functions as does glutamate in excitation of the NMDA receptor; thus, pyridoxine protects brain neuron cells from excess stimulation by lowering both glutamate and homocysteine. Experimental and clinical data are both very clear that B6 has analgesic and anti-seizure benefits that are independent from its support of glutamic acid decarboxylase's conversion of glutamate to GABA; pyridoxine functions both peripherally and centrally, with those central locations including the spinal cord, deep brain structures such as the thalamus (where it inhibits neuron firing), and cortex.[126] The conversion of pyridoxine to its intermediate form pyridoxine-5-phosphate requires magnesium (Mg), as expected with phosphorylation reactions; further, magnesium modulates/reduces calcium entry following NMDAr activation, and thus magnesium has a dual effect in reducing excess NMDAr activation. Conversion of pyridoxine-5-phosphate to the fully activated form pyridoxal-5-phosphate (P5P or PLP) requires an oxidase enzyme which requires riboflavin (vitamin B2); thus, as expected, some patients respond to B6 supplementation optimally/only when B2 is coadministered.[127] Not surprisingly, some patients respond better—or exclusively—to administration of the active P5P when pyridoxine previously failed to provide efficacy[128]; thus, what should be obvious by now is that pyridoxine therapy cannot be considered inefficacious until B2, and Mg have been used concomitantly with it and/or until P5P has been used. Furthermore, pyridoxine doses for adults should be in the range of 50-250 (up to 500) mg/d while determining the proper dosage (ie, response to treatment); multiyear use of extremely high doses, e.g., 2,000 mg/d, can cause sensory nerve damage (dorsal root

---

[122] Morrell MJ. Folic Acid and Epilepsy. *Epilepsy Curr.* 2002 Mar;2(2):31-34
[123] "Patients who consume antiepileptic drugs are susceptible to high levels of homocysteine and low levels of folate in the blood." Paknahad et al. Effects of Common Anti-epileptic Drugs on the Serum Levels of Homocysteine and Folic Acid. *Int J Prev Med.* 2012 Mar;3(Suppl 1):S186-90
[124] Lea et al. Effects of vitamin supplementation and MTHFR (C677T) genotype on homocysteine-lowering and migraine disability. *Pharmacogenet Genomics.* 2009 Jun:422-8
[125] Menon et al. Effects of dietary folate intake on migraine disability and frequency. *Headache.* 2015 Feb;55(2):301-9
[126] Zimmerman M, Bartoszyk GD, Bonke D, et al. Antinociceptive properties of pyridoxine. Neurophysiological and behavioral findings. *Ann N Y Acad Sci.* 1990;585:219-30
[127] Folkers et al. Enzymology of the response of the carpal tunnel syndrome to riboflavin and to combined riboflavin and pyridoxine. *Proc Natl Acad Sci.* 1984 Nov;81(22):7076-8
[128] "We present a female infant with seizures responsive to pyridoxal phosphate but that are resistant to pyridoxine. ... It is suggested that in addition to glutamic acid decarboxylase abnormality, the path from the absorption, transportation, phosphorylation, and oxidation of pyridoxine to pyridoxal phosphate in this patient might be defective. It should be considered whether pyridoxal phosphate can be the drug of choice instead of pyridoxine in treating patients suspected of pyridoxine-dependent epilepsy to reduce failure rate and further delay in seizure control." Kuo MF, Wang HS. Pyridoxal phosphate-responsive epilepsy with resistance to pyridoxine. *Pediatr Neurol.* 2002 Feb;26(2):146-7

ganglionopathy[129]) but this is not a concern when P5P is used (because P5P is considered nontoxic relative to pyridoxine[130]) lower doses are used for shorter periods of time, especially with professional supervision and a modicum of awareness and common sense. Studies using B6 in the treatment of premenstrual syndrome have used doses of 50-500 mg/d; however, regarding dosing, we all need to appreciate the differences between clinical trials (short-term studies with close supervision), use of B6 in epilepsy (high doses are warranted to prevent death and brain damage), and long-term unsupervised use (more likely—although still generally unlikely—to result in adverse effects). Of additional note regarding dosing is the observation that pyridoxine is commonly administered in doses 5-10 mg/kg/d for infants (up to 50 mg/kg/d of P5P[131]) and of 100-500 mg/d for children and adults with the seizure disorder "pyridoxine-dependent epilepsy" (of note: to correct the unfortunate error in the naming of this condition, the name should have been "pyridoxine-responsive epilepsy" or "epilepsy of pyridoxine dependency"). Patients with B6-dependent epilepsy have a gene defect that causes accumulation of a natural substance that blocks the function of P5P; this coenzyme inhibition is overpowered via daily megadosing of B6.[132]

- Effects of pyridoxine supplementation on severity, frequency and duration of migraine attacks in migraine patients with aura. (Iran J Neurol. 2015 Apr[133]): "This double-blind randomized clinical trial study was conducted on 66 patients with migraine with aura (MA)... Patients were randomly allocated to receive either pyridoxine supplements (80 mg pyridoxine per day) or placebo.... Pyridoxine supplementation led to a significant decrease in headache severity, attacks duration, and HDR (headache diary results) compared with placebo, but was not effective on the frequency of migraine attacks. CONCLUSION: Pyridoxine supplementation in patients with MA was effective on headache severity, attacks duration and HDR, but did not affect the frequency of migraine attacks."

- Antiseizure benefit of pyridoxine administration in an infant with normal blood levels of pyridoxine and deficient brain levels of GABA (Neuropediatrics. 1992 Oct[134]): "In an infant with typical pyridoxine-dependent seizures, CSF GABA level, was determined before treatment with pyridoxine. Before onset of treatment, level of GABA in CSF was highly lowered (16 pmol/ml), pyridoxine level in serum was within normal range. Immediately after application of 80 mg pyridoxine fits stopped and the EEG was without seizure activity. The data substantiate previous findings in brain tissue from a patient with pyridoxine-dependent seizures. They are proof of a disturbed GABA metabolism in pyridoxine dependent seizures."

- Pyridoxine deficiency is extremely common in adult patients with severe epilepsy. (Epilepsy Behav. 2015 Nov[135]): "An 8-year-old girl treated at our facility for superrefractory status epilepticus was found to have a low pyridoxine level at 5microg/L. After starting pyridoxine supplementation, improvement in the EEG for a 24-hour period was seen. ... All but six [of 81] patients admitted for status epilepticus [SE] had low normal or undetectable pyridoxine levels. A selective pyridoxine deficiency was seen in 94% of patients with status epilepticus (compared to 39.4% in the outpatients) which leads us to believe that there is a relationship between status epilepticus and pyridoxine levels." Very clearly, all seizure/epilepsy patients must be tested for vitamin B6 deficiency and/or treated empirically with pyridoxine; relatedly and very importantly, several anti-seizure medications cause deficiency of folic acid and/or vitamin D—deficiency of either can promote seizure. Thus, testing for serum pyridoxine, homocysteine, and 25-OH-vitamin D should be mandatory in all seizure/epilepsy patients; empiric nutritional treatment is safe and provides collateral benefits.

- Pyridoxine lowers serum/blood glutamate levels (Am J Clin Nutr 1992 Apr[136]): "Initially, the plasma PLP concentration of the subjects was 45 ± 2 nmol/L ... and after 7 d of oral supplementation with 27 mg PN-HC1 it reached 377 nmol/L. This represented an 8.5-fold increase from the initial concentration (P < 0.0001).

---

[129] Baxter P. Pyridoxine-dependent seizures: a clinical and biochemical conundrum. *Biochim Biophys Acta.* 2003 Apr 11;1647(1-2):36-41
[130] Lewis PJ. Pain in the hand and wrist. Pyridoxine supplements may help patients with carpal tunnel syndrome. *BMJ.* 1995 Jun 10;310(6993):1534
[131] Wang HS et al. Pyridoxal phosphate is better than pyridoxine for controlling idiopathic intractable epilepsy. Arch Dis Child. 2005 May;90(5):512-5
[132] The gene defect leads to reduced activity of antiquitin which would leads to accumulation of L-alpha-aminoadipic semialdehyde (L-AASA) and its reciprocal L-alpha-piperideine 6-carboxylate (P6C), the latter of which inhibits P5P. Mills et al. Genotypic and phenotypic spectrum of pyridoxine-dependent epilepsy (ALDH7A1 deficiency). *Brain.* 2010 Jul;133(Pt 7):2148-59
[133] Sadeghi et al. Effects of pyridoxine supplementation on severity, frequency and duration of migraine attacks in migraine patients with aura. *Iran J Neurol.* 2015 Apr 4;14:74-80
[134] Kurlemann et al. Disturbance of GABA metabolism in pyridoxine-dependent seizures. *Neuropediatrics.* 1992 Oct;23(5):257-9
[135] Dave et al. Pyridoxine deficiency in adult patients with status epilepticus. Epilepsy Behav. 2015 Nov;52(Pt A):154-8
[136] Kang-Yoon SA, Kirksey A. Relation of short-term pyridoxine-HCl supplementation to plasma vitamin B-6 vitamers and amino acid concentrations in young women. *Am J Clin Nutr.* 1992 Apr;55(4):865-72

PLP concentration remained essentially unchanged as long as PN supplementation was continued; after 14 d of vitamin supplementation the plasma PLP concentration was 429 ± 16 nmol/L. The increase in plasma PLP concentrations of individuals ranged from 400% to 1400% after supplementation. The concentration of plasma glutamic acid decreased 31% in the supplemented group after 7d of supplementation and 47% after 14 d of supplementation compared with the unsupplemented group." This very important study shows that supplementation with 27mg per day of synthetic pyridoxine hydrochloride raised blood levels of the pyridoxine and the active phosphorylated form P5P—pyridoxal-5-phosphate. Important for patients with migraine, chronic pain and seizures is the fact that serum glutamate levels were reduced by 31%. The authors somewhat erroneously describe 27 mg as a "large oral dose"; most nutritional doctors comfortably and frequently use doses of 50 to 250 to 500 mg per day of pyridoxine. Also noteworthy is that 2 of 10 of the patients showed moderate elevations—not reductions—in serum glutamate. Magnesium (typical adult dose is 600 mg/d) should always be supplemented when vitamin B6 is used. Also of high importance is the fact that—although this study showed clear safety and effectiveness for lowering blood levels of glutamate—levels of glutamate that surround the brain in the CSF were not measured.

- Vitamin administration improves the analgesic efficacy of pharmacotherapy with diclofenac following knee surgery (*Drug Res* 2013 Jun[137]): "Forty eight patients programmed to total knee arthroplasty with a pain level =7 in a 1-10 cm visual analogue scale were allocated to receive a single intramuscular injection of sodium diclofenac (75 mg) alone or combined with thiamine (100 mg), pyridoxine (100 mg) and cyanocobalamin (5 mg), and the pain level was evaluated during 12 h post-injection. Diclofenac+B vitamins mixture showed a superior analgesic effect during the assessed period and also a better assessment of the pain relief perception by patients than diclofenac alone."

- Pyridoxine (vitamin B6) alleviates neuropsychiatric aspects of premenstrual syndrome (*J R Coll Gen Pract.* 1989 Sep[138]): "A randomized double-blind crossover trial was conducted to study the effects of pyridoxine (vitamin B6) at a dose of 50 mg per day on symptoms characteristic of the premenstrual syndrome. ...In these women a significant beneficial effect (P less than 0.05) of pyridoxine was observed on emotional type symptoms (depression, irritability and tiredness)."

* Vitamin D3 (cholecalciferol): All patients with persistent pain must be tested for non-optimal vitamin D status and empirically treated with vitamin D3 to optimize serum vitamin D.[139,140] Failure to assess and correct vitamin D deficiency and implement effective correction in patients with persistent pain is medical-professional negligence; the data is very clear on the induction of chronic and often debilitating pain by vitamin D deficiency and the merits of vitamin D supplementation in alleviating pain. Several case reports have documented the effectiveness of vitamin D supplementation in the treatment and prevention of migraine.[141,142] Vitamin D is

Excess vitamin D
> 100 ng/mL (250 nmol/L) with hypercalcemia

Optimal range
50 - 100 ng/mL (125 - 250 nmol/L)

Insufficiency range
< 20- 40 ng/mL (50 - 100 nmol/L)

Deficiency
< 20 ng/mL (50 nmol/L)

Image © 2004-2015 by Vasquez A in "**Functional Inflammology, volume 1**" published 2014 and "**Inflammation Mastery, 4th Edition**" published 2015. See InflammationMastery.com/reprints for Dr Vasquez's original paper Vasquez et al. The clinical importance of vitamin D (cholecalciferol): a paradigm shift with implications for all healthcare providers. *Altern Ther.Health Med.* 2004 Sep-Oct

---

[137] Magaña-Villa et al. B-vitamin mixture improves the analgesic effect of diclofenac in patients with osteoarthritis: a double blind study. *Drug Res.* 2013 Jun;63(6):289-92
[138] Doll H, Brown S, Thurston A, Vessey M. Pyridoxine (vitamin B6) and the premenstrual syndrome: a randomized crossover trial. *J R Coll Gen Pract.* 1989 Sep;39(326):364-8
[139] Moore D, Wahl R, Levy P. Hypovitaminosis D presenting as diffuse myalgia in a 22-year-old woman: a case report. *J Emerg Med.* 2014 Jun;46(6):e155-8
[140] "The findings suggest a role of low vitamin D levels for heightened central sensitivity, particularly augmented pain processing upon mechanical stimulation in chronic pain patients." von Känel et al. Vitamin D and central hypersensitivity in patients with chronic pain. *Pain Med.* 2014 Sep;15(9):1609-18
[141] "Therapeutic replacement with vitamin D and calcium resulted in a dramatic reduction in the frequency and duration of their migraine headaches." Thys-Jacobs S. Alleviation of migraines with therapeutic vitamin D and calcium. *Headache.* 1994 Nov-Dec;34(10):590-2
[142] "These observations suggest that vitamin D and calcium therapy should be considered in the treatment of migraine headaches." Thys-Jacobs S. Vitamin D and calcium in menstrual migraine. *Headache.* 1994 Oct;34(9):544-6

anti-inflammatory (including reductions in glial activation) and immunomodulatory[143] and also modulates vascular tone by reducing intracellular hypercalcinosis.[144] Reasonable replacement doses are 2,000 IU per day for children and 4,000 IU per day for adults; monitoring serum calcium ensures safety. Optimal vitamin D status correlates with serum 25(OH)D levels of 50-100 ng/mL, or 125-250 nmol/L—see our review article for more details[145]; levels greater than 100 ng/mL are unnecessary and increase the risk of hypercalcemia.

- 5-Hydroxytryptophan (5-HTP)—typical dose is 50-300 mg per day in divided doses: 5-Hydroxytryptophan (5-HTP) is a natural constituent of the human body and is also found in some plants and is thus available as a nutritional supplement. Altered serotonin metabolism has been observed in headache patients, and this observation serves to support the use of selective serotonin reuptake inhibitors (SSRIs) in headache patients, while supplementation with 5-HTP increases serotonin levels naturally. Among various types of headache, migraine would be expected to show the best response to 5-HTP because the conversion of serotonin to melatonin would extend the benefits of serotonin-mediated analgesia to include the protection of mitochondrial function, an important benefit of melatonin.

> **Tryptophan/serotonin insufficiency & glial activation: common in migraine, chronic pain, and fibromyalgia**
>
> "The recently shown **high prevalence of migraine in the population of fibromyalgia sufferers**, suggests a common ground shared by fibromyalgia and migraine. Migraine has been demonstrated to be characterized by a defect in the serotonergic and adrenergic systems. A parallel dramatic failure of serotonergic systems and a defect of adrenergic transmission have been evidenced to affect fibromyalgia sufferers, too."
>
> Nicolodi M, Sicuteri F. Fibromyalgia and migraine, two faces of the same mechanism. Serotonin as the common clue. *Adv Exp Med Biol*. 1996;398:373-9

  Conversion of serotonin to melatonin occurs in non-migraine headaches but is of lesser therapeutic importance since these disorders (cluster headaches excepted) are not associated directly with mitochondrial dysfunction. 5-HTP has a better safety and efficacy profile than does the drug methysergide in the treatment of migraine according to a study of 124 adults and children with migraine.[146]

- Vitamin E supplementation with mixed tocopherols 400-1,200 IU/d: As noted previously, free radicals and prostaglandins contribute to migraine pathophysiology; both are reduced by administration of supplemental doses of vitamin E.[147] Vitamin E—known chemically as tocopherol—is present in several forms, arguably the most important of which is the gamma form; most nutritionally competent clinicians generally recommend that vitamin E supplementation contain approximately 40% gamma tocopherol. Another form of vitamin E with more specificity for enhancing/protecting mitochondrial function is known as tocopherol succinate; a combination of various tocopherols is reasonable and is likely to produce enhanced therapeutic efficacy. Isoprostanes are lipid-derrived mediators produced in direct proportion to free radical burden; isoprostanes directly trigger pain and their production is inhibited by antioxidant protection, especially with vitamin E.

    - Vitamin E for the treatment of menstrual migraine. (*Med Sci Monit.* 2009 Jan[148]): "During a placebo-controlled double-blinded trial, 72 women with menstrual migraine received placebo (identical in appearance to vitamin E) daily for five days, two days before to three days after menstruation for two cycles followed by a one-month wash-out and one vitamin E softgel (400 IU) daily for five days in the next two cycles. ... There were statistically significant differences in the pain severity and functional disability scales between the placebo and the vitamin E treatments. Vitamin E effect was also superior to placebo regarding photophobia, phonophobia, and nausea. CONCLUSIONS: Vitamin E is effective in relieving symptoms due to menstrual migraine."

- Combination nutritional supplementation: Patients with migraine show multiple abnormalities in metabolism, inflammation, and oxidative stress; as expected therefore, multi-nutrient supplementation shows benefits via numerous mechanisms.

---

[143] Timms et al. Circulating MMP9, vitamin D and variation in the TIMP-1 response with VDR genotype. *QJM*. 2002 Dec;95(12):787-96
[144] Vasquez A. Intracellular Hypercalcinosis. *Naturopathy Digest* 2006 September naturopathydigest.com/archives/2006/sep/vasquez.php
[145] Vasquez et al. The Clinical Importance of Vitamin D (Cholecalciferol). *Alternative Therapies in Health Med* 2004; 10: 28-37. ichnfm.academia.edu/AlexVasquez
[146] "The most beneficial effect of 5-HTP appears to be felt with regard to the intensity and duration rather than the frequency of the attacks... These results suggest that 5-HTP could be a treatment of choice in the prophylaxis of migraine." Titus et al. 5-Hydroxytryptophan versus methysergide in prophylaxis of migraine. *Eur Neurol*.1986; 25:327-329
[147] Shaik MM, Gan SH. Vitamin supplementation as possible prophylactic treatment against migraine with aura and menstrual migraine. *Biomed Res Int*. 2015;2015:469529
[148] Ziaei S, Kazemnejad A, Sedighi A. The effect of vitamin E on the treatment of menstrual migraine. *Med Sci Monit*. 2009 Jan;15(1):CR16-9

Chapter 5.1—Functional Inflammology Protocol for Metabolic Inflammation: Migraine and Fibromyalgia

- <u>Alleviation of migraine symptoms with a supplement containing riboflavin, magnesium, Q10, and other nutrients. (*J Headache Pain*. 2015 Dec[149])</u>: "130 adult migraineurs (age 18 - 65 years) with ≥ three migraine attacks per month were randomized into two treatment groups: dietary supplementation or placebo in a double-blind fashion." The product contained "400 mg riboflavin (vitamin B2), 600 mg magnesium, 150 mg coenzyme Q10 along with a multivitamin/trace elements combination per 4 capsules. The amount of additional multivitamin/trace elements per 4 capsules is as follows: 750 mcg vitamin A, 200 mg vitamin C, 134 mg, vitamin E, 5 mg thiamin, 20 mg niacin, 5 mg vitamin B6, 6 mcg vitamin B12, 400 mcg folic acid, 5 mcg vitamin D, 10 mg pantothenic acid, 165 mcg biotin, 0.8 mg iron, 5 mg zinc, 2 mg manganese, 0.5 mg copper, 30 mcg chromium, 60 mcg molybdenum, 50 mcg selenium, 5 mg bioflavonoids." "Migraine days per month declined from 6.2 days during the baseline period to 4.4 days at the end of the treatment with the supplement and from 6.2.days to 5.2 days in the placebo group. The intensity of migraine pain was significantly reduced in the supplement group compared to placebo. The sum score of the HIT-6 questionnaire was reduced by 4.8 points from 61.9 to 57.1 compared to 2 points in the placebo-group. The evaluation of efficacy by the patient was better in the supplementation group compared to placebo."

- <u>Use of a pine bark extract and antioxidant vitamin combination product as therapy for migraine refractory to pharmacologic medication. (*Headache*. 2006 May[150])</u>: "Twelve patients with a long-term history of migraine with and without aura who had failed to respond to multiple treatments with beta-blockers, antidepressants, anticonvulsants, and 5-hydroxytryptamine receptor agonists were selected for the study. They were treated with 10 capsules of an antioxidant formulation of 120 mg pine bark extract, 60 mg vitamin C, and 30 IU vitamin E in each capsule daily for 3 months. ... There was a significant mean improvement in migraine disability assessment (MIDAS) score of 50.6% for the 3-month treatment period compared with the 3 months prior to baseline. The treatment was also associated with significant reductions in number of headache days and headache severity score. Mean number of headache days was reduced from 44.4 days at baseline to 26.0 days after 3 months' therapy and mean headache severity was reduced from 7.5 of 10 to 5.5. CONCLUSION: These data suggest that the antioxidant therapy used in this study may be beneficial in the treatment of migraine possibly reducing headache frequency and severity."

Infections & Dysbiosis Patients with migraine have a higher-than-average prevalence of gastric infection with *H. pylori*, and significant symptomatic improvement is obtained following eradication of *H pylori* in these patients, according to two studies[151,152] and refuted by two others.[153,154] As usual, gastrointestinal dysbiosis should be assessed and corrected on a *per patient* (rather than *per disease*) basis—see Chapter 4 of *Integrative Rheumatology / Functional Inflammology* for details and interventions. For patients with recalcitrant headaches (and for those seeking comprehensive whole-patient health care), assessment of digestion, absorption, and gastrointestinal microecology is a reasonable component of evaluation that can help guide treatment. Although *Helicobacter pylori* is a common inhabitant of the human gastrointestinal tract (found in more than 50% of Americans over age 50), immunologic responses to the organism can range from nonreactive on one end of the spectrum to diverse diseases like chronic gastritis, chronic urticaria, autoimmune thrombocytopenia, or reactive arthritis on the more severe and systemic end of the spectrum. Thus, the host-microbe relationship is of greater significance than the identity and microbiological characteristics of the microbe. As a Gram-negative bacterium that produces endotoxin (LPS), the obvious means by which *H pylori* can contribute to migraine is via LTR-4 mediated systemic inflammation and microglial activation.

---

[149] Gaul et al. Improvement of migraine symptoms with a proprietary supplement containing riboflavin, magnesium and Q10: a randomized, placebo-controlled, double-blind, multicenter trial. *J Headache Pain*. 2015 Dec;16:516

[150] Chayasirisobhon S. Use of a pine bark extract and antioxidant vitamin combination product as therapy for migraine in patients refractory to pharmacologic medication. *Headache*. 2006 May;46(5):788-93

[151] "H. pylori is common in subjects with migraine. Bacterium eradication causes a significant decrease in attacks of migraine. The reduction of vasoactive substances produced during infection may be the pathogenetic mechanism underlying the phenomenon." Gasbarrini et al. Beneficial effects of Helicobacter pylori eradication on migraine. *Hepatogastroenterology*. 1998 May-Jun;45(21):765-70

[152] "Helicobacter pylori should be examined in migranous patients and eradication of the infection may be helpful for the treatment of the disease." Tunca et al. Is Helicobacter pylori infection a risk factor for migraine? A case-control study. *Acta Neurol Belg*. 2004 Dec;104(4):161-4

[153] "Our study suggests that chronic Helicobacter pylori infection is not more frequent in patients with migraine than in controls and that infection does not modify clinical features of the disease." Pinessi et al. Chronic Helicobacter pylori infection and migraine: a case-control study. *Headache*. 2000 Nov-Dec;40(10):836-9

[154] "In conclusion, our results do not support any specific correlation between Hp infection and migraine." Ciancarelli et al. Helicobacter pylori infection and migraine. *Cephalalgia*. 2002 Apr;22(3):222-5

Nutritional Immunomodulation Migraine is most essentially viewed as a combination of mitochondrial dysfunction and glial activation—those two primary components are fundamental, consistent, and explanatory of nearly all other permutations— resulting in physiologic fragility and sensitivity toward (intolerance of) otherwise minor "subclinical" or "subsymptomatic" stressors such as dietary, hormonal, and environmental stressors. With regard to immunohyperresponsiveness—specifically in this case the glial activation, treatments to reduce neuroinflammation should be implemented, such as vitamin D, anti-inflammatory polyphenolics, resveratrol, the n3 fatty acids EPA and DHA, etc. For patients with overt systemic inflammation, allergy, and/or autoimmunity, a more comprehensive antiinflammatory protocol can and should be implemented, especially 1) identifying and removing the inflammatory triggers, and 2) promoting immunotolerance with the complete nutritional immunomodulation protocol outlined in Chapter 4, Section 3. Elevated homocysteine can promote microglial activation, and reducing homocysteine levels with nutritional supplementation is indicated for hyperhomocysteinemic migraineurs as described in protocol component #5: *Style of Living and Special Considerations*.

Dysmetabolism & Mitochondrial Dysfunction We must appreciate that migraine is a multifaceted phenomenon with neuroemotional, structural, allergic-immunologic-inflammatory, and mitochondrial components. With regard to the latter, we can understand migraine from the perspective of defects in the mitochondrial electron transport chain (ETC), namely **NADH-dehydrogenase**, **citrate synthase** and **cytochrome-c-oxidase**; defects in **NADH-cytochrome-c-reductase** appear to be specific to migraine with aura.[155] Thus, not surprisingly, nutrients which are intimately involved with these steps of the ETC have shown impressive efficacy in the treatment and prevention of migraine via the Le Chatelier principle which states—from the perspective of orthomolecular nutrition—that metabolic defects can be compensated for by the administration of supraphysiologic quantities of nutrients to push defective pathways toward completion; this is a main component of Linus Pauling's orthomolecular concept[156] (ie, that using the right molecules in the right amounts can optimize health by optimizing biochemistry and metabolism, as most recently and authoritatively reviewed by his colleague Bruce Ames in a masterful review.[157] In the case of migraine, we are bypassing or compensating for defects in mitochondrial function by supplying supraphysiologic doses of the nutrients involved in those pathways with the end result being enhancement/restoration/normalization/optimization of mitochondrial function. In the world of nutritional medicine, we would probably not be familiar with these concepts were it not for the independent and synergistic works of Roger Williams, Linus Pauling, Jeff Bland, and Bruce Ames; to these men do we owe gratitude for our understanding of this phenomenon and its clinical application. Furthermore however, we must also appreciate that nutrients have numerous functions and *affect* and *effect* numerous (not singular) pathways and processes, such that a "mitochondrial nutrient" may exert its action via a *non-mitochondrial* effect, as mentioned per nutrient in the itemized section that follows:

> **The evidence of mitochondrial dysfunction in migraine is strong, consistent and irrefutable**
> 1. Biochemical evidence: High intracellular calcium, excessive production of free radicals, low activity of superoxide dismutase, activation of cytochrome-c oxidase and nitric oxide, high levels of lactate and pyruvate, and low ratios of phosphocreatine-inorganic phosphate and "deficient oxidative phosphorylation, which ultimately causes energy failure in neurons and astrocytes, thus triggering migraine mechanisms, including spreading depression."
> 2. Cellular, histologic evidence: Muscle biopsy shows ragged red fibers, accumulation of giant mitochondria with paracrystalline inclusions
> 3. Genetic evidence: Various mitochondrial DNA polymorphisms/mutations have been demonstrated
> 4. Therapeutic evidence: "Several agents that have a positive effect on mitochondrial metabolism have shown to be effective in the treatment of migraines. The agents include riboflavin (B2), coenzyme Q10, magnesium, niacin, carnitine, topiramate, and lipoic acid."
>
> Yorns WR Jr, Hardison HH. Mitochondrial dysfunction in migraine. *Semin Pediatr Neurol*. 2013 Sep;20:188-93

- CoQ10—100-400 mg per day: CoQ10 supplementation significantly reduces migraine headache frequency, duration, and intensity.[158] As shown in the diagram below, CoQ10 shuttles electrons from Complex 1 to Complex 2, and from "Complex 2" to Complex 3. Thus, CoQ10 supplementation helps to bypass defects in the electron transport chain (ETC) of mitochondria to promote, preserve, and protect optimal cellular energy/ATP

---

[155] "NADH-dehydrogenase, citrate synthase and cytochrome-c-oxidase activities in both patient groups were significantly lower than in controls, while NADH-cytochrome-c-reductase activity was reduced in migraine with aura." Sangiorgi et al. Abnormal platelet mitochondrial function in migraine with and without aura. *Cephalalgia* 1994 Feb; 21-3
[156] Pauling L. Orthomolecular psychiatry. Varying concentrations of substances normally present in human body may control mental disease. *Science*. 1968 Apr 19;160:265-71
[157] Ames et al. High-dose vitamin therapy stimulates variant enzymes with decreased coenzyme binding affinity (increased K(m)). *Am J Clin Nutr*. 2002 Apr;75(4):616-58
[158] Rozen et al. Open label trial of coenzyme Q10 as a migraine preventive. *Cephalalgia* 2002;22(2):137-41

production. Furthermore, research by Folkers et al[159] strongly suggests that CoQ10 has an anti-allergy and immunomodulatory role; thus the anti-migraine benefits of CoQ10 may be mediated via immunomodulation in addition to enhancement of mitochondrial function.

- <u>Riboflavin (vitamin B2)—50-400 mg per day</u>: Flavin adenine dinucleotide (FAD) is required at Complex 2 of the ETC. High-dose vitamin B2 shows "high efficacy, excellent tolerability, and low cost" in the prevention of migraine headaches.[160] The standard dose for riboflavin in the treatment of migraine is 400 mg taken orally each morning; identical or lower doses can be used in children and/or when riboflavin is used with other nutrients such as CoQ10, magnesium, etc. Riboflavin is safe and effective for long-term use in children and adults[161]; riboflavin is also effective in the long-term treatment of the migraine variant condition known as cyclic vomiting syndrome (CVS) in children.[162] So-called "B vitamins" such as riboflavin—as with zinc and iron—should always be taken with food to avoid the nausea that commonly occurs when vitamins are taken on an empty stomach; other vitamins such as vitamins A, D, and E are generally well tolerated on an empty stomach.
    - <u>Riboflavin prophylaxis in pediatric and adolescent migraine. (J Headache Pain. 2009 Oct[163])</u>: "This retrospective study reports on our experience of using riboflavin for migraine prophylaxis in 41 pediatric and adolescent patients, who received 200 or 400 mg/day single oral dose of riboflavin for 3, 4 or 6 months. ... In conclusion, riboflavin seems to be a well-tolerated, effective, and low-cost prophylactic treatment in children and adolescents suffering from migraine."
- <u>Acetyl-L-Carnitine—1,000-2,000 mg taken twice daily between meals</u>: The amino acid L-carnitine is necessary for fatty acid transport into the mitochondria for oxidative metabolism and energy production. Deficiency of or metabolic inability to use carnitine can precipitate or perpetuate migraine headaches, which can be alleviated by carnitine supplementation.[164] As a natural component of the human diet, carnitine has a wide safety margin, and supplemental doses of 2-4 g/d are commonly used; acetyl-L-carnitine is generally the preferred form. Carnitine and acetyl-carnitine provide best benefit when combined with other nutrients, especially CoQ10, magnesium, and lipoic acid. A study in 2015 using 3 grams per day of acetyl-carnitine as monotherapy showed no benefit in the treatment of migraine[165]; one possibility is that 3 grams per day of acetyl-carnitine as monotherapy is an excessive dose. A study using magnesium 500 mg/d and/or carnitine 500 mg/d showed benefit in all groups.[166]
- <u>Lipoic acid—300 mg twice-thrice daily</u>: Lipoic (thiotic) acid is an essential component of the pyruvate dehydrogenase complex, is a potent antioxidant, and it also inhibits NFkB-mediated inflammation. A placebo-controlled clinical trial showed benefit of lipoic acid (600 mg/d) supplementation in migraine patients.[167]
- <u>Melatonin—5-20 mg at night</u>: As mentioned previously, melatonin is a potent protector of mitochondrial function, which has been demonstrated in experimental models of mitochondrial inhibition by bacterial endotoxin. As a powerful antioxidant, melatonin scavenges oxygen and nitrogen-based reactants generated in mitochondria and thereby limits the loss of the intramitochondrial glutathione; this prevents mitochondrial protein and DNA damage. Melatonin increases the activity of Complexes 1 and 4 of the ETC, promoting mitochondrial ATP synthesis under various physiological/experimental conditions.[168]
- <u>Niacin, including inositol hexaniacinate and niacinamide—dose varies per type used</u>: High-dose niacin alleviates migraine headaches and headaches of various etiologies, whether administered orally,

---

[159] Ye CQ, Folkers K, et al. A modified determination of coenzyme Q10 in human blood and CoQ10 blood levels in diverse patients with allergies. *Biofactors*. 1988 Dec;1:303-6
[160] Schoenen J, Jacquy J, Lenaerts M. Effectiveness of high-dose riboflavin in migraine prophylaxis. A randomized controlled trial. *Neurology* 1998;50(2):466-70
[161] Sherwood M, Goldman RD. Effectiveness of riboflavin in pediatric migraine prevention. *Can Fam Physician*. 2014 Mar;60(3):244-6
[162] "They received prophylactic monotherapy with riboflavin for at least 12 months. Excellent response and tolerability was observed." Martinez-Esteve Melnikova et al. Riboflavin in cyclic vomiting syndrome: efficacy in three children. *Eur J Pediatr*. 2015 Jul 31. [Epub ahead of print]
[163] Condò et al. Riboflavin prophylaxis in pediatric and adolescent migraine. *J Headache Pain*. 2009 Oct;10(5):361-5
[164] Kabbouche et al. Carnitine palmityltransferase II (CPT2) deficiency and migraine headache: two case reports. *Headache*. 2003 May;43(5):490-5
[165] "After a four-week run-in-phase, 72 participants were randomized to receive either placebo or 3 g acetyl-l-carnitine for 12 weeks. ...In this triple-blind crossover study no differences were found in headache outcomes between acetyl-l-carnitine and placebo. Our results do not provide evidence of benefit for efficacy of acetyl-l-carnitine as prophylactic treatment for migraine." Hagen et al. Acetyl-l-carnitine versus placebo for migraine prophylaxis: A randomized, triple-blind, crossover study. *Cephalalgia*. 2015 Oct;35(11):987-95
[166] "In this clinical trial, 133 migrainous patients were randomly assigned into three intervention groups: magnesium oxide (500 mg/day), L-carnitine (500 mg/day), and Mg-L-carnitine (500 mg/day magnesium and 500 mg/day L-carnitine), and a control group. .. Oral supplementation with magnesium oxide and L-carnitine and concurrent supplementation of Mg-L-carnitine besides routine treatments could be effective in migraine prophylaxis; however, larger trials are needed to confirm these preliminary findings." Tarighat Esfanjani et al. The effects of magnesium, L-carnitine, and concurrent magnesium-L-carnitine supplementation in migraine prophylaxis. *Biol Trace Elem Res*. 2012 Dec;150(1-3):42-8
[167] Magis D, Ambrosini A, et al. A randomized double-blind placebo-controlled trial of thioctic acid in migraine prophylaxis. *Headache*. 2007 Jan;47(1):52-7
[168] León J, Acuña-Castroviejo D, Escames G, Tan DX, Reiter RJ. Melatonin mitigates mitochondrial malfunction. *J Pineal Res*. 2005 Jan;38(1):1-9

intramuscularly, or intravenously; niacin can also be used to halt acute migraine attacks.[169,170] Niacinamide adenine dinucleotide (NADH) is an essential component of the first stage (Complex 1) of the ETC, a step that is commonly defective in migraine patients. High-dose niacin facilitates this step and thus enhances energy production. Another anti-migraine benefit of high-dose niacin is its sparing effect on tryptophan, allowing its conversion to serotonin. Niacin also has a vasodilating action and may thereby address the vasculogenic component of headache. Efficacious oral doses of niacin can range from 300 to 1500 mg/d; lower doses are used for children. High-dose niacin, particularly in time-released tablets, presents some risk for hepatic damage, and thus safer forms of niacin such as plain niacin, slow-release niacin (e.g., Niaspan®), and inositol hexaniacinate are preferred; niacinamide and NADH might also be efficacious but neither has vasodilating actions provided by the other forms of niacin. Doses of niacin exceeding 500 to 1000 mg/d are probably unnecessary in headache patients if other treatments such as coenzyme Q10 (CoQ10), vitamin D, and fatty acids are being used; before implementing high-dose niacin, patients should be selected, informed, and monitored appropriately.

- Oxygen: One-hundred percent oxygen delivered by facial mask at 8 L/min for 10 minutes can help abort an attack of cluster headache. Oxygen is the required electron and proton acceptor of the mitochondrial electron transport chain (ETC) for ATP production; thus, supraphysiologic oxygen, like supraphysiologic doses of mitochondria-specific nutrients, generally improves mitochondrial energy (ATP) production. The model of pain sensitization in so-called "chronic pain syndromes"—specifically migraine, cluster headache, fibromyalgia, myofascial pain syndrome and complex regional pain syndrome (CRPS)—that I have developed includes a tripartite vicious cycle of mitochondrial dysfunction, glial activation, and neuronal hyperexcitation, all of which promote brain inflammation, central sensitization, and perpetuation and amplification of pain. Strong support for this model comes from both its biochemical and physiologic rationale as well as the efficacy of corresponding treatments that address each of the main components: mitochondrial dysfunction, glial activation, neuronal hyperexcitation, brain inflammation.

    Appreciation of the efficacy of oxygen therapy—whether as normobaric (more available and affordable) or hyperbaric (more effective and more expensive; hyperbaric oxygen therapy [HBOT])—in various pain states invites us to revisit the naturopathic profession's hierarchy of therapeutics. Symptomatic therapy clearly has a role in patient care but, in order to avoid repeated use of urgent/emergency care and the creation of unnecessary medical dependency, repeated acute care or even "maintenance therapy" should never replace treatment of the underlying cause(s). Patients need to receive treatment aimed at the underlying pathophysiology so that health is optimized and patients are moved toward better general well-being and disease-specific health. Oxygen therapy is abortive of pain and shows some contribution to breaking the vicious cycles of mitochondrial impairment (immediate treatment of headaches, current-prospective treatment of fibromyalgia and CRPS) but this therapy does not address the other aspects of mitochondrial dysfunction (e.g., CoQ10 deficiency) and does not address the cause (e.g., small intestine bacterial overgrowth [SIBO] in fibromyalgia) and therefore should remain as supplemental, abortive/acute, and adjunctive therapy not as the foundation of therapy. Obviously, hyperbaric therapy makes more money for doctors/clinics and—(except/including) when patients buy their own home hyperbaric units—will therefore receive more press and more endorsement than therapies that are curative and empowering (ie, autonomous, without medical dependence). Given that most patients with persistent inflammation and pain are antioxidant deficient and therefore at increased risk for oxygen toxicity (including subclinical damage), antioxidant repletion should occur prior to oxygen therapy; the idea of administering supraphysiologic doses of oxygen to pull more protons and electrons through an *already damaged* and *pro-inflammatory* and *ROS-generating* electron transport chain is not meritorious, although some will defend it—weakly—on the theoretic basis of hormesis.

    - High-flow oxygen therapy for all types of headache (*Am J Emerg Med* 2012 Nov[171]): "We performed a prospective, randomized, double-blinded, placebo-controlled trial of patients presenting to the ED with a chief complaint of headache. The patients were randomized to receive either 100% oxygen via nonrebreather mask at 15 L/min or the placebo treatment of room air via nonrebreather mask for 15 minutes in total. ... A total of 204 patients agreed to participate in the study and were randomized to the oxygen

---

[169] Velling DA, Dodick DW, Muir JJ. Sustained-release niacin for prevention of migraine headache. *Mayo Clin Proc*. 2003 Jun;78(6):770-1
[170] Prousky J, Seely D. The treatment of migraines and tension-type headaches with intravenous and oral niacin (nicotinic acid). *Nutr J*. 2005 Jan 26;4:3
[171] Ozkurt et al. Efficacy of high-flow oxygen therapy in all types of headache: a prospective, randomized, placebo-controlled trial. *Am J Emerg Med*. 2012 Nov;30(9):1760-4

(102 patients) and placebo (102 patients) groups. Patient headache types included tension (47%), migraine (27%), undifferentiated (25%), and cluster (1%). Patients who received oxygen therapy reported significant improvement in visual analog scale scores at all points when compared with placebo: 22 mm vs 11 mm at 15 minutes, 29 mm vs 13 mm at 30 minutes, and 55 mm vs 45 mm at 60 minutes. ... In addition to its role in the treatment of cluster headache, high-flow oxygen therapy may provide an effective treatment of all types of headaches in the ED setting.

- Oxygen (normobaric/hyperbaric) for migraine (*Headache* 1995 Apr[172]): "The purpose of this study was to compare the effects of hyperbaric oxygen and normobaric oxygen in migraine. Twenty migraineurs were divided randomly into two groups and studied in a hyperbaric chamber during a typical headache attack. ... One group received 100% oxygen at 1 atmosphere of pressure (normobaric) while the other received 100% oxygen at 2 atmospheres of pressure (hyperbaric). One of the 10 patients in the normobaric group achieved significant relief of headache symptoms, while 9 of 10 in the hyperbaric group found relief. Based on a chi-square test, this difference is significant at the $P < .005$ level. Those patients who did not find significant relief from normobaric oxygen were given hyperbaric oxygen as above. All nine found significant relief. The results suggest that hyperbaric (but not normobaric) oxygen may be useful in the abortive management of migraine headache. Possibilities for the mechanism of this effect, in addition to vasoconstriction, include an increase in the rate of energy-producing and neurotransmitter-related metabolic reactions in the brain which require molecular oxygen."

- Hyperbaric oxygen in the treatment of migraine with aura. (*Headache* 1998 Feb[173]): "Female subjects with confirmed migraine were randomly assigned to begin with either the control (100% oxygen, no pressure) or hyperbaric treatment (100% oxygen, pressure). ... Results suggest that hyperbaric oxygen treatment reduces migraine headache pain..."

- Minimal prophylactic/preventive effect of hyperbaric oxygen therapy on migraine. (*Cephalalgia* 2004 Aug[174]): Not surprisingly, this study found that prophylactic administration of oxygen was generally inefficacious in reducing future migraine attacks; while patients with migraine always have a basal level of mitochondrial dysfunction, oxygen administration does not prevent future attacks. Oxygen is of main value in the treatment of active migraine (and other headache) attacks.

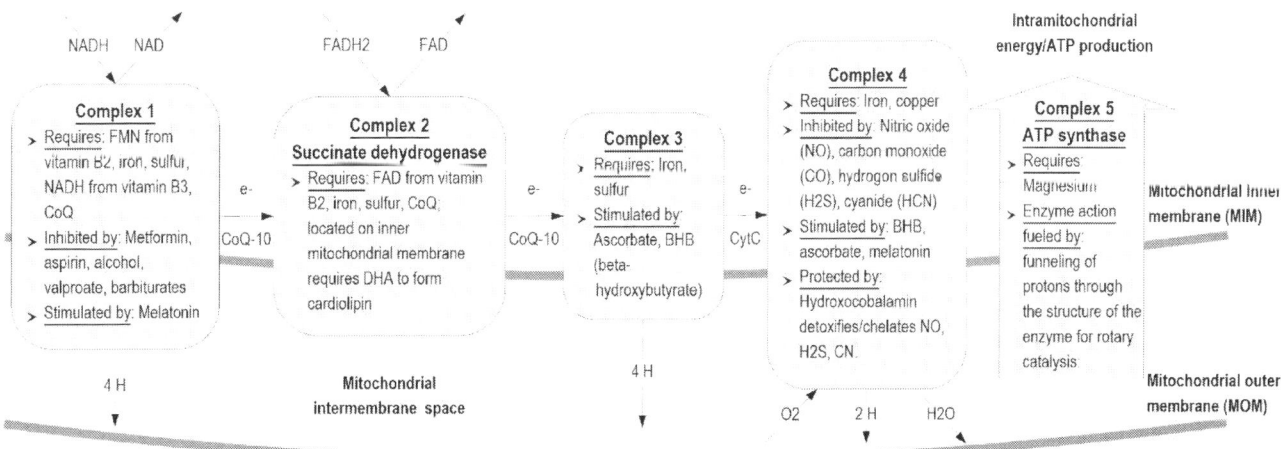

Mitochondrial ETC: Schematic diagram of the electron transport chain, function of which is wholly dependent upon niacin, riboflavin, CoQ10, iron, copper, and—to a lesser extent—vitamin C. Beyond showing where nutrients function in the ETC, this diagram also shows where drugs inhibit ETC function while endogenous substances such as melatonin and beta-hydroxybutyrate stimulate the ETC at complexes 1 and 4 (melatonin) and 3 and 4 (BHB), respectively. For a thorough review of mitochondrial nutrition and mitochondrial medicine, see videos at https://vimeo.com/ondemand/mitochondrialmedicine.

---

[172] Myers DE, Myers RA. A preliminary report on hyperbaric oxygen in the relief of migraine headache. *Headache*. 1995 Apr;35(4):197-9
[173] Wilson et al. Hyperbaric oxygen in the treatment of migraine with aura. *Headache*. 1998 Feb;38(2):112-5
[174] Eftedal et al. A randomized, double blind study of the prophylactic effect of hyperbaric oxygen therapy on migraine. *Cephalalgia*. 2004 Aug:639-44

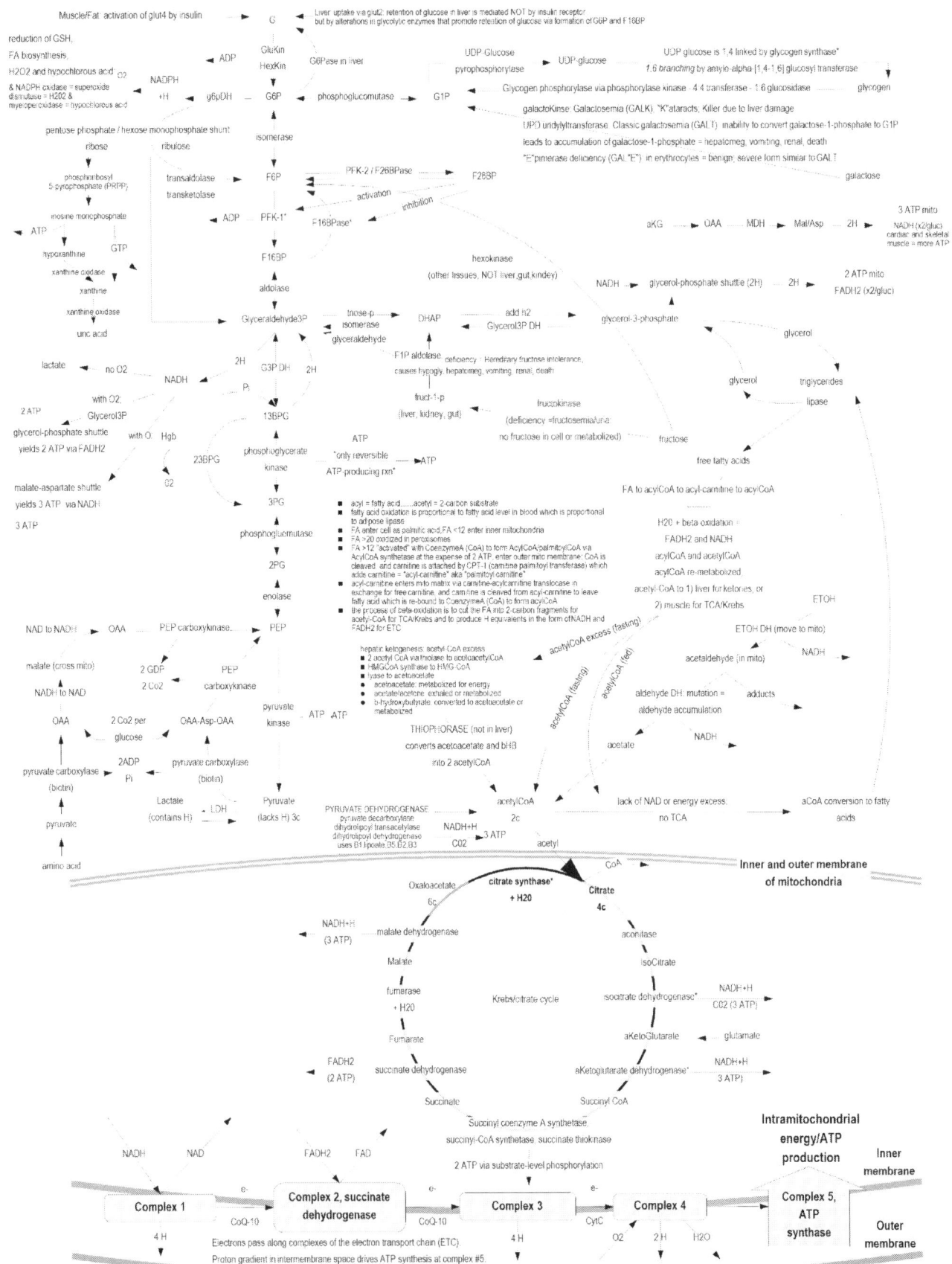

Schematic overview of glycolytic pathways, pyruvate dehydrogenase complex, the Krebs/citrate cycle, and the electron transport chain: Items in bold/red are those which are commonly defective in patients with migraine headaches (and fibromyalgia)—consult your biochemistry text as needed for details and definitions. Note from DrV: The origin of this diagram is from my first year of medical school, during which I created this diagram for my study notes and could recite it from memory.

Style of Living, Special Considerations, Surgical & Somatic/Spinal Treatments

- Style of living—lifestyle optimization: Patients with cluster headaches show a greater percentage of increased work-related stress, self-employment, tobacco smoking, and alcohol use or abuse. These concerns should be addressed per patient as indicated. Lifestyle factors such the standard American diet, overconsumption of caffeine and alcohol, and use of tobacco can result in mitochondrial impairment through various mechanisms, not the least of which are nutrient (especially magnesium) deficiency and accumulation of cyanide (from tobacco smoke), a known mitochondrial poison.

- Self-knowledge—psychological exploration of emotional tension: If someone has chronic "tension headaches" then the question becomes *"Why does this person have chronic tension?"* Generally, the answer is a combination of magnesium deficiency along with some underlying emotional issue(s). If someone feels compelled to maintain a static posture all day without taking a necessary and healthful break to stretch and relax, this suggests that they are over-focusing on their work at the expense of taking care of their body and health—this is not a sign of health, and it suggests an underlying compulsion or dissociation. Patients with cluster headaches show a greater percentage of increased work-related stress, self-employment, tobacco smoking and alcohol use and abuse.[175] Address as indicated.

> **Osteopathic treatment should include manual musculoskeletal medicine and nutritional interventions**
>
> "In contrast to the description of the osteopathic medical profession by the American Osteopathic Association, namely, "doctors of osteopathic medicine, or D.O.s, apply the philosophy of treating the whole person to the prevention, diagnosis and treatment of illness, disease and injury," [the authors of the article in question] essentially reviewed only pharmacologic treatment. ... It is hoped that future reviews in this journal can include a more balanced survey of the literature, inclusive of non-pharmacologic and "holistic" interventions that are consistent with osteopathic philosophy."
>
> **Vasquez A**. Interventions Need to be Consistent with Osteopathic Philosophy. [Letter] *JAOA: Journal of the American Osteopathic Association* 2006 Sep
> http://jaoa.org/cgi/content/full/106/9/528

- Stress reduction, relaxation and biofeedback: Biofeedback is proven effective in the prevention of chronic headaches, including pediatric migraine.[176,177] Relaxation and stress management are more effective than drug treatment with metoprolol for pediatric migraine.[178]

- Somatic treatments—cervical myofascial trigger points: Check the upper cervical spine musculature (especially suboccipital muscles and sternocleidomastoid) for the characteristic manifestations of myofascial trigger points (MFTP, see Chapter 3): palpable nodule, twitch response, and elicitation of referred pain with deep palpation/provocation. Myofascial trigger points are much more common in patients with migraine[179] than in non-headache controls, and they are an important cervicogenic contribution to chronic headaches.[180,181] If located, the MFTP can be effectively treated with in-office/at-home post-isometric stretching and exercises.[182]

> **Post-isometric stretching is of high value in the treatment of cervical MFTP that contribute to headache**
>
> If deep palpation of the upper cervical musculature produces referred pain to the head or face, then you know the patient has MFTP, and you can then treat them with at-home and in-office stretching and exercises. Generally their "chronic pain" will be greatly reduced within 3-10 days. Better results will be obtained with alkalinization and concomitant supplementation with magnesium, vitamin D, and fish oil.

- Somatic treatments—spinal manipulation: Myofascial, arthrogenic, and dyskinetic contributions to headache are significant, and these cervicogenic problems can be addressed with manual spinal and myofascial manipulation. Spinal manipulation can help alleviate headaches with efficacy comparable to commonly used

---

[175] Manzoni GC. Cluster headache and lifestyle: remarks on a population of 374 male patients. *Cephalalgia* 1999 Mar;19(2):88-94
[176] Scharff L, Marcus DA, Masek BJ. A controlled study of minimal-contact thermal biofeedback treatment in children with migraine. *J Pediatr Psychol*.2002; 27:109 –119
[177] "Feedback training was accompanied by significant reduction of cortical excitability. This was probably responsible for the clinical efficacy of the training; a significant reduction of days with migraine and other headache parameters was observed." Siniatchkin et al. Self-regulation of slow cortical potentials in children with migraine: an exploratory study. *Appl Psychophysiol Biofeedback*.2000; 25:13 –32
[178] "The overall results of the study showed that relaxation training combined with stress management training was significantly more effective in reducing the headache index than treatment with the betablocker metoprolol." Sartory et al. A comparison of psychological and pharmacological treatment of pediatric migraine. *BehavResTher*1998;36:1155-1170
[179] "Trigger points were found in 92 (93.9%) migraineurs and in nine (29%) controls (P < 0.0001). The number of individual migraine trigger points varied from zero to 14, and was found to be related to both the frequency of migraine attacks, and the duration of the disease." Calandre et al. Trigger point evaluation in migraine patients: an indication of peripheral sensitization linked to migraine predisposition? *Eur J Neurol*. 2006 Mar;13(3):244-9
[180] "Myofascial trigger points can refer pain to the head and face in the cervical region, thus contributing to cervicogenic headache." Borg-Stein J. Cervical myofascial pain and headache. *Curr Pain Headache Rep*. 2002 Aug;6(4):324-30
[181] "Because treating myofascial problems may be the only way to offer complete relief from certain types of headache, clinicians must learn to diagnose and manage trigger points in neck, shoulder, and head muscles." Davidoff RA. Trigger points and myofascial pain: toward understanding how they affect headaches. *Cephalalgia*. 1998 Sep;18(7):436-48
[182] Lewit K, Simons DG. Myofascial pain: relief by post-isometric relaxation. *Arch Phys Med Rehabil* 1984 Aug;65(8):452-6. This is a classic "must read" article.

first-line prophylactic prescription medications.[183,184,185,186] Spinal manipulation should be performed only by professionals with graduate and postgraduate training in relevant spinal biomechanics, patient assessment, and manipulative technique.[187,188,189,190]

- Somatic treatments—acupuncture: Acupuncture is effective symptomatic treatment for migraine and tension headaches, with cost-effectiveness comparable to standard medical treatment.[191,192,193] Acupuncture is known to affect regional blood flow, and the neurophysiological mechanisms involved include increased release of endogenous analgesics such as endomorphin-1, beta-endorphin, enkephalin, and serotonin.[194] In a head-to-head study of acupuncture versus drug treatment with metoprolol, "2 of 59 patients randomized to acupuncture withdrew prematurely from the study compared to 18 of 55 randomized to metoprolol … The proportion of responders was 61% for acupuncture and 49% for metoprolol. Both physicians and patients reported fewer adverse effects in the acupuncture group."[195] While pain relief is an important benefit, it should not be the primary goal in treatment if an underlying physiological or biochemical disturbance (including nutritional deficiencies or imbalances) can be corrected.

- Somatic treatment—exercise: When people have pain, they are disinclined to move and exercise; however, lack of movement and exercise promotes the continuation of pain via several mechanisms including allowing the formation of myofascial adhesions and contractions, "untraining" of vestibulocerebellar circuits that are necessary for neuromuscular coordination, which is important for reducing musculoskeletal microtrauma that results from uncoordinated/dyscoordinated movements. Exercise and movement are necessary for maintaining myofascial elasticity; all three of these (exercise, movement/stretching, elasticity) support flow of blood, nutrients, oxygen, interstitial fluid, lymph throughout tissues to maintain cellular metabolism and remove wastes and inflammatory mediators. Movement also reduces pain directly via proprioceptive/sensory inhibition of nociception; stated simply: *movement sensation blocks pain reception*.
    - Intensive dynamic training for females with chronic neck/shoulder pain. (*Clin Rehabil*. 1998 Jun[196]): In a clinical research study with 77 women who suffered from chronic neck and shoulder pain, women who performed their exercises **three times per week for 5 sets of 20 repetitions** had better results than those who performed their exercises three times per week for 1 set of 20 repetitions. *More exercise gives better results—faster and more complete relief of pain.*

- Special supplementation for hyperhomocysteinaemia/hyperhomocysteinuria: Reasonable doses are listed; clinicians will have to combine and "dose to effect" per patient. All of these treatments are well-accepted as safe and generally effective. Homocysteine contributes to the increased cardiovascular and stroke risk seen in patients with migraine, while also contributing directly to neuroinflammation and NMDA receptor activation resulting in—identically as with glutamate—increased intracellular calcium and neurotoxicity.[197]
    - Folinic acid or methylfolate 2-5 mg/d: Use in combination with other vitamins, especially vitamin B12, in the form of hydroxocobalamin, adenosylcobalamin, or methylcobalamin—cyanocobalamin is obviously to be avoided because of its clinically relevant content of cyanide.
    - Vitamin B12 >2,000 mcg per day orally, or 1-2 mg per week by injection: Use in the form of hydroxocobalamin, adenosylcobalamin, or methylcobalamin—cyanocobalamin is obviously to be avoided because of its clinically relevant content of cyanide.

---

[183] Bronfort et al. Efficacy of spinal manipulation for chronic headache: a systematic review. *J Manipulative Physiol Ther* 2001 Sep;24(7):457-66
[184] Tuchin PJ, Pollard H, Bonello R. A randomized controlled trial of chiropractic spinal manipulative therapy for migraine. *J Manipulative Physiol Ther*. 2000 Feb;23(2):91-5
[185] "SMT appears to have a better effect than massage for cervicogenic headache. It also appears that SMT has an effect comparable to commonly used first-line prophylactic prescription medications for tension-type headache and migraine headache." Bronfort et al. Efficacy of spinal manipulation for chronic headache: a systematic review. *J Manipulative Physiol Ther* 2001 Sep;24(7):457-66
[186] "The average response of the treatment group (n = 83) showed statistically significant improvement in migraine frequency (P < .005), duration (P < .01), disability, and medication use…" Tuchin et al. A randomized controlled trial of chiropractic spinal manipulative therapy for migraine. *J Manipulative Physiol Ther*. 2000 Feb;23(2):91-5
[187] Kirk CR, Lawrence DJ, Valvo NL. *States Manual of Spinal, Pelvic, and Extravertebral Technics. Second Edition*. Lombard, Illinois: National College of Chiropractic; 1985
[188] Kimberly PE. *Outline of Osteopathic Manipulative Procedures. The Kimberly Manual 2006*. Kirksville College of Osteopathic Medicine. Walsworth Publishing
[189] Bergmann TF, Peterson DH, Lawrence DJ. *Chiropractic Technique*. New York; Churchill Livingstone: 1993
[190] Gatterman MI. *Chiropractic Management of Spine Related Disorders*. Baltimore; Williams and Wilkins: 1990
[191] Endres et al. Acupuncture for tension-type headache. *J Headache Pain*. 2007 Oct; 8(5): 306–314
[192] Vickers et al. Acupuncture of chronic headache disorders in primary care: randomised controlled trial and economic analysis. *Health Technol Assess*. 2004 Nov;8(48):iii, 1-35
[193] Wonderling D, et al. Cost effectiveness analysis of a randomised trial of acupuncture for chronic headache in primary care. *BMJ*. 2004 Mar 27;328(7442):747
[194] Cabyoglu MT, Ergene N, Tan U. The mechanism of acupuncture and clinical applications. *Int J Neurosci*. 2006 Feb;116(2):115-25
[195] Streng et al. Effectiveness and tolerability of acupuncture compared with metoprolol in migraine prophylaxis. *Headache*. 2006 Nov-Dec;46(10):1492-502
[196] "… pain scores were only significantly improved in the intensive group at 12 months follow-up." Randlov et al. Intensive dynamic training for females with chronic neck/shoulder pain. *Clin Rehabil*. 1998 Jun:200-10
[197] Abushik et al. The role of NMDA and mGluR5 receptors in calcium mobilization and neurotoxicity of homocysteine in trigeminal and cortical neurons and glial cells. *J Neurochem*. 2014 Apr;129(2):264-74

Chapter 5.1—Functional Inflammology Protocol for Metabolic Inflammation: Migraine and Fibromyalgia

- Vitamin B6, pyridoxine 50-250 mg/d: The phosphorylated form (P5P) can also be used; when the HCL form is used, additional attention must be given to magnesium status/supplementation and urinary alkalinization.
- Riboflavin 20-400 mg/d: Small doses of 2 mg/d have been shown to significantly reduce homocysteine levels, and doses of 400 mg/d are common and well-tolerated in the treatment of migraine.
- Thyroid optimization: Hypothyroidism causes elevated homocysteine and promotes insulin resistance[198] and should be treated appropriately per Chapter 1.
- NAC 600 mg per day and upward to 500-1,500 mg thrice daily: Doses of NAC 4,800 mg/d have been used with success and safety in the treatment of SLE.
- Avoidance of homocysteine-elevating factors: High coffee intake (>5 cups per day), ethanol, tobacco smoking, and medications/treatments (such as methotrexate, metformin, niacin and fibrate drugs); fish oil can raise homocysteine levels in some patients. Metformin is well-known to cause malabsorption of vitamin B12 and to thereby exacerbate "diabetic neuropathy" and promote depression and dementia/psychosis.
- Choline, phosphatidylcholine, lecithin (approximately 2.6 g choline/d): Each TBS (tablespoon, approximately 15 mL) of lecithin contains 275 mg of choline; thus, if the goal is to get to 2.6 g choline, one would need to use 10 TBS (150 mL) per day of granulated lecithin.
- Betaine, trimethylglycine 6–12 g/day: Effects are weak/modest; likely more relevant for patients taking drugs such as metformin and fibrates that promote loss of betaine in urine.

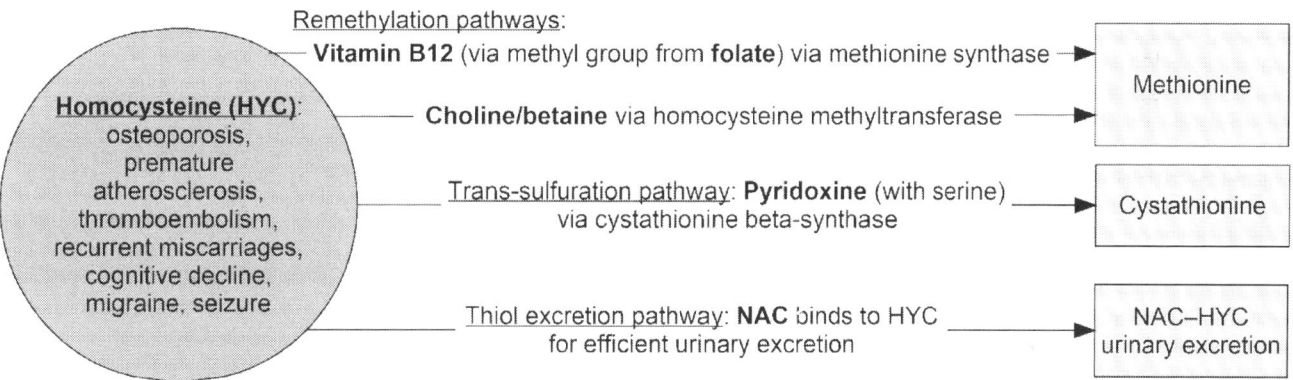

Lowering homocysteine (HYC) via nutritional supplementation: Folate gives methyl group to cobalamin (vitamin B12) to convert HYC via methionine synthase to methionine; choline/betaine can remethylate homocysteine via homocysteine methyltransferase to form methionine. Pyridoxine promotes conversion of HYC via cystathionine beta-synthase to cystathionine. The amino acid N-acetyl-cysteine (NAC) binds to HYC for efficient renal excretion of NAC-HYC.[199]

- Special supplementation—Feverfew (*Tanacetum parthenium*) as monotherapy or with ginger, willow: Results of numerous studies support the use of feverfew for the safe and cost-effective treatment and prevention of migraine headaches. Feverfew has several mechanisms of action including antithrombosis and inhibition of NFkB. Feverfew products are generally concentrated to 0.2% to 0.7% parthenolide, and a reasonable starting dose is 250 mcg/d of parthenolide; lower doses can be used within a context of multicomponent treatment. Feverfew can be used alone, with other nutrients, or with other botanical medicines. The combination of ginger and feverfew has shown efficacy for halting incipient migraine attacks when started within 2 hours of pain onset.[200] Similarly, the combination of feverfew and willow extract was shown to be remarkably safe and effective in preventing and reducing migraine.[201]

---

[198] Yang N et al. Novel Clinical Evidence of an Association between Homocysteine and Insulin Resistance in Patients with Hypothyroidism or Subclinical Hypothyroidism. *PLoS One*. 2015 May 4;10(5):e0125922
[199] "NAC intravenous administration induces an efficient and rapid reduction of plasma thiols, particularly of Hcy; our data support the hypothesis that NAC displaces thiols from their binding protein sites and forms, in excess of plasma NAC, mixed disulphides (NAC-Hcy) with an high renal clearance." Ventura et al. N-Acetyl-cysteine reduces homocysteine plasma levels after single intravenous administration by increasing thiols urinary excretion. *Pharmacol Res*. 1999 Oct;40(4):345-50
[200] Cady RK, Schreiber CP, Beach ME, Hart CC. Gelstat Migraine (sublingually administered feverfew and ginger compound) for acute treatment of migraine when administered during the mild pain phase. *Med Sci Monit*. 2005 Sep;11(9):PI65-9
[201] Shrivastava R, Pechadre JC, John GW. Tanacetum parthenium and Salix alba (Mig-RL) combination in migraine prophylaxis. *Clin Drug Investig*. 2006;26(5):287-96

- Special supplementation—Butterbur (*Petasites hybridus*): Butterbur/*Petasites* has consistently shown excellent efficacy in the prophylaxis of migraine, with frequency reductions of ~60%.[202,203] Adult doses used in clinical trials have been variantly described per product as "two capsules 25 mg BID" for a total of 100 mg and "Petasites extract 75 mg bid, Petasites extract 50 mg bid" with the higher doses showing greater efficacy. Anderson et al[204] noted that antimigraine and anti-allergy benefits are mediated via sesquiterpene esters of petasin and furanopetasin which reduce leukotriene biosynthesis, inhibit cyclooxygenase (COX-1 and COX-2), ameliorate activation of p38 mitogen–activated protein kinase stress signaling, and reduce NFkB activation in rat microglial cells; partial blockade of calcium channels has also been noted. Rare (< 0.01%) cases of liver damage (acute hepatitis) and liver failure have been reported; Anderson et al (op cit) per their comparative *in vitro* studies recommend limiting the petasin content (<17%) to improve the hepatobiliary safety profile. Differently, Utterback et al[205] attributed the hepatotoxicity to pyrrolizidine alkaloids (PAs) and stated that products should be virtually PA-free, with less than 0.08 ppm PA; their review concluded that butterbur is safe for antimigraine treatment in children (>6yo) and adults. Contraindications include hypersensitivity/allergy to butterbur or any of the related Asteraceae plants: ragweed, marigolds, daisies, and chrysanthemums.
- Special supplementation—botanical medicines with anti-inflammatory and/or GABA agonist effects: Numerous botanical medicines have proven anti-inflammatory and analgesic benefits. *Zingiber officinale* (ginger) demonstrates multiple antiinflammatory mechanisms and is commonly consumed as food, juice and as a nutritional supplement in the form of pills/powder. Components of ginger reduce production of the leukotriene LTB4 by inhibiting 5-lipoxygenase and reduce production of the prostaglandin PG-E2 by inhibiting cyclooxygenase.[206,207] With its dual reduction in the formation of inflammation-promoting prostaglandins and leukotrienes, ginger has been shown to safely reduce musculoskeletal pain in general[208,209] and to provide relief from osteoarthritis of the knees and migraine headaches.[210] The traditional Chinese herbal medicine *Scutellaria baicalensis* contains, among other active phytochemicals, baicalein, which is anti-inflammatory, neuroprotective, and protective/therapeutic against persistent pain and neuroinflammation[211]; baicalein, oroxylin A, and skullcapflavone II bind the benzodiazepine site of GABA-A receptors with a Ki value of 13.1, 14.6 and 0.36 micromol/L, respectively.[212] Since Ki value increases with decreasing affinity (ie, the lower the value, the stronger the ligand-receptor interaction), we note that skullcapflavone has more GABA receptor affinity than does the more "famous" baicalein. Botanical medicines that have proven clinical benefit via—at least in large part—their activation of the GABA/benzodiazepine receptor include *Matricaria recutita* (Chamomile), *Melissa officinalis* (lemon balm), *Passiflora incarnata* (passionflower), *Piper methysticum* (kava), *Scutellaria lateriflora* (skullcap), *Valeriana species* (valerian), and *Withania somnifera* (ashwagandha).[213]
- Special supplementation—intranasal capsaicin: Capsaicin is the "hot" spicy component of hot chili peppers; when applied to the skin or mucus membranes, it damages pain-sensing nerves and thereby reduces the sensation of pain after an initial exacerbation of pain. Intranasal capsaicin (300 mcg/100 microliters) is remarkably well studied in the treatment and prevention of cluster headaches, beginning with the first report published by Sicuteri et al[214] in 1989. Treatment of active cluster headache with intranasal capsaicin (compared

---

[202] Grossmann M, Schmidramsl H. An extract of Petasites hybridus is effective in the prophylaxis of migraine. *Int J Clin Pharmacol Ther*. 2000 Sep;38(9):430-5. See also Grossman W, Schmidramsl H. An extract of Petasites hybridus is effective in the prophylaxis of migraine. *Altern Med Rev*. 2001 Jun;6(3):303-10
[203] Lipton et al. Petasites hybridus root (butterbur) is an effective preventive treatment for migraine. *Neurology*. 2004 Dec 28;63(12):2240-4
[204] Anderson et al. Toxicogenomics applied to cultures of human hepatocytes enabled an identification of novel petasites hybridus extracts for the treatment of migraine with improved hepatobiliary safety. *Toxicol Sci*. 2009 Dec;112(2):507-20
[205] Utterback et al. Butterbur extract: prophylactic treatment for childhood migraines. *Complement Ther Clin Pract*. 2014 Feb;20(1):61-4
[206] Kiuchi et al. Inhibition of prostaglandin and leukotriene biosynthesis by gingerols and diarylheptanoids. *Chem Pharm Bull* (Tokyo) 1992 Feb;40(2):387-91
[207] Tjendraputra et al. Effect of ginger constituents and synthetic analogues on cyclooxygenase-2 enzyme in intact cells. *Bioorg Chem* 2001 Jun;29(3):156-63
[208] Srivastava KC, Mustafa T. Ginger (Zingiber officinale) in rheumatism and musculoskeletal disorders. *Med Hypotheses*. 1992 Dec;39(4):342-8
[209] Srivastava KC, Mustafa T. Ginger (Zingiber officinale) and rheumatic disorders. *Med Hypotheses*. 1989 May;29(1):25-8
[210] "It is proposed that administration of ginger may exert abortive and prophylactic effects in migraine headache without any side-effects." Mustafa T, Srivastava KC. Ginger (Zingiber officinale) in migraine headache. *J Ethnopharmacol*. 1990 Jul;29(3):267-73
[211] "Baicalein (BE), isolated from the traditional Chinese herbal medicine Scutellaria baicalensis Georgi (or Huang Qin), has been demonstrated to have anti-inflammatory and neuroprotective effects. ...Intrathecal and oral administration of BE at different doses could alleviate the mechanical allodynia in CIBP rats. Intrathecal 100 μg BE could inhibit the production of IL-6 and TNF-α in the spinal cord of CIBP rats. ...The analgesic effect of BE may be associated with the inhibition of the expression of the inflammatory cytokines IL-6 and TNF-α and through the activation of p-p38 and p-JNK MAPK signals in the spinal cord." Hu et al. The Analgesic and Antineuroinflammatory Effect of Baicalein in Cancer-Induced Bone Pain. *Evid Based Complement Alternat Med*. 2015;2015:973524
[212] "A benzodiazepine binding assay directed separation led to the identification of 3 flavones baicalein (1), oroxylin A (2), and skullcapflavone II (3) from the water extract of Scutellaria baicalensis root. Compounds 1, 2, and 3 interacted with the benzodiazepine binding site of GABAA receptors with a Ki value of 13.1, 14.6 and 0.36 micromol/L, respectively." Liao et al. Benzodiazepine binding site-interactive flavones from Scutellaria baicalensis root. *Planta Med*. 1998 Aug;64(6):571-7
[213] Sarris et al. Plant-based medicines for anxiety disorders, part 2: a review of clinical studies with supporting preclinical evidence. *CNS Drugs*. 2013 Apr;27(4):301-19
[214] Sicuteri et al. Beneficial effect of capsaicin application to the nasal mucosa in cluster headache. *Clin J Pain*. 1989;5(1):49-53

Chapter 5.1—Functional Inflammology Protocol for Metabolic Inflammation: Migraine and Fibromyalgia

with placebo) reduced severity after 7 days of treatment.[215] In a small controlled clinical trial, patients stated that intranasal capsaicin alleviated chronic migraine suffering by 50% to 80%.[216] Burning pain, sneezing, and increased nasal secretions induced by topical capsaicin application are intense for the first few applications but decrease over time, generally within a week or so; clinical benefits generally begin on the eighth day of consecutive treatment. Episodic cluster headache patients appear to benefit more than do chronic cluster headache patients. Cluster headaches are typically unilateral, and capsaicin should be applied to the nostril on the same side as the head pain.[217]

- Surgical closure of patent foramen ovale: While the prevalence of patent foramen ovale in the general adult population is approximately 20% to 30%, patients with migraine—especially migraine with aura—show a higher prevalence (55–65%) of this physiological cardiopulmonary shunt. Surgical closure of a patent foramen ovale can provide relief from headache in many migraine patients.[218,219] If the cardiopulmonary shunt is severe enough to result in reduced blood oxygenation, it can exacerbate the already reduced energy production caused by the aforementioned mitochondrial dysfunction. More likely, bypassing the lungs results in failure of pulmonary degradation of proinflammatory mediators. The lungs inactivate proinflammatory mediators such as prostaglandins E1, E2, and F2-alpha, all of the leukotrienes, and norepinephrine (30% reduction). Thus, surgical closure of the patent foramen ovale stops inflammatory mediators from bypassing the lungs and may provide a systemic anti-inflammatory benefit that reduces migraine severity.

Endocrine Imbalances All hormones have either pro-inflammatory or anti-inflammatory effects. As such, patients with pain and inflammation are candidates for complete hormonal evaluation, as reviewed in Chapter 4, Section 5. Melatonin's utility in migraine—as in fibromyalgia—is more likely related to its antioxidant and mitochondrial-supportive effects than to a true "hormonal" effect. Thyroid status must be optimized in all migraine patients.

- Melatonin—3-20 mg at night: Melatonin is a hormone made in the pineal gland from the neurotransmitter serotonin which is derived from the amino acid tryptophan and 5-hydroxytryptophan. Melatonin levels are low in patients with migraine and cluster headache. According to case reports and studies with small numbers of patients, 10 mg of melatonin taken at night relieves cluster headaches in approximately 50% of patients, with results beginning 3 to 5 days after the start of treatment and continuing for the duration of treatment.[220,221] Clinical trials using melatonin in migraine patients have shown consistently positive results, with a significant number of patients becoming completely migraine-free.[222] Melatonin is generally administered at night in doses ranging from 3 to 10 mg, although studies in cancer patients have safely used doses as high as 20 to 40 mg and have shown antitumor and pro-survival benefits. Melatonin has antioxidant and immunomodulatory actions in addition to its ability to preserve mitochondrial function, which is particularly relevant to migraine and cluster headaches. According to small studies and case reports with small numbers of patients, 10 mg of melatonin taken at night relieves cluster headaches in approximately 50% of patients.[223,224]
- Testosterone: Testosterone has antiinflammatory/immunomodulatory actions, and as such has clinical utility in migraine, cluster headache, rheumatoid arthritis, and fibromyalgia via reduction/modulation of peripheral inflammation and the brain inflammation and microglial activation that promote central sensitization to pain. In this study[225], following subcutaneous testosterone implant, "Improvement in headache severity was noted by 92% of patients and the mean level of improvement was statistically significant (3.3 on a 5 point scale). ... Seventy-four percent of patients reported a headache severity score of '0' (none) on testosterone implant therapy

---

[215] Marks et al. A double-blind placebo-controlled trial of intranasal capsaicin for cluster headache. *Cephalalgia*. 1993 Apr;13(2):114-6
[216] Fusco et al. Repeated intranasal capsaicin applications to treat chronic migraine. *Br J Anaesth*. 2003 Jun;90(6):812
[217] Fusco et al. Preventative effect of repeated nasal applications of capsaicin in cluster headache. *Pain*. 1994 Dec;59(3):321-5
[218] Rigatelli et al. Primary patent foramen ovale closure to relieve severe migraine. *Ann Intern Med*. 2006 Mar 21;144(6):458-60
[219] Dubiel et al. Migraine Headache Relief after Percutaneous Transcatheter Closure of Interatrial Communications. *J Interv Cardiol*. 2007 Dec 18; [Epub ahead of print]
[220] Leone et al. Melatonin versus placebo in the prophylaxis of cluster headache: a double-blind pilot study with parallel groups. *Cephalalgia* 1996 Nov;16(7):494-6
[221] "Melatonin levels have been found to be decreased in cluster headache patients. ... We report two chronic cluster headache patients who had both daytime and nocturnal attacks that were alleviated with melatonin." Peres MF, Rozen TD. Melatonin in the preventive treatment of chronic cluster headache. *Cephalalgia*. 2001 Dec;21(10):993-5
[222] Vogler et al. Role of melatonin in the pathophysiology of migraine: implications for treatment. *CNS Drugs*. 2006;20(5):343-50
[223] "Five of the 10 treated patients were responders whose attack frequency declined 3-5 days after treatment, and they experienced no further attacks until melatonin was discontinued." Leone et al. Melatonin versus placebo in the prophylaxis of cluster headache: a double-blind pilot study with parallel groups. *Cephalalgia* 1996 Nov;16(7):494-6
[224] "Melatonin levels have been found to be decreased in cluster headache patients. ... We report two chronic cluster headache patients who had both daytime and nocturnal attacks that were alleviated with melatonin." Peres MF, Rozen TD. Melatonin in the preventive treatment of chronic cluster headache. *Cephalalgia*. 2001 Dec;21(10):993-5
[225] Glaser et al. Testosterone pellet implants and migraine headaches: a pilot study. *Maturitas*. 2012 Apr;71(4):385-8

for the 3-month treatment period. Continuous testosterone was effective therapy in reducing the severity of migraine headaches in both pre- and post-menopausal women." In another study, "Seven male and 2 female patients, seen between July 2004 and February 2005, and between the ages of 32 and 56, are reported with histories of treatment resistant cluster headaches accompanied by borderline low or low serum testosterone levels. The patients failed to respond to individually tailored medical regimens, including melatonin doses of 12 mg a day or higher, high flow oxygen, maximally tolerated verapamil, antiepileptic agents, and parenteral serotonin agonists. Seven of the 9 patients met 2004 International Classification for the Diagnosis of Headache criteria for chronic cluster headaches; the other 2 patients had episodic cluster headaches of several months duration. After neurological and physical examination all patients had laboratory investigations including fasting lipid panel, PSA (where indicated), LH, FSH, and testosterone levels (both free and total). All 9 patients demonstrated either abnormally low or low, normal testosterone levels. After supplementation with either pure testosterone in 5 of 7 male patients or combination testosterone/estrogen therapy in both female patients, the patients achieved cluster headache freedom for the first 24 hours. Four male chronic cluster patients, all with abnormally low testosterone levels, achieved remission."[226]

Xenobiotic Accumulation/Detoxification Persistent organic pollutants promote inflammation generally and mitochondrial dysfunction specifically. Treat as described throughout this text and as reviewed in Chapter 2 and also in Chapter 4, Section 7.

---

[226] Stillman MJ. Testosterone replacement therapy for treatment refractory cluster headache. *Headache*. 2006 Jun;46(6):925-33

Fibromyalgia (FM, FMS, FMD) & Complex Regional Pain Syndrome (CRPS)

Introduction:

Fibromyalgia (FM)—also referred to as fibromyalgia syndrome (FMS) but more properly referred to as fibromyalgia disease (FMD)—is an organic clinical entity that has remained enigmatic to the medical profession despite the consistent publication of research that delineates its cause and its effective treatments. This chapter summarizes clinical assessments, treatments, and essential background information that—when properly applied—should provide empowering knowledge for clinicians and for the patients suffering with this condition. For any disease, we as scientists and clinicians must establish a model of the disease so that we can start with *some shared idea about the disease itself* and have some common language and understanding so that we can engage in meaningful conversations and valuable research. The model of FM that I have been the first and only clinician researcher to construct (starting in 2008) is that FM starts from small intestine bacterial overgrowth, and that the absorbed microbial molecules lead to the pain-amplifying central sensitization and fatigue-inducing mitochondrial dysfunction that clearly characterize this condition; variations of and deviations from the classic model of any disease will undoubtedly be encountered, but they can always be traced back to this classic model.

Fibromyalgia is uniquely exemplary in its clouded diagnostic criteria and etiology; I argue in this review and the associated videos and related articles that the diagnostic criteria for FM have been intentionally clouded and deconstructed via influence of drug companies that want the FM label to be applied to as many patients as possible for the longest duration possible and that the etiology and very nature of the disorder has been clouded and hijacked by medical/science writers who are heavily paid by drug companies that sell the FDA-approved drugs for fibromyalgia and which directly profit from a population of confused doctors and helpless patients which create a multi-billion dollar international drug/medical market.

After considerable literature review and personal deliberation, I have decided to include information about complex regional pain syndrome (CRPS) in this section that is otherwise exclusive to fibromyalgia. At the time of this writing, I am impressed that these are two variants of essentially the same pathophysiology, just as migraine and cluster headaches are variants on the same theme. With all due respect for the differences and distinctions between these conditions, I appreciate that their pathophysiologic similarities are of greater importance for understanding and treating both disorders. Both FM and CRPS share 1) central sensitization, 2) peripheral sensitization, 3) neuroinflammation, 4) neurogenic inflammation, 5) increased glutaminergic/NMDA-mediated central and peripheral neurotransmission, 6) mitochondrial dysfunction, 7) increased intestinal permeability, and 8) gastrointestinal dysbiosis. Relative to my study of migraine and fibromyalgia which has a history of publications over many years, I have only recently begun to study CRPS, but my impression is that it is a spinal cord (SC)-specific regional activation of glia and neurons and failure of regional neuroinhibition; somewhat analogous to a spinal cord migraine/seizure or at least the spinal cord variant of fibromyalgia. I agree with Littlejohn[227] that both FM and CRPS have a component of neurogenic inflammation; I think the role of neurogenic inflammation in CRPS is stronger than it is in FM, also—perhaps obviously—more focal, limited to a region of spinal cord segments. While the pathophysiologic similarities among FM and CRPS suggest that effective nutritional therapeutics for FM will be at least partly efficacious for CRPS, we generally do not have such proof due to lack of studies in CRPS; however, the risk:benefit ratio of these highly safe and frequently efficacious interventions clearly favors empiric implementation followed by monitoring of therapeutic response in a condition notorious for its debilitating severity and therapeutic recalcitrance.

Introduction and Clinical Presentation

- <u>Overview</u>: Fibromyalgia (FM) is commonly described as an "idiopathic" (of unknown origin) syndrome principally characterized by widespread body pain and numerous myofascial tender points at specific locations. FM is most common in women 20-50 years of age, and the condition often presents with associated complaints of fatigue, headaches, subjective numbness, altered sleep patterns, and gastrointestinal disturbances. FM in children and adolescents presents similarly to FM in adults except for the comparatively higher prevalence of sleep disturbance and the finding of fewer tender points in children.[228] Until recently,

---

[227] Littlejohn G. Neurogenic neuroinflammation in fibromyalgia and complex regional pain syndrome. *Nat Rev Rheumatol*. 2015 Nov;11(11):639-48. At the time this citation is added in 2015 Dec, I (DrV) had recently submitted a reply to *Nat Rev Rheumatol*; my reply will be posted at ichnfm.academia.edu/AlexVasquez as soon as possible.
[228] Siegel DM, Janeway D, Baum J. Fibromyalgia syndrome in children and adolescents: clinical features at presentation and status at follow-up. *Pediatrics* 1998;101:377-82

fibromyalgia was considered a *diagnosis of exclusion* after infection, autoimmunity, or other primary causes of widespread pain were excluded by clinical and laboratory assessment. However, current criteria base the diagnosis on positive findings of chronic, widespread musculoskeletal pain in characteristic locations; these criteria will be described below. Fibromyalgia shares several clinical, demographic, and pathologic features with chronic fatigue syndrome (CFS) and irritable bowel syndrome (IBS); the reason for these overlaps is not generally understood by most clinicians and researchers but will be made plain in this writing.

- The common medical view—scientifically inaccurate, financially leveraged: The prevailing medical view, expressed by most medical doctors and the authors of widely cited articles, is that fibromyalgia is idiopathic—*of unknown origin*—with strong neuropsychogenic (*neuro*=nerves and brain, *psyche*=mind, *genic*=origin) influences (in other words, "It's all in your head.") and that, since the underlying causes of the condition have not been identified, the best therapeutic approach is symptom suppression via perpetual pharmacotherapy with adjunctive use of psychotherapy and limited exercise.[229,230,231] This prevailing medical view is unscientific (not based on science) and counterscientific (ignores and contradicts published and validated research), unethical (fails to provide effective treatment when such treatment is available; condemns patients to medicalization and suffering), and commercially leveraged (diagnostic criteria revision and many review articles discussing treatment are sponsored by drug companies; medical profession benefits financially by having many long-term drug-dependent patients).

- Fibromyalgia—in its original "1990-based" description—is a disease, not a syndrome: The term *syndrome* connotes that a cluster of symptoms is of a nonorganic, psychogenic, or idiopathic nature, whereas *disease* validates the organic and pathophysiological nature of an illness. This author advocates the use of *disease* rather than *syndrome* when describing fibromyalgia in appreciation of the real, organic, biochemical, and histopathological (*histo*=cells and tissues, *pathological*=disease) findings which clearly indicate that fibromyalgia is a specific disease entity and not simply a psychogenic or enigmatic cluster of symptoms. If fibromyalgia is a real, organic clinical entity (as will be documented here), then the appropriate designation is *fibromyalgia disease* (FMD) rather than *fibromyalgia syndrome* (FMS). For consistency and clarity within this section, the general term "fibromyalgia" will be used. Relatedly, the term "irritable bowel syndrome" (IBS) is also a misnomer that confuses professionals as well as the general public into thinking that the condition does not have identified causes and (nonpharmaceutical) treatments; despite promulgations to the contrary, the cause of IBS is well-known[232], and effective treatment is readily available. When in 2010 the American College of Rheumatology (ACR) mistakenly allowed the diagnostic criteria for FM to be hijacked to leverage more drug sales—facilitate the making of the diagnosis, broaden the number of people affected, and impair the "cure" of and escape from the disease—the legitimate meaning of the diagnosis of course became less specific and much broader and therefore less meaningful. Per the original diagnostic criteria published in 1990, fibromyalgia is a legitimate functional disease and should be described and labeled as such; per the ridiculous diagnostic criteria published in 2010, fibromyalgia is a symptom cluster of aches, pains, and other signs and symptoms that can accompany virtually any other disease. Classic legitimate "FM" or "FMD" is what is described in this section (per the 1990 criteria), not simply the pain and symptom cluster (FMS) per the 2010 criteria. The idea that any and all patients with widespread pain should qualify for a diagnosis of fibromyalgia is ludicrous, and the 2010 ACR criteria for fibromyalgia are an obfuscating disservice to patients, doctors, and researchers while only benefiting the drug companies that can now sell their expensive and ineffective drugs *more quickly* to a *larger audience* for a *longer duration*. Patients develop true *primary* FM/FMD per the description that follows; patients can develop "fibromyalgia-like syndrome" or "fibromyalgia syndrome"—ie, FMS—by many means, including head injury or major trauma—these cases are more legitimately titled *secondary* hyperalgesic/allodynic central sensitization and/or myofascial pain syndromes.

- Prevalence, symptoms, and clinical findings: Fibromyalgia is one of the most common chronic pain conditions, affecting an estimated 10 million people in the U.S. and an estimated 3-6% of people world-wide.[233]

---

[229] Chakrabarty S, Zoorob R. Fibromyalgia. *Am Fam Physician*. 2007 Jul 15;76(2):247-54
[230] Tierney ML. McPhee SJ, Papadakis MA (eds). *Current Medical Diagnosis and Treatment 2006, 45th Edition*. New York: Lange Medical Books, pages 820-821
[231] Simms RW. Nonarticular soft tissue disorders. In Andreoli TE, Carpenter CCJ, Griggs RC, Benjamin IJ (eds). *Cecil Essentials of Medicine. 7th Edition*. Elsevier 2007:851-2
[232] Lin HC. Small intestinal bacterial overgrowth: a framework for understanding irritable bowel syndrome. *JAMA*. 2004 Aug 18;292(7):852-8
[233] "Fibromyalgia is one of the most common chronic pain conditions. The disorder affects an estimated 10 million people in the U.S. and an estimated 3-6% of the world population." National Fibromyalgia Association. fmaware.org/PageServera6cc.html?pagename=fibromyalgia_affected Accessed Sept 2012.

Approximately 10% of affected patients have severe symptoms resulting in partial or total disability. Affected patients report chronic aches, pains, and stiffness with a proclivity for localization near the neck, shoulders, low back, and hips. Pain and fatigue are typically exacerbated following physical exertion or psychological stress. Associated manifestations include fatigue, sleep disorders (including insomnia, unrefreshing sleep, and objective abnormalities such as an increase in stage 1 sleep, a reduction in delta sleep, and alpha-delta sleep anomaly), subjective numbness, headaches, and gastrointestinal disturbances consistent with a clinical diagnosis of irritable bowel syndrome (IBS). Clinical findings shared between FM and IBS include abdominal pain and discomfort, changed frequency of stool, diarrhea and/or constipation, abdominal bloating/distention/gas and flatulence, dyspepsia/heartburn, headaches especially migraine-type headaches), fatigue, myalgias, restless leg syndrome, anxiety, and depression. The **high prevalence (>50%) of migraine-type headaches in FM patients** suggests an underlying pathogenesis shared between cephalgia (*ceph*=head, *algia*=pain) and widespread myalgia (*myo*=muscle, *algia*=pain); one of the established and most likely causative abnormalities shared between migraine and FM is impaired mitochondrial function, which will be explained in greater detail later in this publication. Cognitive symptoms such as "brain fog" ("fibro-fog") and difficulty with memory and word retrieval, as well as **environmental intolerance (EI) and multiple chemical sensitivity (MCS)**, are seen in both FM and CFS[234]; again, this overlap of shared symptoms suggests a common etiopathogenesis (*etio*=cause, *patho*=disease, *genesis*=initiation). Routine physical examination and laboratory findings are generally normal, with the exception the physical examination finding of fibromyalgia tender points (described and diagrammed below in the section on Diagnosis per the 1990 diagnostic criteria).

	Objective "organic" (ie, real) abnormalities in fibromyalgia dispel the myth that the condition is psychogenic
1.	Histologic and functional abnormalities in muscle tissue: Disorganization of actin filaments, accumulation of lipofuscin bodies consistent with premature muscle aging, increased DNA fragmentation, and focal areas of chronic muscle contraction, reduced perfusion of muscle tissue during exercise (i.e., reduced blood flow to muscles).
2.	Mitochondrial defects: Accumulation of glycogen (muscle sugar) and lipid (fat) indicate that intracellular energy production is impaired and that the cells are unable to efficiently convert fuel sources into energy in the form of ATP, adenosine triphosphate, which is the basic fuel source for cellular metabolism. Also noted are significant reductions in the number of mitochondria, reduced activity of important enzymes such as 3-hydroxy-CoA dehydrogenase, citrate synthase, and cytochrome oxidase. Nutritional deficiencies, such as CoQ-10 deficiency, promote mitochondrial dysfunction, thus leading to mitochondrial destruction (mitophagy) which ultimately results in reduced numbers of mitochondria and perpetuates and aggravates muscle fatigue, pain, and neurocognitive dysfunction (i.e., brain fog, difficulty thinking, depression).
3.	Oxidative stress: Increased oxidative stress results from mitochondrial dysfunction and nutrient depletion.
4.	Neuroendocrine abnormalities: Hypothalamic-pituitary-adrenal (HPA) disturbance indicates impaired function of the brain and endocrine system.
5.	Elevated brain glutamate, homocysteine, and interlukin-8: See in FM, also CFS and CRPS.
6.	Low-grade immune activation: Increased cytokine production indicates a pro-inflammatory state.
7.	Bacterial overgrowth in the intestines: FM patients have excess/overgrowth of bacteria in their intestines, referred to as SIBO (small intestine bacterial overgrowth); CRPS patients have an abnormal pattern of microbial growth.
8.	Increased intestinal permeability, "leaky gut": Generally this indicates a gastrointestinal disorder, including but not limited to microbial imbalance (dysbiosis) or overt infection. Leaky gut is seen in both FM and CRPS.
9.	High prevalence of vitamin D deficiency: Common in the general population but more common in patients with chronic pain; vitamin D deficiency causes chronic pain, depression/anxiety, and low-grade inflammation—all of these problems are seen in patients with fibromyalgia.
10.	Low blood levels of L-tryptophan: FM patients have low levels of the amino acid tryptophan in their blood, despite adequate dietary intake. The most likely explanation for the deficiency of tryptophan is destruction of tryptophan by bacterial enzyme action. Several intestinal bacteria produce the enzyme tryptophanase, which destroys the amino acid tryptophan. Bacterial overgrowth results in more tryptophanase, resulting in tryptophan deficiency. Deficiency of tryptophan results in deficiencies of the hormones serotonin and melatonin, which result in anxiety, depression, food/sugar cravings, unrestful sleep, and mitochondrial dysfunction, since deficiency of melatonin causes reduced mitochondrial energy-production efficiency.

---

[234] Brown MM, Jason LA. Functioning in individuals with chronic fatigue syndrome: increased impairment with co-occurring multiple chemical sensitivity and fibromyalgia. *Dyn Med*. 2007 May 31;6:6 dynamic-med.com/content/6/1/6

Pathophysiology of Pain and Mitochondrial Dysfunction in Fibromyalgia

- Patients with FM demonstrate multiple biochemical abnormalities, centering on mitochondrial dysfunction, oxidative stress, and increased pain perception: Ultrastructural and biochemical abnormalities appear to be more pathologically significant and clinically relevant than the noted histological changes in skeletal muscle biopsy samples. Importantly, **the biochemical abnormalities *are the cause* of the histologic/tissue abnormalities**. Numerous **mitochondrial enzyme defects are seen**, including reduced activity of 3-hydroxy-CoA dehydrogenase, citrate synthase, and cytochrome oxidase. Levels of free magnesium are reduced by 31%, and levels of complexed ATP-magnesium are reduced by 12% in muscle from FM patients compared with levels seen in healthy controls; these biochemical and bioenergetic defects contribute to rapid-onset fatigue and muscle pain. From a neurophysiological perspective, *magnesium deficiency* can promote hypersensitivity to pain due to a reduction in the partial blockade of N-methyl-D-aspartate (NMDA) neurotransmitter receptor sites.[235] Reduced perfusion of muscle tissue during exercise results in relative tissue hypoxia, reduced muscle healing after the microtrauma of exercise, and promotion of muscle soreness due to accumulation of L-lactate (lactic acid).[236] **Increased oxidative stress** is also seen in FM patients,[237] providing additional objective evidence of the systemic, organic, and non-psychogenic nature of the illness. Evidence of hypothalamic-pituitary-adrenal disturbance and **increased cytokine production (particularly interleukin-8, which promotes sympathetic pain, and interleukin-6, which induces hyperalgesia [increased perception of pain], fatigue, and depression**[238]) further characterize the systemic and organic nature of this condition and are well documented in the research literature. **The majority of fibromyalgia patients demonstrate laboratory evidence of bacterial overgrowth in the small bowel**[239], and the details and important implications of this will be discussed below. **Vitamin D deficiency**—a recognized cause of chronic widespread pain as well as depression, muscle fatigue, and chronic low-grade inflammation—is also common in fibromyalgia patients.[240,241] FM patients have **significantly elevated blood levels of pentosidine**, which is an advanced glycation end-product (AGE) and marker of oxidative stress and glycosylation (sugar-protein binding); AGEs promote chronic inflammation and nociceptive sensitization leading to chronic pain.[242] Another AGE very similar to pentosidine, **carboxy-methyl-lysine (CML) is found in higher levels in the blood and muscle of FM patients**[243]; both pentosidine and CML cause expedited "muscle aging" and promote chronic pain and inflammation. These objective abnormalities of biochemical, histological, nutritional, and microbiological/gastrointestinal status force clinicians to appreciate the valid and organic nature of fibromyalgia. As previously stated, this evidence refutes promulgations espoused within standard allopathic/pharmaceutical medicine that fibromyalgia is an idiopathic condition warranting lifelong medicalization with expensive and potentially hazardous analgesic and antidepressant drugs; the focus on drug treatment to mask/suppress the pain of fibromyalgia detours doctors and patients away from focusing on the legitimate and validated causes of fibromyalgia and physiology-based (rather than pharmacology-based) means for alleviating the suffering and pain that these patients experience. CRPS patients also show evidence of biochemical and metabolic abnormalities, including increased oxidative stress and impaired mitochondrial function.[244]
    - Chronic widespread pain: increased glutamate and lactate concentrations in the trapezius muscle and plasma. (*Clin J Pain.* 2014 May[245]): "Chronic widespread pain (CWP), including fibromyalgia syndrome (FM), is associated with prominent negative consequences. CWP has been associated with alterations in

---

[235] Park JH, Niermann KJ, Olsen N. Evidence for metabolic abnormalities in the muscles of patients with fibromyalgia. *Curr Rheumatol Rep.* 2000 Apr;2(2):131-40
[236] Elvin et al. Decreased muscle blood flow in fibromyalgia patients during standardised muscle exercise. *Eur J Pain.* 2006 Feb;10(2):137-44
[237] Altindag O, Celik H. Total antioxidant capacity and the severity of the pain in patients with fibromyalgia. *Redox Rep.* 2006;11(3):131-5
[238] Wallace DJ, Linker-Israeli M, Hallegua D, et al. Cytokines play an aetiopathogenic role in fibromyalgia. *Rheumatology* (Oxford). 2001 Jul;40(7):743-9
[239] Pimentel et al. A link between irritable bowel syndrome and fibromyalgia may be related to findings on lactulose breath testing. *Ann Rheum Dis.* 2004 Apr;63(4):450-2
[240] Huisman AM, White KP, Algra A, et al. Vitamin D levels in women with systemic lupus erythematosus and fibromyalgia. *J Rheumatol.* 2001 Nov;28(11):2535-9
[241] Armstrong DJ, Meenagh GK, Bickle I, et al. Vitamin D deficiency is associated with anxiety and depression in fibromyalgia. *Clin Rheumatol.* 2007 Apr;26(4):551-4
[242] Hein G, Franke S. Are advanced glycation end-product-modified proteins of pathogenetic importance in fibromyalgia? *Rheumatology* (Oxford). 2002 Oct;41(10):1163-7
[243] "In the interstitial connective tissue of fibromyalgic muscles we found a more intensive staining of the AGE CML, activated NF-kappaB, and also higher CML levels in the serum of these patients compared to the controls. RAGE was only present in FM muscle." Rüster M, Franke S, Späth M, Pongratz DE, Stein G, Hein GE. Detection of elevated N epsilon-carboxymethyllysine levels in muscular tissue and in serum of patients with fibromyalgia. *Scand J Rheumatol.* 2005 Nov-Dec;34(6):460-3
[244] "Recent evidence demonstrates that oxidative stress is associated with clinical symptoms in patients with CRPS-I. ... This review summarises the effect of oxidative stress and mitochondrial dysfunction in the pathogenesis of CRPS." Taha R, Blaise GA. Update on the pathogenesis of complex regional pain syndrome: role of oxidative stress. *Can J Anaesth.* 2012 Sep;59(9):875-81
[245] Gerdle et al. Chronic widespread pain: increased glutamate and lactate concentrations in the trapezius muscle and plasma. *Clin J Pain.* 2014 May;30(5):409-20. See also "Significantly higher interstitial concentrations of pyruvate and lactate were found in patients with fibromyalgia- syndrome. The multivariate regression analyses of group membership and pressure pain thresholds of the trapezius confirmed the importance of pyruvate and lactate." Gerdle et al. Increased interstitial concentrations of pyruvate and lactate in the trapezius muscle of patients with fibromyalgia: a microdialysis study. *J Rehabil Med.* 2010 Jul;42(7):679-87

the central processing of nociception. ...CWP patients had significantly increased interstitial muscle and plasma concentrations of lactate and glutamate. No significant differences existed in blood flow between CWP and CON [controls]. The interstitial concentrations-but not the plasma levels-of glutamate and lactate correlated significantly with aspects of pain such as pressure pain thresholds of the trapezius and tibialis anterior and the mean pain intensity in CWP but not in CON." Elevated lactate correlates with and suggests the presence of impaired energy/ATP production, consistent with mitochondrial dysfunction, while elevated glutamate levels is expected to promote enhanced pain reception, given that glutamate is the classic excitatory neurotransmitter (activator of glutamate receptors in general and the NMDA receptor in particular). The combination of elevated lactate (e.g., muscle impairment leading to muscle pain) and elevated glutamate (e.g., enhanced sensitivity to pain) is the perfect recipe and explanation for muscle-generated pain with enhanced sensitivity to pain that might otherwise be well tolerated. In addition to lactate and glutamate, patients with chronic muscle pain also show tissue-specific elevations of pyruvate (again indicating impaired energy/ATP production in muscles, leading to easy fatigability and increased achiness/pain) and elevated serotonin[246]; although we generally think of serotonin as having a relaxing and analgesic effect in the *central* nervous system, serotonin in the *periphery* appears to promote pain perception and sensitization while also promoting inflammation.[247]

- Patients with FM demonstrate abnormalities in muscle tissue and mitochondrial function: Muscle biopsies from patients with fibromyalgia show numerous histological, ultrastructural, and biochemical abnormalities, including defects in mitochondrial structure and function, reduced numbers of capillaries in skeletal muscle (leading to reduced blood supply to muscles), thickened capillary endothelium (thicker vessel walls), and ragged red fibers consistent with the development of **mitochondrial myopathy** (*myo*=muscle, *pathos*=disease). The histological finding of "rubber-band morphology" with reticular threads connecting neighboring cells in muscle biopsies of FM patients is associated with prolonged contractions in adjacent/neighboring muscle fibers; these abnormalities result in and perpetuate a low-energy state within myocytes (*myo*=muscle, *cytes*=cells).[248] Other studies have shown disorganization of actin filaments, accumulation of lipofuscin (cellular debris) consistent with premature muscle aging, accumulation of glycogen and lipid accumulation consistent with **mitochondrial impairment**, increased DNA fragmentation, **significant reductions in the number of mitochondria**, and focal areas of chronic muscle contraction.[249] These histological abnormalities are important and establish the fact that **fibromyalgia is a** *disease of metabolic dysfunction* rather than an *emotional disorder of psychogenic origin*; therefore, attributing the pain and fatigue of fibromyalgia to a mental-psychological cause or a central nervous system disorder such as central sensitization is unscientific and illogical. Patients with CRPS also show evidence of impaired oxygen diffusion and mitochondrial impairment[250], and Tan et al[251] specifically noted that mitochondrial ETC "complex II activity in the CRPS I patients was significantly lower."

---

[246] "Several studies clearly showed elevated levels of serotonin, glutamate, lactate, and pyruvate in localized chronic myalgias and may be potential biomarkers." Gerdle et al. Chronic musculoskeletal pain: review of mechanisms and biochemical biomarkers as assessed by the microdialysis technique. *J Pain Res*. 2014 Jun 12;7:313-26
[247] "5-HT, acting in combination with other inflammatory mediators, may ectopically excite and sensitize afferent nerve fibers, thus contributing to peripheral sensitization and hyperalgesia in inflammation and nerve injury." Sommer C. Serotonin in pain and analgesia in the periphery. *Mol Neurobiol*. 2004 Oct:117-25. See also: "5-HT sensitizes afferent nerve fibers, thus contributing to hyperalgesia in inflammation and nerve injury." Sommer C. Is serotonin hyperalgesic or analgesic? *Curr Pain Headache Rep*. 2006 Apr; 101-6
[248] Olsen NJ, Park JH. Skeletal muscle abnormalities in patients with fibromyalgia. *Am J Med Sci*. 1998 Jun;315(6):351-8
[249] Sprott H, Salemi S, Gay RE, et al. Increased DNA fragmentation and ultrastructural changes in fibromyalgic muscle fibres. *Ann Rheum Dis*. 2004 Mar;63(3):245-51
[250] "The mean venous oxygen saturation (S(v)O(2)) value (94.3% ± 4.0%) of the affected limb was significantly higher than S(v)O(2) values found in healthy subjects (77.5% ± 9.8%) pointing to a severely decreased oxygen diffusion or utilization within the affected limb. ... Ultrastructural investigations of soleus skeletal muscle capillaries revealed thickened endothelial cells and thickened basement membranes. Muscle capillary densities were decreased in comparison with literature data. High venous oxygen saturation levels were partially explained by impaired diffusion of oxygen due to thickened basement membrane and decreased capillary density. ...The abnormal skeletal muscle findings points to severe disuse but only partially explain the impaired diffusion of oxygen; mitochondrial dysfunction seems a likely explanation in addition." Tan et al. Impaired oxygen utilization in skeletal muscle of CRPS I patients. *J Surg Res*. 2012 Mar;173(1):145-52
[251] "We observed that mitochondria obtained from CRPS I muscle tissue displayed reduced mitochondrial ATP production and substrate oxidation rates in comparison to control muscle tissue. Moreover, we observed reactive oxygen species evoked damage to mitochondrial proteins and reduced MnSOD levels." Tan et al. Mitochondrial dysfunction in muscle tissue of complex regional pain syndrome type I patients. *Eur J Pain*. 2011 Aug;15(7):708-15

CONTROL

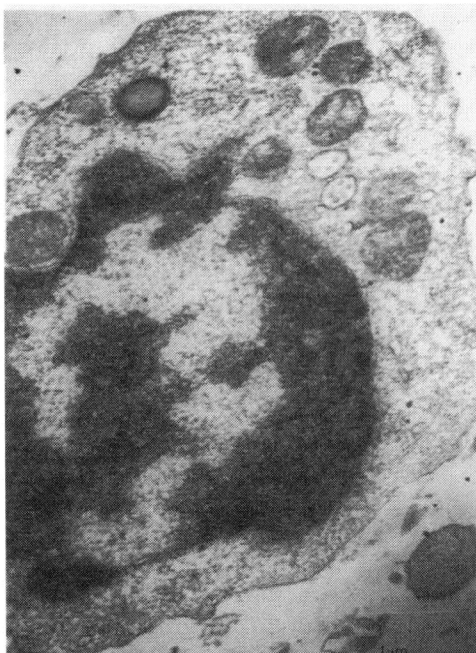

PATIENT

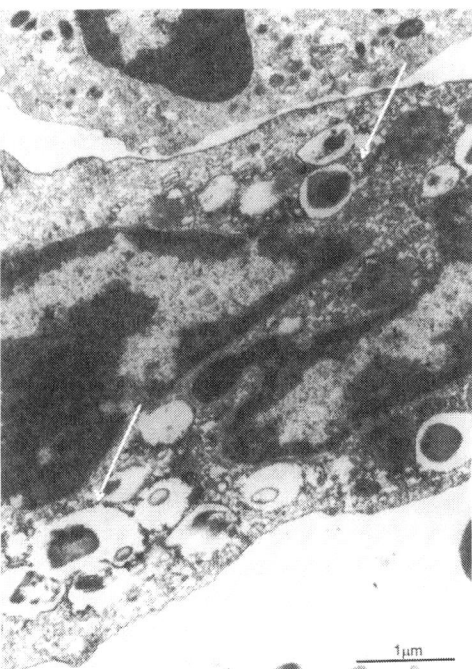

Blood cells in FM patients show mitochondrial destruction (mitophagy), smaller size and lower number of mitochondria: Structure of blood mononuclear cells (BMCs, cells of the immune system) from FM patients. The healthy/control BMCs show mitochondria with a normal structure. Autophagosomes (indicated by arrows), where mitochondria are destroyed (the process of mitophagy [*mito*=mitochondria, *phagy*=consumption], are noted in the BMCs of patients with FM. [Bar = 1 micrometer]. This open-access image is respectfully attributed to the brilliant research published by these researchers Cordero MD, De Miguel M, Moreno Fernández AM, Carmona López IM, Garrido Maraver J, Cotán D, Gómez Izquierdo L, Bonal P, Campa F, Bullon P, Navas P, Sánchez Alcázar JA. Mitochondrial dysfunction and mitophagy activation in blood mononuclear cells of fibromyalgia patients. *Arthritis Res Ther.* 2010;12(1):R17 arthritis-research.com/content/12/1/R17

Mitophagy: The body's inherent mechanism for the destruction of dysfunctional mitochondria
Concept: Autophagic destruction of mitochondria is termed "mitophagy" and is the body's inherent mechanism for eliminating superfluous or dysfunctional mitochondria; this generally has a protective and life-sustaining effect. However, in the case of fibromyalgia wherein the mitochondrial dysfunction is persistent, prolonged mitophagy contributes to failure of adequate energy production and thereby contributes to clinical manifestations of fatigue, dyscognition, and impaired exercise/activity performance. Further, the consistent documentation of significant mitophagy in patients with fibromyalgia proves the biological/organic/real/pathophysiologic character of the illness and refutes the pharmacocentric paradigm which holds that the condition is of psychogenic or neurologic origin and thus to be treated with so-called "antidepressants" and/or analgesic drugs, respectively. • "The removal of damaged mitochondria that could contribute to cellular dysfunction or death is achieved through process of mitochondrial autophagy, i.e. mitophagy." Novak I. *Antioxid Redox Signal.* 2011 • "Mitochondrial number and health are regulated by mitophagy, a process by which excessive or damaged mitochondria are subjected to autophagic degradation." Rambold. *Cell Cycle.* 2011 • **"Autophagy can be beneficial for the cells by eliminating dysfunctional mitochondria, but massive autophagy can promote cell injury and may contribute to the pathophysiology of FM."** Cordero. *Arthritis Res Ther.* 2010

- <u>Pain in fibromyalgia originates peripherally and is amplified centrally</u>: The pain of fibromyalgia originates from the muscles[252] secondary to stimulation by oxidative and inflammatory mediators and is excessively amplified in the brain and spinal cord; another possible peripheral contribution to pain inputs is degeneration of nerve fibers in the skin.[253] To risk redundancy for clarity: **FM pain originates *peripherally* in the muscles**

---

[252] "Results of these studies suggest that FM pain is associated with widespread primary and secondary cutaneous hyperalgesia, which are dynamically maintained by tonic impulse input from deep tissues and likely by brain-to-spinal cord facilitation. Enhanced somatic pains are accompanied by mechanical hyperalgesia and allodynia in FM patients as compared with healthy controls. FM pain is likely to be at least partially maintained by peripheral impulse input from deep tissues. This conclusion is supported by results of several studies showing that injection of local anesthetics into painful muscles normalizes somatic hyperalgesia in FM patients." Staud R. Is it all central sensitization? Role of peripheral tissue nociception in chronic musculoskeletal pain. *Curr Rheumatol Rep.* 2010 Dec;12(6):448-54

[253] "The study's instruments comprised the Michigan Neuropathy Screening Instrument (MNSI), the Utah Early Neuropathy Scale (UENS), distal-leg neurodiagnostic skin biopsies, plus autonomic-function testing (AFT). We found that 41% of skin biopsies from subjects with fibromyalgia vs 3% of biopsies from control subjects were diagnostic for small-fiber polyneuropathy (SFPN), and MNSI and UENS scores were higher in patients with fibromyalgia than in control subjects (all $P \leq 0.001$). Abnormal AFTs were equally prevalent, suggesting that fibromyalgia-associated SFPN is primarily somatic. Blood tests from subjects with fibromyalgia and SFPN-diagnostic skin biopsies provided insights into causes. All glucose tolerance tests were normal, but 8 subjects had dysimmune markers, 2 had hepatitis C serologies, and 1 family had apparent genetic causality. These

Chapter 5.1—Functional Inflammology Protocol for Metabolic Inflammation: Migraine and Fibromyalgia

(and likely in the skin as well, at least in some patients) and is amplified *centrally* in the spinal cord and brain. Following reception and amplification of the original muscular pain, the peripheral "receptive field" grows both in size and intensity/hypersensitivity to include the skin, so that various skin inputs are perceived as pain; the two main types of dysfunctional pain sensitivity/sensations are allodynia (reception of nonpainful stimuli as pain) and hyperalgesia (extended duration and increased intensity of pain).

- Enhanced central pain processing of fibromyalgia patients is maintained by muscle afferent input (*Pain.* 2009 Sep[254]): "Lidocaine injections increased local pain thresholds and decreased remote secondary heat hyperalgesia in FM patients, emphasizing the important role of peripheral impulse input in maintaining central sensitization in this chronic pain syndrome; similar to other persistent pain conditions such as irritable bowel syndrome and complex regional pain syndrome."

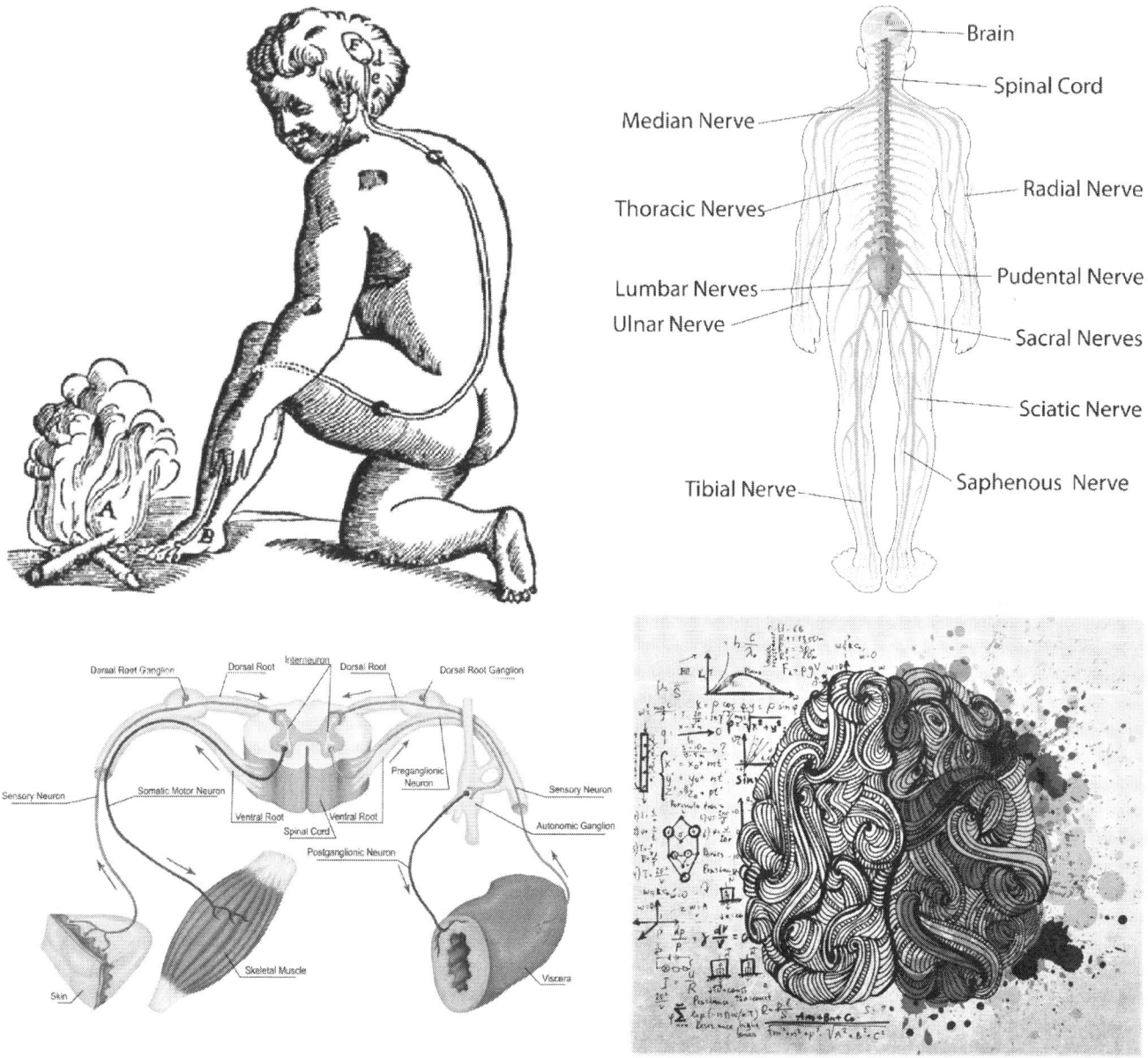

More than 400 years of the history and development of neuroanatomy and neurophysiology represented in four images: These four images in sequence represent the history and development of the fields of neuroanatomy and neurophysiology, ❶ starting with the drawing by Descartes in the 1600s, ❷ the tracing of nerves throughout the body, ❸ what might be called the

---

findings suggest that some patients with chronic pain labeled as fibromyalgia have unrecognized SFPN, a distinct disease that can be tested for objectively and sometimes treated definitively." Oaklander et al. Objective evidence that small-fiber polyneuropathy underlies some illnesses currently labeled as fibromyalgia. *Pain.* 2013 Nov;154(11):2310-6
[254] Staud et al. Enhanced central pain processing of fibromyalgia patients is maintained by muscle afferent input. *Pain.* 2009 Sep;145(1-2):96-104

start of the "functional neurology revolution" with the work of Melzack and Wall in 1965, and ❹ culminating with today's understanding of neurophysiology and neuropathophysiology as highly dynamic and interactive processes, far advanced from the simple models of sequential reception, perception, and interpretation.

1. *Image upper left*—**Peripheral reception and central perception of pain as simple linear connectivity**: The famous historical image from Rene Descartes (French polymath, 1596-1650) was an important beginning in society's understanding that pain had a neurophysiologic basis; in this model, pain sensations were received peripherally, transmitted by nerves to the brain where they were perceived and acted upon (e.g., reflexive withdrawal from painful stimuli).

2. *Image upper right*—**Nervous system as simple anatomic connectivity**: Advances in anatomy and neuroanatomy later provided more details about neurologic pathways and synapses (connections between nerves), but the overall model remained very mechanical, resembling electric circuitry (e.g., wiring) transmitting signals which were simply received in the brain for interpretation. Important within this obvious, simple, and overly simplistic model of neuroanatomy is the facile assumption of "stimulus-response specificity"—the idea that specific sensory receptors receive only one type of sensory input such as temperature or light touch or vibration; what we appreciate instead is that receptors, nerves, and their connections within the spinal cord, brainstem, and brain can communicate a wide range of sensory inputs in ways that commonly defy anatomic expectations. As stated by Wall[255], "Obviously anatomy does not predict physiology." Wall goes on to dismantle "the specificity theory" with the observations that "nociceptors become sensitized by prolonged or repeated noxious stimulation so that their threshold drops and they are excited by normally innocuous stimuli" and "one should be cautious in attributing one and only one function to a particular cell ...the degree of convergence is under control of other afferents and descending systems and is also dependent on the activity of segmental interneurons."

3. *Image lower left*—**Interconnections of nerves allows for "spill over" of sensory inputs and dynamic changes in perception, including amplification and misinterpretation**: Research published by Melzack and Wall[256] in 1965 is commonly credited with revolutionizing our view of pain processing by changing the paradigm from a static model of linear connectivity (e.g., sensory receptors to peripheral nerve to spinal cord to the brain) to one of dynamic interconnectivity, with interactive interconnections and opportunities for inhibition, amplification and misinterpretation at nearly every level. The perception of pain is neither simple, nor solely anatomy-based, nor static; it is a dynamic process reflecting the sum total interplay of peripheral reception (e.g., modifiable by inflammation), nerve transmission (e.g., modifiable by nutritional status and glutamate levels), reception and "intermixing" (e.g., pre-synaptic inhibition/facilitation, post-synaptic inhibition/facilitation) in the spinal cord, brain stem, subcortical structures such as the thalamus, and the "conscious" brain cortex. Furthermore and very importantly, we know that sensory nerves and sensory nerve endings do much more than simply receive information; in addition to sensing various types of stimuli through various types of sensors in nerve endings, sensory nerves directly influence the peripheral environment by releasing inflammatory mediators via a process called "neurogenic inflammation." Neurogenic (*neuro*—nerve, *genic*—generated, originating) inflammation again forces us to expand our perception of the nervous system and its components; in neurogenic inflammation, stimulated sensory nerves respond to painful stimuli by releasing inflammatory mediators into the self-same environment, thereby adding an inflammatory component to whatever is perceived as pain-inducing. One can think of this as providing a survival advantage in the short-term, as the inflammatory response generated by the release of these pro-inflammatory mediators released by nerves will serve to prepare the area for infiltration by immune cells to remove and repair damage and repel any infectious agents that might have entered with the injury. However, when chronically sustained in so-called "chronic diseases", neurogenic inflammation contributes to tissue injury, as we see in complex regional pain syndrome (CRPS). Indeed, CRPS might be considered a prototype of the effects of neurogenic inflammation. In a review published in 2015, Littlejohn[257] writes, "Neurogenic inflammation—comprising tissue swelling, vasomotor changes and marked allodynia—also contributes substantially to the clinical features of CRPS."; he goes on to mention the clinical sequelae that have a strong or dominant origination from neurogenic inflammation, including bone marrow edema, osteopenia, visceral pain, hyperalgesia (increased intensity and duration of pain sensitivity), allodynia (misperception of sensory input as pain), abnormal hair and nail growth (including absence of same), rashes, sweating, vasodilation (causing heat and redness), vasoconstriction (causing cold and blueness), skin ulceration and fibrosis of skin and joints. Littlejohn goes on to describe how neurogenic inflammation can promote itself, as nerve-released inflammatory mediators cause tissue damage that leads to pain, the self-same sensory nerves respond to the self-generated pain by causing more neurogenic inflammation; the proper term for this concept and phenomenon is "*neurogenic* neurogenic inflammation" or "neurogenic inflammation-induced neurogenic inflammation", a vicious cycle. I (DrV[258]) disagree with Littlejohn that this vicious cycle is at the core of fibromyalgia, although I agree that it could and likely does play a contributing part. Lastly and also included in the review by Littlejohn, the idea of "neurogenic neuroinflammation" likely has merit based on our current understanding of neuropathophysiology; especially in the spinal cord and brain, neuronal activity triggers microglial inflammation, which thereby inflames the nearby nerves for a vicious cycle of neurogenic neuroinflammation. This neurogenic neuroinflammation is almost certain to play a major role in CRPS, importantly but to a lesser degree in FM (chronic) and migraine (periodic). In the context of the brain and spinal cord,

---

[255] Wall PD. The gate control theory of pain mechanisms. A re-examination and re-statement. *Brain*. 1978 Mar;101(1):1-18
[256] Melzack R, Wall PD. Pain mechanisms: a new theory. *Science*. 1965 Nov 19;150(3699):971–979
[257] Littlejohn G. Neurogenic neuroinflammation in fibromyalgia and complex regional pain syndrome. *Nat Rev Rheumatol*. 2015 Nov;11(11):639-48
[258] At the time this citation is added in 2015 Dec, I (DrV) had recently submitted a reply to the article by Littlejohn (Neurogenic neuroinflammation in fibromyalgia and complex regional pain syndrome. *Nat Rev Rheumatol*. 2015 Nov) pointing out some of the bitemporal hemianopsia of the aforecited article; my reply will be published either by *Nat Rev Rheumatol* or elsewhere and will be posted at ichnfm.academia.edu/AlexVasquez as soon as possible. In reply to the previously cited article, Cordeo—a preeminent researcher in the field of fibromyalgia's mitochondrial dysfunction and its remediation by the vitamin-like substance coenzyme Q10 (CoQ10)—noted that neurogenic neuroinflammation includes activation of the NLRP3 inflammasome, which he suggests could be a therapeutic target (e.g., by CoQ10). "Interestingly, inflammasome activation in the CNS primarily occurs in microglia and macrophages. Microglia have been highly studied as key contributors to pathological and chronic pain mechanisms, and are involved in hyperalgesia and allodynia in both fibromyalgia and chronic fatigue syndrome, as well as pain in CRPS. Microglial inflammasome activation promotes the recruitment of peripheral innate immune cells (macrophages) and adaptive immune cells (T cells and B cells), as well as further activating nearby glial cells." Cordero MD. The inflammasome in fibromyalgia and CRPS: a microglial hypothesis? *Nat Rev Rheumatol*. 2015 Nov;11(11):630. See also Cordero MD et al. NLRP3 inflammasome is activated in fibromyalgia: the effect of coenzyme Q10. *Antioxid Redox Signal*. 2014 Mar 10;20(8):1169-80

neurogenic neuroinflammation contains an important pro-inflammatory contributor that is phenotypically ready to perceive, amplify, and exacerbate neuroinflammation caused by increased neuroactivity—the microglial cells; thus, neurogenic (in this context, including the microglia and astrocytes as components of the brain and spinal cord) neuroinflammation would be expected to participate in seizure disorders and vaccine-induced encephalomyelitis.

4. **Image lower right—The nervous system (represented by artistic brain image) is now appreciated as dynamic and interactive receiver and processor of sensory information**: In modern times, pain processing is appreciated as a dynamic, complex, and interactive process at every level, from ❶ peripheral reception of stimuli (e.g., in the skin or muscles), to the ❷ spinal cord, to the ❸ brainstem, to the ❹ subcortical structures especially the thalamus, to the ❺ cortex. Generally appreciated is that much "spill-over", "misinterpretation", inhibition and amplification" can occur in the spinal cord, brainstem, and structures of the brain, so that the initial perception of pain—in the muscles for example in the case of FM—is amplified at multiple levels and "spills over" to be perceived as skin pain, resulting in allodynia (misinterpretation of light touch as pain) and hyperalgesia (pain perception is amplified in intensity and duration). The brain is constantly adapting to input; for example the brain forms neuron-neuron connections in various patterns to produce memory. When the brain is constantly receiving messages of pain, the brain "rewires" the neuron-neuron interconnections to increase pain processing—what might be called a "pain memory"—in a way that facilitates the perception of pain, leading to enhanced pain perception, e.g., more pain felt by the patient.

Diagnosis

- Clinical criteria—description and contrast of the 1990 criteria and the 2010 criteria: Per guidelines published in 1990 by the American College of Rheumatology (ACR), a diagnosis of fibromyalgia can be made in a patient with inexplicable, widespread myofascial pain of at least 3 months' duration; *inexplicable* denotes normalcy of routine laboratory and physical examination findings and failure to find an alternate explanation or diagnosis, while *widespread* denotes bilateral pain above and below the waist not attributable to trauma or rheumatic disease and with pain at 11 of 18 classic tender point locations (see illustration below).

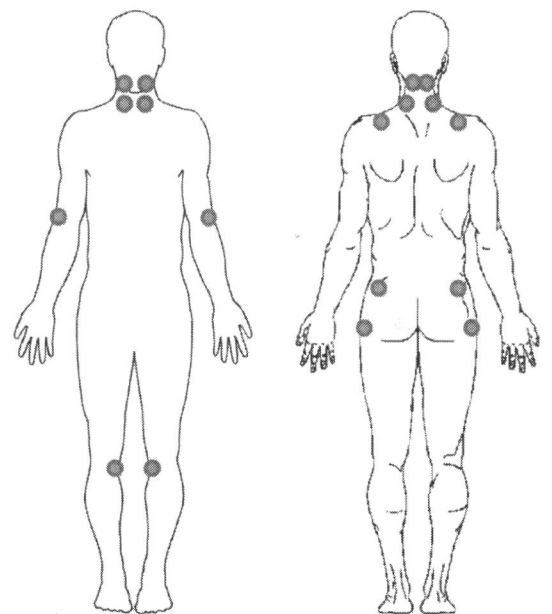

Illustration of the 9 paired locations of FM tender points:
Pain, on digital palpation, must be present in at least 11 of the following 18 tender point sites:
1. Occiput: at the suboccipital muscle insertions.
2. Low cervical: at the anterior aspects of the intertransverse spaces at C5-C7.
3. Trapezius: at the midpoint of the upper border.
4. Supraspinatus: at origins, above the scapula spine near the medial border.
5. Second rib: upper lateral to the second costochondral junction.
6. Lateral epicondyle: 2 cm distal to the epicondyles.
7. Gluteal: in upper outer quadrants of buttocks in anterior fold of muscle.
8. Greater trochanter: posterior to the trochanteric prominence.
9. Knee: at the medial fat pad proximal to the joint line.

Per 1990 ACR guidelines, the diagnosis of FM is supported when at least 11 out of 18 of these locations are painful. Digital palpation should be performed with an approximate force of 4 kg (9 lbs). A tender point has to be painful at palpation, not just "tender."[259]

FM tender points are assessed bilaterally at 9 paired sites: (sub)occiput (below the head at the neckline), low cervical spine (lower neck), trapezius and supraspinatus (two of the shoulder muscles), second rib (anterior, near costosternal [rib-breastbone] junction), lateral epicondyle, gluteal region, greater trochanter, and medial fat pad of the knees. Tender points are provoked by the clinician's application of approximately 9 pounds of fingertip pressure, which is sufficient to cause blanching of the clinician's nail bed. The tender points of fibromyalgia are distinguished from myofascial trigger points (MFTP, described by Travell[260]) and strain-counterstrain tender points (described in the osteopathic literature by Jones[261]). Pain must have been consistent

---

[259] The American College of Rheumatology 1990 Criteria for the Classification of Fibromyalgia. nfra.net/Diagnost.htm Accessed Nov 2011
[260] Simons DG, Travell JG, Simons LS. *Travell & Simons' Myofascial Pain and Dysfunction. The Trigger Point Manual*. Baltimore: Lippincott Williams & Wilkins; 1999
[261] Jones L, Kusunose R, Goering E. *Jones Strain-Counterstrain*. Carlsbad, Jones Strain Counterstrain Incorporated, 1995. [ISBN 0964513544]

for at least three months and must not be attributable to another (obvious) cause. In contrast to MFTP, which are located toward the center of the muscle fiber and which refer pain and show spontaneous electrocontractile activity[262], tender points of fibromyalgia are located near the tendinous insertions of muscle to bone and cause local pain only, without pain referral or contractile activity.

The new 2010 ACR guidelines for the diagnosis and assessment of FM[263] are significantly different from the 1990 guidelines. These new guidelines are mostly illogical and appear to have been structured to broaden the definition of fibromyalgia, to allow patients and nonphysicians to make the diagnosis, and generally to increase the patient population available for a diagnosis of fibromyalgia, thereby increasing the sales of drugs. Very curiously, the authors state that one of their objectives was to create criteria that "do not require a tender point examination"; at first, this seems odd and clinically inconsistent considering that the tender point examination ❶ takes only about 60 seconds to perform, ❷ is noninvasive, ❸ was previously the standard by which the diagnosis was made, and ❹ is reasonable and responsible—physical examination of patients with pain is a reasonable standard of care. Oddly, the authors of the new guidelines note several "important problems" with the 1990 ACR criteria, such as "Patients who improved or whose symptoms and tender points decreased could fail to satisfy the ACR 1990 classification definition" and "there was little variation in symptoms among fibromyalgia patients." Clinicians should note that these so-called "problems" *are not problems at all* because patients who improve and thus no longer meet diagnostic criteria should not be considered to have an active disease/diagnosis, and that high-quality clinical criteria should indeed result in the specific definition of clinical disorder and thus in a well-defined cohort of patients; correcting these "problems" results in patients being diagnosed for longer periods of time (more *long-term* patients) and also results in more patients being diagnosed with fibromyalgia (more *total* patients). Perhaps even more curious is the fact that development of these new guidelines was sponsored by Lilly Research Laboratories—the front page of the article states "these criteria were developed with support from the study sponsor, Lilly Research Laboratories", which is the "research" section of Eli Lilly and Company, one of the world's largest drug companies and the manufacturer of duloxetine/Cymbalta® which is one of the only drugs approved by the US Food and Drug Administration (FDA) for the treatment of fibromyalgia.[264] Among patients labeled with fibromyalgia, the new criteria increase the percentage of patients diagnosable by criteria from 75% to 88%; whether the motivation to expand the patient population diagnosed with fibromyalgia is altruistic or financially motivated is subject to debate. The new criteria rely on a summation of two tallies—"widespread pain index" (WPI) and "symptom severity" (SS, parts 1 and 2)—with the diagnosis being supported by either **"WPI >7 and SS >5"** or **"WPI 3–6 and SS >9"**.

A descriptive video of the distinctions between the 1990 and 2010 diagnostic criteria is available for free from www.InflammationMastery.com/pain.

---

[262] Hubbard DR, Berkoff GM. Myofascial trigger points show spontaneous needle EMG activity. *Spine*. 1993 Oct 1;18(13):1803-7
[263] Wolfe F, et al. The ACR preliminary diagnostic criteria for fibromyalgia and measurement of symptom severity. *Arthritis Care Res*. 2010;62(5):600-10
[264] lilly.com/research/Pages/research.aspx and newsroom.lilly.com/ReleaseDetail.cfm?releaseid=316740 Accessed January 2012

Chapter 5.1—Functional Inflammology Protocol for Metabolic Inflammation: Migraine and Fibromyalgia

Widespread pain index (WPI): Each positive location receives one point (max = 19)

1. Shoulder girdle, left
2. Shoulder girdle, right
3. Upper arm, left
4. Upper arm, right
5. Lower arm, left
6. Lower arm, right
7. Hip (buttock/trochanter), left
8. Hip (buttock/trochanter), right
9. Upper leg, left
10. Upper leg, right
11. Lower leg, left
12. Lower leg, right
13. Jaw, left
14. Jaw, right
15. Chest (sternum area)
16. Abdomen
17. Neck
18. Upper back
19. Lower back

Symptom severity (SS)—part 1: Each of these three problems is quantified with the following scale (max = 9):
0 none: no problem **1 mild**: intermittent or mild problems
2 moderate: often present, considerable problems **3 severe**: continuous, life-disturbing problems

0 1 2 3 Fatigue   0 1 2 3 Waking unrefreshed   0 1 2 3 Cognitive symptoms

Symptom severity (SS)—part 2: The clinician considers the patient's "somatic symptoms in general" (listed below) and applies the following scale (max = 3):
0 no symptoms 1 few symptoms
2 a moderate number of symptoms 3 a great deal of symptoms

0 1 2 3 muscle pain
0 1 2 3 irritable bowel syndrome
0 1 2 3 fatigue/tiredness
0 1 2 3 thinking or remembering problems
0 1 2 3 muscle weakness
0 1 2 3 headache
0 1 2 3 pain/cramps in the abdomen
0 1 2 3 numbness/tingling
0 1 2 3 dizziness
0 1 2 3 insomnia
0 1 2 3 depression
0 1 2 3 constipation
0 1 2 3 pain in the upper abdomen
0 1 2 3 nausea
0 1 2 3 nervousness
0 1 2 3 chest pain
0 1 2 3 blurred vision
0 1 2 3 fever
0 1 2 3 diarrhea
0 1 2 3 dry mouth

0 1 2 3 itching
0 1 2 3 wheezing
0 1 2 3 Raynaud's phenomenon
0 1 2 3 hives/welts
0 1 2 3 ringing in ears
0 1 2 3 vomiting
0 1 2 3 heartburn
0 1 2 3 oral ulcers
0 1 2 3 loss of/change in taste
0 1 2 3 seizures
0 1 2 3 dry eyes
0 1 2 3 shortness of breath
0 1 2 3 loss of appetite
0 1 2 3 skin rash
0 1 2 3 sun sensitivity
0 1 2 3 hearing difficulties
0 1 2 3 easy bruising
0 1 2 3 hair loss
0 1 2 3 frequent urination
0 1 2 3 painful urination
0 1 2 3 bladder spasms

Tally points from above; patients may be diagnosed with FM if "**WPI >7 and SS >5**" or "**WPI 3–6 and SS >9**".
WPI = _____
SS1 + SS2 = _____

2010 Fibromyalgia diagnostic criteria—summary, chart for clinical use, discussion: Per the 2010 diagnostic criteria[265], patients may be diagnosed with fibromyalgia if "WPI >7 and SS >5" or "WPI 3–6 and SS >9". Of note, these new criteria reflect a major departure from the former criteria published and codified in 1990; of additional note, publication of these new criteria received sponsorship from a drug company which has an FDA-approved drug for this condition—this represents a massive conflict of interest, full manifestation of which would have the same for-profit entity influencing the criteria used by doctors for the diagnosis and then providing (i.e., selling for profit) one of the only approved drug treatment options. I have discussed these conundrums in video format at Vimeo.com/ICHNFM and Vimeo.com/DrVasquez.

- <u>Clinical profile and findings on common laboratory tests</u>: New-onset fibromyalgia is unlikely over age 50, and the condition never causes fever, significant weight loss, or other objective signs of acute or subacute illness. Hypothyroidism is common and can produce widespread myofascial pain along with depression and other complaints, resulting in a clinical picture that closely resembles FM; thus, a complete thyroid evaluation (detailed later) is essential during the initial evaluation of any fibromyalgia-like condition. Common rheumatic

---

[265] Wolfe F, et al. The ACR preliminary diagnostic criteria for fibromyalgia and measurement of symptom severity. *Arthritis Care Res.* 2010;62(5):600-10

conditions such as rheumatoid arthritis (RA) and systemic lupus erythematosus (SLE) are excluded by the lack of other clinical manifestations (e.g., joint pain and swelling) and the lack of positive laboratory findings such as anti-cyclic citrullinated protein (CCP) antibodies and antinuclear antibodies (ANA), respectively. C-reactive protein (CRP) and erythrocyte sedimentation rate (ESR) are normal in FM patients; abnormalities with these or other common laboratory assessments suggest inflammatory disease, infection, or other concomitant illness. Hypophosphatemia (a low level of the electrolyte phosphate in the blood) can cause bone pain and muscle weakness; this condition is easily excluded by demonstration of normal serum phosphate level.

Standard Medical Treatment for Fibromyalgia
- Overview: Mild exercise, "patient education", and the use of pain-relieving drugs are mainstays of standard medical treatment delivered by most allopathic medical doctors (MDs), and osteopathic medical doctors (DOs) may add manual musculoskeletal treatments to enhance the benefits of drugs.[266] These interventions are only partially effective and offer no hope of actually curing the disease; thus, medical treatment relegates patients to a future of drug dependency, potential adverse effects (some of which can be fatal), and therapeutic inefficacy insofar as none of these treatments addresses the underlying cause of the disorder.
    - Amitriptyline: For many years, the most widely used drug for symptomatic treatment of fibromyalgia was amitriptyline (a tricyclic antidepressant), which has been used "off label"—without approval from the FDA—for this application. In the treatment of FM, the drug has low efficacy and high potential for adverse effects; up to 20% of patients suffer from weight gain, constipation, orthostatic hypotension, and/or agitation as a side-effect of the drug. Only 25% to 30% of fibromyalgia patients experience clinically significant improvement with amitriptyline.[267] According to recent research in rats, administration of amitriptyline causes deficiency of CoQ10, impaired mitochondrial function, reduced ATP/energy production, and increased oxidative stress and free radical damage[268]; all of these drug-induced problems (discussed in detail later in this paper) are expected to worsen the pain and suffering experienced by FM patients. Thus, the use of amitriptyline cannot be considered to be consistent with the practice of good medicine due to its low efficacy and unacceptable risks for adverse effects.
    - Pregabalin: In 2007, the United States Food and Drug Administration (US FDA) approved pregabalin (Lyrica® sold/marketed by Pfizer) for symptomatic treatment of fibromyalgia[269]; however, because the drug does not address the primary cause(s) of the disease, patients must continue treatment indefinitely. Adverse effects of pregabalin include dizziness, sleepiness, blurred vision, weight gain, dry mouth, swelling of hands and feet, impairment of motor function, and problems with concentration and attention. Pregabalin when given at the recommended dose of 150-225 mg twice per day for fibromyalgia costs $94-190 per month (pricing in 2013).

      > **Suicide and depression risk warning for pregabalin/Lyrica from the US FDA**
      >
      > "Antiepileptic drugs (AEDs), including Lyrica, increase the risk of suicidal thoughts or behavior in patients taking these drugs for any indication. Patients treated with any AED for any indication should be monitored for the emergence or worsening of depression, suicidal thoughts or behavior, and/or any unusual changes in mood or behavior."
      >
      > fda.gov/Safety/MedWatch/SafetyInformation/Safety-RelatedDrugLabelingChanges/ucm154524.htm

    - Duloxetine: In 2008, the FDA announced duloxetine (Cymbalta® sold/marketed by Lilly) as the second approved drug for the treatment of fibromyalgia. Ironically, many physicians consider any "approved" drug to have scientific substantiation; however, in the case of duloxetine (as well as pregabalin) the exact mechanism of action is unknown[270] although duloxetine appears to inhibit reuptake of norepinephrine and serotonin, thereby increasing the action of these neurotransmitters in the synaptic cleft. Adverse effects from duloxetine include nausea, dry mouth, sleepiness, constipation, decreased appetite, and increased

---

[266] Gamber RG, Shores JH, Russo DP, Jimenez C, Rubin BR. Osteopathic manipulative treatment in conjunction with medication relieves pain associated with fibromyalgia syndrome: results of a randomized clinical pilot project. *J Am Osteopath Assoc.* 2002 Jun;102(6):321-5 jaoa.org/content/102/6/321.full.pdf
[267] Leventhal LJ. Management of fibromyalgia. *Ann Intern Med.* 1999 Dec 7;131(11):850-8
[268] "Amitriptyline is a tricyclic antidepressant commonly prescribed for the treatment of several neuropathic and inflammatory illnesses. We have already reported that amitriptyline has cytotoxic effect in human cell cultures, increasing oxidative stress, and decreasing growth rate and mitochondrial activity." Bautista-Ferrufino MR, Cordero MD, Sánchez-Alcázar JA, et al. Amitriptyline induces coenzyme Q deficiency and oxidative damage in mouse lung and liver. *Toxicol Lett.* 2011 Jul 4;204(1):32-7
[269] FDA Approves First Drug for Treating Fibromyalgia. fda.gov/bbs/topics/NEWS/2007/NEW01656.html
[270] "exact mechanism of action unknown; inhibits norepinephrine and serotonin reuptake" https://online.epocrates.com; "Both Lyrica and Cymbalta reduce pain and improve function in people with fibromyalgia. While those with fibromyalgia have been shown to experience pain differently from other people, the mechanism by which these drugs produce their effects is unknown. fda.gov/ForConsumers/ConsumerUpdates/ucm107802.htm. Accessed January 2012

Chapter 5.1—Functional Inflammology Protocol for Metabolic Inflammation: Migraine and Fibromyalgia

sweating; **duloxetine can also increase the risk of suicidal thinking and behavior and for this reason the drug carries a black box warning on the container.** Duloxetine can cause serious and fatal adverse effects including the following: worsening depression and suicidality, serotonin syndrome, neuroleptic malignant syndrome, seizures, and Stevens-Johnson syndrome. Duloxetine given at the recommended dose of 60 mg per day for FM costs $170 per month (pricing in 2013).

Suicide and depression risk warning for duloxetine/Cymbalta
"WARNING: Suicidality and Antidepressant Drugs: Antidepressants increased the risk compared to placebo of suicidal thinking and behavior (suicidality) in children, adolescents, and young adults in short-term studies of major depressive disorder (MDD) and other psychiatric disorders. Anyone considering the use of Cymbalta or any other antidepressant in a child, adolescent, or young adult must balance this risk with the clinical need. Short-term studies did not show an increase in the risk of suicidality with antidepressants compared to placebo in adults beyond age 24; there was a reduction in risk with antidepressants compared to placebo in adults aged 65 and older. Depression and certain other psychiatric disorders are themselves associated with increases in the risk of suicide. Patients of all ages who are started on antidepressant therapy should be monitored appropriately and observed closely for clinical worsening, suicidality, or unusual changes in behavior. Families and caregivers should be advised of the need for close observation and communication with the prescriber. Cymbalta is not approved for use in pediatric patients."
pi.lilly.com/us/cymbalta-pi.pdf

- Milnacipran: Approved for the treatment of FM by the US FDA in 2009, milnacipran (Savella® sold/marketed by Forest Pharmaceuticals) inhibits norepinephrine and serotonin reuptake, i.e., it potentiates (increases the effect of) the neurotransmitters norepinephrine and serotonin, both of which decrease the experience of pain and elevate mood. Of course, other non-drug treatments (such as nutrients and dietary optimization) can have the same effect, but most medical doctors have no training in nondrug

Suicide and depression risk warning for milnacipran/Savella
"Savella is a selective serotonin and norepinephrine reuptake inhibitor (SNRI), similar to some drugs used for the treatment of depression and other psychiatric disorders. Antidepressants increased the risk compared to placebo of suicidal thinking and behavior (suicidality) in children, adolescents, and young adults in short-term studies of major depressive disorder (MDD) and other psychiatric disorders."
frx.com/pi/Savella_pi.pdf, linked as "Full Prescribing Information" from savella.com/important-risk-information.aspx

treatments[271,272,273,274] and thus habitually turn to drugs as the one-and-only answer to the patients' problems[275], especially when these are sanctified by FDA/government approval. Nondrug treatments that enhance serotonergic and noradrenergic neurotransmission include exercise, relaxation, massage, and nutritional supplementation with omega-3 fatty acids (as found in fish oil), nutritional supplementation in general and vitamin D supplementation in particular. Adverse effects associated with use of milnacipran include seizures, suicidality, depression, worsening hypomania/mania, Stevens-Johnson syndrome (which is a medical emergency that can be fatal), serotonin syndrome, neuroleptic malignant syndrome, hypertensive (elevated blood pressure) crisis, tachycardia (rapid heart rate), hyponatremia (low sodium in the blood, which can occasionally result in permanent brain damage), abnormal bleeding (due to abnormal platelet function), glaucoma, and liver toxicity.[276] Treatment of fibromyalgia is the only FDA-approved use of this medication, which when used at the recommended dose of 50 mg twice daily costs $144 per month (pricing in April 2013).

- Cyclobenzaprine, Tramadol, and acetaminophen: Cyclobenzaprine (a muscle-relaxing drug), Tramadol (a non-typical opioid, centrally-acting narcotic analgesic) and acetaminophen (centrally acting analgesic),

---

[271] "Internal medicine interns' perceive nutrition counseling as a priority, but lack the confidence and knowledge to effectively provide adequate nutrition education." Vetter ML, Herring SJ, Sood M, Shah NR, Kalet AL. What do resident physicians know about nutrition? An evaluation of attitudes, self-perceived proficiency and knowledge. *J Am Coll Nutr*. 2008 Apr;27(2):287-98 ncbi.nlm.nih.gov/pmc/articles/PMC2779722/

[272] "The amount of nutrition education that medical students receive continues to be inadequate." Adams KM, Kohlmeier M, Zeisel SH. Nutrition education in U.S. medical schools: latest update of a national survey. *Acad Med*. 2010 Sep;85(9):1537-42

[273] "Scientific advances on the relationship of dietary substances to the cellular mechanisms of disease occur with regularity and frequency. Yet, despite the prevalence of nutritional disorders in clinical medicine and increasing scientific evidence on the significance of dietary modification to disease prevention, present day practitioners of medicine are typically untrained in the relationship of diet to health and disease." Halsted CH. The relevance of clinical nutrition education and role models to the practice of medicine. *Eur J Clin Nutr*. 1999 May;53 Suppl 2:S29-34

[274] Vasquez A. Interventions need to be consistent with osteopathic philosophy. *J Am Osteopath Assoc*. 2006 Sep;106(9):528-9 jaoa.org/content/106/9/528.full.pdf

[275] Ely et al. Analysis of questions asked by family doctors regarding patient care. *BMJ*. 1999 Aug 7;319(7206):358-61 ncbi.nlm.nih.gov/pmc/articles/PMC28191/

[276] https://online.epocrates.com/noFrame/showPage.do?method=drugs&MonographId=4950 Accessed April 2012.

show low efficacy and have little research supporting their use in the treatment of FM; these drugs also carry important risks for adverse effects, and they do not favorably alter the course of the disease over the long-term.[277] Per recent information from the American College of Rheumatology (ACR), treatment of FM with opioid drugs "may cause greater pain sensitivity or make pain persist."[278]

- Exercise: Low-intensity aerobic exercise may initially exacerbate symptoms but can result in very modest mental and physical improvement. Exercise alone cannot cure FM.
- Cognitive-behavioral therapy (CBT): Cognitive-behavioral therapy helps patients deal with and adapt to the impact of the illness. Therapy alone cannot cure FM.
- Patient (mis)education in standard medicine: "Patient education" from a *medical* perspective generally means telling patients that ❶ they will probably have the condition forever, ❷ they will not immediately die from it, ❸ they need to take it seriously (i.e., comply with medical treatment), and ❹ they need to rely on drugs for alleviation of symptoms since no cause of the condition is known and therefore no direct treatment is available. From the medical perspective, these communications are considered "helpful" and "reassuring"; however, part of the effect that is created is **dependency** ("You need these drugs from me."), **passivity** ("There's nothing you can do about this, so don't even try to think for yourself or seek 'alternative' treatments."), and **co-victimization** ("We are both victims of our ignorance; I am in this with you in that we are both blind and dependent on drug management."). In the examples that follow, I will review and summarize patient educational materials from major medical groups; for efficiency, I will use quotes followed by my comments in *italics*:
- Press release from the American Pain Society "Fibromyalgia Has Central Nervous System Origins", written by an author heavily funded by drug companies[279]:
    - "Fibromyalgia is the second most common rheumatic disorder behind osteoarthritis and, though still widely misunderstood, is now considered to be a lifelong central nervous system disorder, which is responsible for amplified pain that shoots through the body in those who suffer from it." — *This lunacy does not require additional refutation: the American Pain Society clearly affiliates with drug companies and wants to promote drug sales and use of their specialty organization; for this, they need to foster the illusion that FM begins in the brain, has no cure, and needs to be treated with drugs for the duration of patients' lives.*
- "Patient Education" from the American College of Rheumatology, Rheumatology.org[280]:
    - "Though there is no cure, medications can relieve symptoms." — *This is a commonly used statement within the medical community from doctors to patients to create passivity and drug/medical dependency.*
    - "There likely are certain genes that can make people more prone to getting fibromyalgia and the other health problems that can occur with it. Genes alone, though, do not cause fibromyalgia." — *These are common statements in the medical community, basically summed as "We don't know what we are doing but your only hope is to depend on us."*
    - "For the person with fibromyalgia, it is as though the "volume control" is turned up too high in the brain's pain processing centers." — *This promotes the concept of "primary central sensitization" (i.e., the brain has defied normal physiology and has somehow [without known cause, by itself] become too sensitive to pain); this "blame the brain" concept is used to leverage drug sales for pain-relieving and anti-depressant drugs as I have recently reviewed in video: youtube.com/watch?v=41opevN87qs*
    - "There is no cure for fibromyalgia. However, symptoms can be treated with both medication and non-drug treatments." — *This is the standard "party line" for the medical profession, whose chief goal is not to cure diseases but rather to drug them indefinitely, thereby creating a perpetual audience for their services and prescriptions. Honorable mention (more accurately: dishonorable mention) is generally given to "lifestyle modification" but is generally done so in a way that provides vague advice for ineffective interventions, thereby **creating the illusion of options** while undercutting any potential for these "options" to actually work.*
    - Non-drug treatments reviewed: relaxation, deep breathing, meditation, sleep, avoidance of nicotine and caffeine, exercise including such miniscule revelations as "take the stairs instead of the elevator, or park further [sic] away from the store", and "education" from other medical and special interest groups. — *These*

---

[277] Goldenberg DL, Burckhardt C, Crofford L. Management of fibromyalgia syndrome. *JAMA*. 2004 Nov 17;292(19):2388-95
[278] rheumatology.org/practice/clinical/patients/diseases_and_conditions/fibromyalgia.asp Accessed April 2012
[279] Fibromyalgia Has Central Nervous System Origins. americanpainsociety.org/about-us/press-room/fibromyalgia-clauw May 16, 2015
[280] rheumatology.org/practice/clinical/patients/diseases_and_conditions/fibromyalgia.asp Accessed March 31, 2012

Chapter 5.1—Functional Inflammology Protocol for Metabolic Inflammation: Migraine and Fibromyalgia

- are all essentially worthless suggestions, but they are effective distractions for patients and doctors so that effective treatments are marginalized and drug/medical dependency is fostered.
  - Prescription drugs are given the primary emphasis in the treatment section.—*Whether drugs are effective or not, the medical profession relies on drugs for its societal position and will therefore reflexively advocate them.*
- Patient education from American Academy of Family Physicians, FamilyDoctor.org[281]:
  - "…your muscles and organs are not being damaged."—*This is false/inaccurate information. Several primary research studies have demonstrated consistently pathologic and biochemical abnormalities in muscle tissue from patients with fibromyalgia; this research has been published in widely available peer-reviewed medical journals. While the muscle damage in FM is not gross or overt myopathy, stating that no damage is occurring is not histologically accurate and supports the drug-friendly model that FM is idiopathic and (neuro)psychiatric.*
  - "This condition is not life-threatening, but it is chronic (ongoing). Although there is no cure,…"—*This is false information because obviously the condition is curable; the statement as it reads produces patient passivity and drug-dependency, which is exactly what the medical profession and the drug industry want.*
  - "There isn't currently a cure for fibromyalgia. Your care will focus on helping you minimize the impact of fibromyalgia on your life and treating your symptoms. Your doctor can prescribe medicine to help with your pain,… The treatment recommendations your doctor makes won't do any good unless you follow them."—*Again, false information promoting passivity and drug-dependency. This is basically drug propaganda, encouraging passivity and compliance on the parts of doctors and patients alike.*
  - Weak recommendations under the guise of "taking an active role in your healthcare" include 1) maintaining a healthy outlook, 2) support groups, 3) **"take medicines exactly as prescribed"**, 4) moderate exercise, 5) stress management, 6) "establish healthy sleep habits", 7) make a routine daily schedule, 8) "make healthy lifestyle choices." —*Most of these recommendations are blatantly passive, vague, and ineffective while fostering drug-dependency.*

The most common pattern in medical books and articles: components of the medical paradigm
1. <u>Diseases are describable yet incomprehensible</u>: The condition generally is described as a complex interplay of genetic factors with numerous environmental factors; generally little or no intellectually worthwhile effort is made to understand these environmental factors, so the ultimate view that medical physicians are taught is the disease is not understood and that drug therapy is the best available treatment.
2. <u>Nearly always the disease is described as chronic and incurable</u>: Even if a cure is known and published in available peer-reviewed research, most medical books and articles conclude that the causes are unknown and that palliative drug treatment is therefore warranted.
3. <u>Characteristics and diagnostic criteria are reviewed</u>: The medical profession is very good at defining and diagnosing problems, but the reliance on drugs often detours from the more effective nondrug treatments. The idea that every disease needs a drug is absurd, but this very same idea is embraced as the axiom of medicine and indeed as proof of its insight and resourcefulness.
4. <u>Drug treatments are emphasized</u>: Drug benefits are inflated and adverse effects are minimized if mentioned at all. Many review articles published in clinical journals are authored by consultants to the drug industry who profit from the sale of the drugs used to treat the condition about which they write; the general conclusion of these articles is that "Medical research is making considerable advances in our understanding of Disease X, which results from a complex interplay of genetic and environmental factors. The appropriate treatment for Disease X is Drug X, then Drug X2…along with generally meaningless lifestyle advice.
5. <u>Nondrug treatments are marginalized</u>: Brief mention is made of diet and lifestyle and other non-drug treatments, but the information is nonspecific, very general, and almost always diluted to the point of inefficacy.
6. <u>Authentic integrative/functional medicine approaches are essentially never mentioned</u>: Generally, the only nondrug treatments that are mentioned are weak or are mentioned so casually that no effective action can be implemented.
7. <u>Hope for the future is always placed back in the hands of "more research" and "drug development"</u>: Often some mention of "hope for the future" is made, generally in the guise of "medical research" and "drug development"; all the while, safe and effective non-drug treatment approaches that go far beyond diet and lifestyle are virtually never mentioned.
Most medical students start medical school in their early 20s, when they are notably young and impressionable, with practically zero life experience other than creating the perfect medical school application (ie, high scores plus some foray into volunteer work to appear socially concerned), and eager to "be a doctor." Medical training is typically 7 years—4 years of school followed by 3 years of hospital-based residency training—during which students and residents are stressed, sleep-deprived, hazed, and fearful of expulsion for any minor infraction. Medical training induces a trance-like state, with several features of Stockholm syndrome, wherein doctors become accustomed in their early training to memorize, pathologize, and conform to expectations. They are eager to do the things that define a doctor's authority and privilege—write prescriptions or recommend procedures/surgery; nutrition—a "nonmedical" treatment that is considered "alternative"—is for "quacks" (nonscientists, nonrationalists, especially those who could not enter a "real" medical school) and dieticians (hierarchically inferior to doctors). With the time pressure (2-7 minutes per patient) and formulary restrictions (zero outpatient nutritional options) of hospital/clinic-based training/practice, practically no *conscientious* medical student/clinician would consider deviation from well ingrained (and intellectually inbred) *expectations*.

---

[281] American Academy of Family Physicians. familydoctor.org/familydoctor/en/diseases-conditions/fibromyalgia.html Accessed March 31, 2012

Functional/Naturopathic Considerations, Assessments, and Interventions

- **Foundational perspectives**: Two fundamental premises of are: (1) ~~chronic~~ sustained *diseases* are manifestations of ~~chronic~~ sustained *dysfunctions*, and (2) dysfunction can result from a wide range of interconnected genotropic (gene-influenced), metabolic, nutritional, microbial, inflammatory, toxic, environmental, and psychological and social influences. **Many of these dysfunctions lie outside the narrow, pathology-based, pharmacocentric (drug-centered) view of standard allopathic medicine.** The functional/naturopathic medicine approach to each individual fibromyalgia patient is based firstly on the presumption that the condition has an underlying primary cause (or several interconnected causes) and that the cause(s) can be identified and addressed—in this manner, the clinical approach is one of positive psychoepistimology (ie, affirmative that we can understand this situation), rather than pathodefetism (ie, "we can't understand this disease") resulting pharmacodependency. The cause(s) may be manifold and multifaceted and may differ among patients with the same diagnostic label. This approach includes the diagnostic and therapeutic considerations of standard medicine but extends far beyond these in assessment, treatment, and understanding. Clinicians should appreciate that as a diagnostic label, fibromyalgia is commonly applied to any patient with chronic, widespread pain and that the current trend to limit diagnostic evaluation in such patients will clearly result in failure to identify and address readily diagnosable and treatable problems that can result in a clinical picture that resembles FM. Clinicians must consider chronic infections (such as with hepatitis C virus, *Borrelia burgdorferi* [the bacteria strongly associated with Lyme disease], *Chlamydia/Chlamydophila pneumoniae*, and the protozoan parasite *Babesia*, which is also associated with Lyme disease and co-infection with *Borrelia burgdorferi*), cancerous conditions such as multiple myeloma and lymphoma, and autoimmune/rheumatic diseases such as polymyositis and polymyalgia rheumatica. A few of the other more exemplary conditions to consider in patients with widespread pain are vitamin D deficiency, hypothyroidism, iron overload, and chronic exposure to and accumulation of xenobiotics—perhaps most importantly mercury and lead.

Common differential diagnoses—conditions that can mimic (or contribute to) fibromyalgia

- <u>Vitamin D deficiency</u>: A clinical picture nearly identical to fibromyalgia—chronic widespread pain, mental depression/anxiety, headaches, low-grade systemic inflammation—can result from vitamin D deficiency.[282] Fibromyalgia patients are commonly deficient in vitamin D, and indeed, **vitamin D deficiency—with its attendant pain, anxiety/depression, and normal lab values on routine laboratory testing—is often misdiagnosed as fibromyalgia**, as reported by Holick.[283] Increased severity of the deficiency correlates with worsening depression and anxiety in these patients.[284] Correction of vitamin D deficiency by administration of vitamin D3 (cholecalciferol) in doses of 5,000-10,000 IU (international units) per day for several months has resulted in a dramatic alleviation of pain; such intervention among patients with low back pain has resulted in cure rates greater than 95%.[285] Other studies with vitamin D3 using doses 400-4,000 IU/day have shown that vitamin D3 supplementation for the correction of vitamin D deficiency alleviates depression and enhances sense of well-being. Vitamin D3 supplementation—or adequate endogenous production from ultraviolet light exposure (approximately 10-30 minutes per day of full-body exposure at midday, near the equator)—to meet physiological requirements of approximately 4,000 IU/day is safe and results in numerous major health benefits.[286,287,288] The only risk associated with vitamin D supplementation is hypercalcemia—too much calcium in the blood, mostly as a result of increased gastrointestinal absorption of calcium; hypercalcemia can cause abdominal pain, bone pain, fatigue, constipation, abnormal heart rhythm (arrhythmia), kidney stones, increased thirst and urination [additional details[289]]. Hypercalcemia caused solely by vitamin D3 supplementation is extremely rare; vitamin D supplementation in the range of 2,000 – 10,000 IU per day for

---

[282] Plotnikoff GA, Quigley JM. Prevalence of severe hypovitaminosis D in patients with persistent, nonspecific musculoskeletal pain. *Mayo Clin Proc*. 2003;78(12):1463-70
[283] Holick MF. Vitamin D: importance in the prevention of cancers, type 1 diabetes, heart disease, and osteoporosis. *Am J Clin Nutr*. 2004 Mar;79(3):362-71
[284] Armstrong DJ, et al. Vitamin D deficiency is associated with anxiety and depression in fibromyalgia. *Clin Rheumatol*. 2007 Apr;26(4):551-4
[285] Al Faraj S, Al Mutairi K. Vitamin D deficiency and chronic low back pain in Saudi Arabia. *Spine*. 2003;28:177-9
[286] Holick MF. Vitamin D: importance in the prevention of cancers, type 1 diabetes, heart disease, and osteoporosis. *Am J Clin Nutr*. 2004 Mar;79(3):362-71
[287] Vieth R. Vitamin D supplementation, 25-hydroxyvitamin D concentrations, and safety. *Am J Clin Nutr*. 1999 May;69(5):842-56
[288] Zittermann A. Vitamin D in preventive medicine: are we ignoring the evidence? *Br J Nutr*. 2003 May;89(5):552-72
[289] Mild hypercalcemia is not necessarily a problem by itself and must be evaluated within the patient's clinical context. When blood levels of calcium (normal range: 8.7-10.4 mg/dL) reach approximately 12.0 mg/dL, patients will start to develop symptoms; with levels of 14 mg/dL or higher, the patient is generally experiencing symptoms and complications and is in need of treatment (initially with administration of intravenous fluids and a loop diuretic such as furosemide).

adults is remarkably safe.[290,291] The main drug-nutrient interaction of relevance to vitamin D supplementation is with the drug hydrochlorothiazide, which is a diuretic drug used for the treatment of high blood pressure; this drug causes calcium retention by the kidney and when combined with vitamin D supplementation may lead to high levels of calcium in the blood (hypercalcemia). *Note from Dr Vasquez: I have only seen this occur one time in my clinical practice in a hypertensive patient taking hydrochlorothiazide who was vitamin D deficient; vitamin D supplementation at 2,000 IU/d caused a mild hypercalcemia within 10 days which was treated simply by discontinuing the vitamin D supplementation (also note that discontinuation of appropriate nutritional supplementation in favor of continuing a symptom-suppressing drug is generally not my preference but in this particular situation it was the best choice).* A group of conditions called granulomatous diseases—which can include lymphoma, sarcoidosis, and Crohn's disease—increase the risk for hypercalcemia; caution and more frequent laboratory monitoring must be employed when using physiological doses of vitamin D3 in patients with these conditions. Diagnosis of vitamin D3 deficiency is simple and is based upon measurement of serum 25-hydroxy vitamin D3 (25[OH]D) levels. Supplementation effectiveness and safety are monitored by measuring 25(OH)D levels and serum calcium, respectively. The two goals with supplementation of vitamin D3 are ❶ safety—avoidance of hypercalcemia or any calcium-related complications, and ❷ efficacy—serum 25[OH]D levels should enter into the optimal range of 50 – 100 ng/mL (125 - 250 nmol/L).

- Functional/metabolic hypothyroidism: Insufficient levels of thyroid hormone lead to an associated clinical condition called hypothyroidism (*hypo*=low, *thyroidism*=thyroid condition). Both mild and overt hypothyroidism are well known in the rheumatology literature as causes of diffuse body pain. As a cause of diffuse muscle pain, mild-moderate hypothyroidism can mimic fibromyalgia; more severe cases of hypothyroidism cause "hypothyroid myopathy" which typically manifests as polymyositis-like disease with proximal muscle weakness and an increased serum level of the enzyme creatine kinase, indicating muscle damage. In its most extreme, hypothyroid myopathy presents as muscle enlargement (pseudohypertrophy); in adults, this condition is called Hoffmann syndrome while in children it is known as Kocher-Debré-Sémélaigne syndrome.[292] Hypothyroidism is well known to cause depression and low-grade systemic inflammation; these are two findings common in FM. Another related problem commonly seen with both FM and hypothyroidism is IBS and small intestinal bacterial overgrowth (SIBO); hypothyroidism causes a slowing of intestinal motility, promoting stasis in the gastrointestinal tract which leads to an overgrowth of bacteria.[293] Detailed thyroid assessment should include measurements of thyroid stimulating hormone (TSH), free T4, free T3, total T3, reverse T3 (rT3), and antithyroid peroxidase (anti-TPO) and antithyroglobulin antibodies. Management of hypothyroidism is detailed in Chapter 1 of *Inflammation Mastery* / *Functional Inflammology* (2014 and later editions).

> **Thyroid hormone production and metabolism**
>
> - **TSH—thyroid stimulating hormone**: Hormone secreted from the anterior pituitary gland to stimulate T4 and T3 production from the thyroid gland.
> - **T4**: The inactive form of thyroid hormone, accounting for about 80% of thyroid gland output. Oddly and contrary to physiology, doctors have been trained to use this inactive form of the hormone despite known problems in conversion to the active T3 form described immediately below.
> - **T3**: The active form of thyroid hormone produced from conversion of T4, accounts for about 20% of thyroid gland output. This is the form of thyroid hormone that is most important, because it is active and ready to stimulate metabolic processes; of note, the brain is unable to convert T4 to T3 and thus when patients are deficient in thyroid hormone and substituted with T4 only, they commonly have suboptimal improvement, especially for brain-specific issues of depression and fatigue.
> - **rT3—reverse T3**: During times of stress and also as a result of some drugs, T4 is preferentially converted to rT3, which is inactive and may actually impair the utilization of active T3. In some people, especially after a period of severe emotional stress, their thyroid hormone metabolism becomes skewed toward rT3 production, perhaps as an adaptive mechanism to conserve energy. However, increased rT3 production results in impaired thyroid hormone function and thereby promotes a clinical picture of hypothyroidism even when gland function is adequate; the problem is the hormone's peripheral metabolism, not its production from the gland.

---

[290] Vieth R. Vitamin D supplementation, 25-hydroxyvitamin D concentrations, and safety. *Am J Clin Nutr*. 1999 May;69(5):842-56
[291] Vasquez et al. Clinical Importance of Vitamin D: Paradigm Shift for All Healthcare Providers. *Altern Ther Health Med* 2004; 10: 28-37 ichnfm.academia.edu/AlexVasquez
[292] Kedlaya D. Hypothyroid Myopathy. emedicine.medscape.com/article/313915-overview Accessed April 2012
[293] Lauritano EC, Bilotta AL, Gabrielli M, et al. Association between hypothyroidism and small intestinal bacterial overgrowth. *J Clin Endocrinol Metab*. 2007 Nov;92(11):4180-4

- Occult infections, especially with *Mycoplasma* species and *Chlamydia/Chlamydophila pneumoniae*: Clinicians are increasingly appreciating the role of occult intracellular infections in the genesis and/or perpetuation of chronic health problems, including some previously perplexing problems such as chronic fatigue syndrome (CFS), inflammatory arthritis, and multiple sclerosis (MS). For chronic *Chlamydophila* (previously *Chlamydia*) *pneumoniae* infection, testing for serum levels of antibodies is useful followed by treatment with antibacterial drugs such as azithromycin and nutritional supplements such as N-acetyl-cysteine (NAC) in appropriately selected patients; for chronic *Mycoplasma* infections, because of the various subspecies involved, polymerase chain reaction (PCR) testing appears to be preferred followed by treatment with doxycycline in adults.
    - Review: *Mycoplasma* blood infection in chronic fatigue and **fibromyalgia** syndromes (*Rheum Int* 2003 Sep[294]): The author notes that "**Chronic fatigue syndrome (CFS) and fibromyalgia syndrome (FMS)** are characterized by a lack of consistent laboratory and clinical abnormalities. Although they are distinguishable as separate syndromes based on established criteria, a great number of patients are diagnosed with both." He goes on to say, "In studies using **polymerase chain reaction [PCR] methods, mycoplasma blood infection has been detected in about 50% of patients with CFS and/or FMS**, including patients with Gulf War illnesses and symptoms that overlap with one or both syndromes. **Such infection is detected in only about 10% of healthy individuals**, significantly less than in patients. Most patients with CFS/FMS who have mycoplasma infection appear to recover and reach their pre-illness state after **long-term antibiotic therapy with doxycycline**, and the infection cannot be detected after recovery. … It is not clear whether mycoplasmas are associated with CFS/FMS as causal agents, cofactors, or opportunistic infections in patients with immune disturbances."
    - Clinical investigation: High prevalence of Mycoplasmal infections in symptomatic (chronic fatigue syndrome) family members of *Mycoplasma*-positive Gulf War illness patients (*Journal of Chronic Fatigue Syndrome* 2003[295]): The authors state, "…a relatively common finding in Gulf War Illness patients is a bacterial infection due to *Mycoplasma* species, we examined military families (149 patients: 42 veterans, 40 spouses, 32 other relatives and 35 children with at least one family complaint of illness) selected from a **group of 110 veterans with Gulf War Illness who tested positive (~41%) for at least one of four *Mycoplasma* species**: *M. fermentans, M. hominis, M. pneumoniae* or *M. genitalium*. Consistent with previous results, over 80% of Gulf War Illness patients who were positive for blood mycoplasmal infections had **only one *Mycoplasma* species, in particular *M. fermentans*** (Odds ratio = 17.9, P <0.001). In healthy control subjects the incidence of mycoplasmal infection was ~8.5% and none were found to have multiple mycoplasmal species."
    - Clinical investigation: Prevalence of antibodies to *Chlamydophila pneumoniae* in persons without clinical evidence of respiratory infection (*Journal of Clinical Pathology* 2002 May[296]): The authors note that "Because there is as yet no standardization of serological criteria for persistent infection, we considered antibody titers of > 1/20 in the IgA fraction, together with **IgG titers of 1/64 to 1/256, to be indicative of persistent infection**." This article supports clinical experience and post-graduate presentations[297] showing that in persons with fatigue and various other chronic health disorders characterized by pain and inflammation (such as chronic inflammatory arthritis[298] or spine[299] inflammation), the finding of IgG antibody levels >1:64 suggests that the patient has a persistent *Chlamydophila pneumoniae* infection which may be alleviated by the administration of—for example—the antibiotic **azithromycin** (adult dose 250 mg every other day due to the drug's long half-life, given for several weeks or months until symptoms are resolved and/or antibody

Chapter 5.1—Functional Inflammology Protocol for Metabolic Inflammation: Migraine and Fibromyalgia

titers are normalized) and **N-acetyl-cysteine** (NAC: 500-1,200 mg 1-3 times per day by mouth between meals). Positive antibody titers (levels) are common because the infection itself is common *as a transient condition*; the issue here is the determination of which patients have a *chronic* and *persistent* low-grade infection. The finding of an elevated antibody titer—that is a level greater than 1:64—indicates the need to consider long-term antimicrobial intervention.

```
Chlamydia pneumoniae IgG      >1:256    High           Neg:<1:16
Chlamydia pneumoniae IgM      <1:10                    Neg:<1:10
```

Elevated titers to *Chlamydia/Chlamydophila pneumoniae* suggesting chronic persistent infection in a 40yo male physician *without pulmonary symptoms* but with a positive history of chronic sinus congestion and low-grade fatigue—improvement with azithromycin and NAC: This patient experienced years of severe psychologic and physiologic stress during a doctorate program and then had an acute upper respiratory illness onset in September 2010 while working in hospital emergency rooms and urgent care clinics; recurrent bouts of upper respiratory illness—attributed to viral infections—persisted for five months until February 2011. By the summer of 2011, the patient was relatively asymptomatic except for persistent sinus congestion and low-grade fatigue. No pulmonary symptoms such as shortness of breath were ever present. Following detection of the elevated antibody titer, the patient started on azithromycin and NAC, which resulted in a short-term (12-hour) exacerbation of symptoms followed by complete and sustained resolution of sinus congestion and improved energy levels and exercise endurance.

- Hemochromatosis and iron overload: Genetic hemochromatosis is a common iron-accumulation disease that causes chronic persistent musculoskeletal pain, even while most routine laboratory tests are normal; thus, the clinical presentation of iron overload may be confused with that of fibromyalgia—both are common conditions commonly presenting with inexplicable (i.e., normal values of routine laboratory tests) nontraumatic musculoskeletal pain. Hemochromatosis is one of the most common hereditary disorders among Caucasians, with a homozygote (two of the same genes, results in more severe disease) frequency of approximately 1 in 200 to 250 persons and a heterozygote (only one affected gene, less severe disease) frequency of approximately 1 in 7 persons. Various other hereditary iron overload disorders affect all races, with the highest prevalence in persons of African descent (as high as 1 in 80 according to some small studies among hospitalized African-American patients).[300,301] Eighty percent of hemochromatosis patients have chronic musculoskeletal pain, which is commonly the earliest or only presenting complaint.[302] In contrast to the clinical presentation of FM, the musculoskeletal manifestations of iron overload are classically arthritic (i.e., in the joints) rather than muscular, with the joints of the hands, wrists, hips, and knees most commonly affected. However, due to the widespread distribution of pain and the normalcy of routine laboratory results, iron overload can mimic fibromyalgia. Given the high population prevalence of iron overload and the high frequency with which it presents with musculoskeletal manifestations, all patients with chronic, nontraumatic musculoskeletal pain must be tested for iron overload. Serum ferritin, which can be used alone or with transferrin saturation, is the best single laboratory test; confirmed results greater than 200 mcg/L in women and 300 mcg/L in men necessitate treatment with diagnostic and therapeutic phlebotomy (frequent "blood donation" is the most effective treatment for chronic iron overload).[303]
- Iron deficiency: Iron is necessary for function of the mitochondrial electron transport chain as well as for formation of the neurotransmitters dopamine and serotonin, both of which can be said to have an analgesic effect. As such, iron deficiency can promote headaches in general and migraine in particular; iron deficiency might also contribute to the clinical presentation of fibromyalgia.[304] Per my previous extensive reviews of the

---

[300] Wurapa RK, Gordeuk VR, Brittenham GM, Khiyami A, Schechter GP, Edwards CQ. Primary iron overload in African Americans. *Am J Med*. 1996 Jul;101(1):9-18
[301] Barton JC, Edwards CQ, Bertoli LF, Shroyer TW, Hudson SL. Iron overload in African Americans. *Am J Med*. 1995 Dec;99(6):616-23
[302] Vasquez A. Musculoskeletal disorders and iron overload disease: comment on the American College of Rheumatology guidelines for the initial evaluation of the adult patient with acute musculoskeletal symptoms. *Arthritis Rheum*. 1996 Oct;39(10):1767-8 Ichnfm.academia.edu/AlexVasquez
[303] Barton JC, McDonnell SM, Adams PC, et al. Management of hemochromatosis. Hemochromatosis Management Working Group. *Ann Intern Med*. 1998 Dec 1;129(11):932-9
[304] "The mean serum ferritin levels in the fibromyalgia and control groups were 27.3 and 43.8 ng/ml, respectively, and the difference was statistically significant. Binary multiple logistic regression analysis with age, body mass index, smoking status and vitamin B12, as well as folic acid and ferritin levels showed that having a serum ferritin level <50 ng/ml caused a 6.5-fold increased risk for FMS." Ortancil et al. Association between serum ferritin level and fibromyalgia syndrome. *Eur J Clin Nutr*. 2010 Mar;64(3):308-12

literature on the topics of iron overload and iron deficiency[305], optimal iron status correlates with serum ferritin values of 40-70 ng/ml. Rarely, a person with what can be described as a defect in the blood-brain barrier transport of iron into the brain will need to have a serum ferritin value of 120 ng/ml in order to promote entry of iron into the brain.

- Accumulation of xenobiotics (including mercury and lead): Xenobiotic (foreign chemical) accumulation may occasionally cause widespread pain resembling fibromyalgia, and xenobiotic detoxification (depuration) can alleviate pain in affected patients. Toxic chemical and toxic metal accumulation is common in humans worldwide and has been well-documented in Americans. Eight percent (8%) of American women of childbearing age have sufficiently high levels of mercury in their blood to increase the risk of health problems such as neurological damage in their children.[306] Americans in general show alarmingly high concentrations and combinations of neurotoxic (nerve-damaging), carcinogenic (cancer-causing), diabetogenic (diabetes-causing), and immunotoxic (immune-poisoning) xenobiotics/toxins.[307] Adverse effects of toxic chemicals (e.g., pesticides, herbicides, solvents, plastics, formaldehyde, petroleum byproducts) and heavy metals (especially lead and mercury) are well described throughout the biomedical literature and have been clinically reviewed by Crinnion.[308,309,310,311] Among toxins with the ability to produce chronic muscle pain, mercury may deserve special recognition given its ubiquitous distribution in the human population and the scientific evidence detailing its numerous adverse effects.[312,313] Whether by metabolic, neurological, or endocrinologic means, occult mercury toxicity may manifest as a syndrome of widespread muscle pain that resembles fibromyalgia.[314] Acrodynia is a subacute peripheral pain syndrome due to mercury toxicity classically seen in children.[315] Acute mercury intoxication can result in severe skeletal muscle damage (rhabdomyolysis).[316] Mercury in organic and inorganic forms interferes with acetylcholine reception and several crucial aspects of the sarcoplasmic reticulum, including calcium-magnesium-ATPase and calcium transport; these adverse effects establish a molecular basis for a *mercurial myopathy* (mercury-induced muscle disease).[317,318] The toxicity of mercury is greatly increased by simultaneous accumulation of lead, elevated levels of which are also common in the U.S. population. Demonstration of high mercury and lead levels in urine following administration of a chelating agent such as dimercaptosuccinic acid (DMSA) can be used to diagnose chronic mercury or lead overload, and orally administered DMSA is also used for treatment.[319,320,321,322] Failure to preadminister a chelating agent prior to measurement of urine mercury renders the test insensitive for chronic accumulation and can thus give the false impression that mercury is not contributory to fibromyalgia, as concluded by Kotter et al.[323] Orally administered selenium, phytochelatins (metal-binding peptides from plants[324]), a high-fiber diet, and potassium citrate can be used to augment mercury excretion.

> **Potential benefits of reducing the body burden of mercury in patients with chronic pain and fatigue**
>
> "We suggest that **metal-driven inflammation** may affect the hypothalamic-pituitary-adrenal axis (HPA axis) and indirectly trigger psychosomatic multisymptoms characterizing **chronic fatigue syndrome, fibromyalgia**, and other diseases of unknown etiology."
>
> Sterzl et al. Mercury and nickel allergy: risk factors in fatigue and autoimmunity. *Neuro Endocrinol Lett.* 1999

---

[305] See Chapter 1 of either *Inflammation Mastery* / *Functional Inflammology* (2014 or later) for the most complete reviews, including assessment, management, and radiographic presentations. See also: Vasquez A. Musculoskeletal disorders and iron overload disease. [Letter] *Arthritis & Rheumatism* 1996; 39:1767-8. Vasquez A. High body iron stores: causes, effects, diagnosis, and treatment. *Nutritional Perspectives* 1994 October
[306] "However, approximately 8% of women had concentrations higher than US Environmental Protection Agency's recommended reference dose (5.8 µg/L), below which exposures are considered to be without adverse effects." Schober et al. Blood mercury levels in US children and women of childbearing age,1999-2000. *JAMA* 2003;289:1667-74
[307] Kristin et al. Chemical Trespass. Pesticide Action Network North America. Available at panna.org. See also: Body Burden: The Pollution in People. ewg.org/ 2006 Feb
[308] Crinnion WJ. Environmental medicine, part 1: the human burden of environmental toxins and their common health effects. *Altern Med Rev.* 2000 Feb;5(1):52-63
[309] Crinnion WJ. Environmental medicine, part 2: health effects of and protection from ubiquitous airborne solvent exposure. *Altern Med Rev.* 2000 Apr;5(2):133-43
[310] Crinnion WJ. Environmental medicine, part 3: long-term effects of chronic low-dose mercury exposure. *Altern Med Rev.* 2000 Jun;5(3):209-23
[311] Crinnion WJ. Environmental medicine, part 4: pesticides - biologically persistent and ubiquitous toxins. *Altern Med Rev.* 2000 Oct;5(5):432-47
[312] Elemental Mercury Vapor Poisoning -- North Carolina, 1988. cdc.gov/mmwr/preview/mmwrhtml/00001499.htm
[313] Shih H, Gartner JC Jr. Weight loss, hypertension, weakness, and limb pain in an 11-year-old boy. *J Pediatr.* 2001 Apr;138(4):566-9
[314] Sterzl I, Prochazkova J, Hrda P, et al. Mercury and nickel allergy: risk factors in fatigue and autoimmunity. *Neuro Endocrinol Lett.* 1999;20:221-8
[315] Padlewska KK. Acrodynia. Last Updated: February 15, 2007 eMedicine emedicine.com/derm/topic592.htm Accessed October 25, 2007
[316] Chugh KS, Singhal PC, Uberoi HS. Rhabdomyolysis and renal failure in acute mercuric chloride poisoning. *Med J Aust.* 1978 Jul 29;2(3):125-6
[317] Chiu VC, Mouring D, Haynes DH. Action of mercurials on the active and passive transport properties of sarcoplasmic reticulum. *J Bioenerg Biomembr.* 1983 Feb;15(1):13-25
[318] Shamoo AE, Maclennan DH, Elderfrawi ME. Differential effects of mercurial compounds on excitable tissues. *Chem Biol Interact.* 1976 Jan;12(1):41-52
[319] Kalra V, et al. Succimer in Symptomatic Lead Poisoning. *Indian Pediatrics* 2002; 39:580-585 indianpediatrics.net/june2002/june-580-585.htm
[320] Bradstreet et al. Case-control study of mercury burden in children with autistic spectrum disorders. *J Am Physicians Surgeons* 2003; 8: 76-79 jpands.org/vol8no3/geier.pdf
[321] Forman et al. A cluster of pediatric metallic mercury exposure cases treated with meso-2,3-dimercaptosuccinic acid (DMSA). *Environ Health Perspect.* 2000 Jun;108(6):575-7
[322] Miller AL. Dimercaptosuccinic acid (DMSA), a non-toxic, water-soluble treatment for heavy metal toxicity. *Altern Med Rev.* 1998 Jun;3(3):199-207
[323] Kotter I, Durk H, Saal JG, et al. Mercury exposure from dental amalgam fillings in the etiology of primary fibromyalgia. *J Rheumatol.* 1995;22:2194-5
[324] Cobbett CS. Phytochelatins and their roles in heavy metal detoxification. *Plant Physiol.* 2000;123:825-32 plantphysiol.org/content/123/3/825

Chapter 5.1—Functional Inflammology Protocol for Metabolic Inflammation: Migraine and Fibromyalgia

- Case report: Therapeutic detoxification to reduce the body burden of lead and mercury in a woman diagnosed with FM leads to complete relief of FM symptoms: This 54-year-old athletic female with healthy diet, lifestyle, and supportive relationship presented with chronic diffuse musculoskeletal pain. Health history was significant for decades of environmental illness/intolerance (EI) and multiple chemical sensitivity (MCS). Family history was positive for maternal temporal (giant cell) arteritis. Physical examination revealed numerous tender points consistent with FM. Stool analysis was unremarkable and unsupportive of either identifiable infection or nonspecific bacterial overgrowth. Laboratory investigations revealed normal results for hsCRP (high-sensitivity c-reactive protein), CK (creatine kinase, a marker of muscle damage), ANA (anti-nuclear antibodies, elevated in many autoimmune diseases such as lupus/SLE), vitamin D, calcium, phosphorus, and comprehensive thyroid evaluation. The patient was then (defensively) referred to an excellent osteopathic medical internist who diagnosed the patient with fibromyalgia. The patient was unsatisfied with the diagnosis of FM and returned to the current author, who then performed urine heavy metal testing provoked with 10 mg per kilogram of dimercaptosuccinic acid (DMSA). Results revealed the highest levels of lead and mercury encountered in the author's practice at that time. As in the accompanying lab results, lead levels were 6x above the reference range and mercury levels were 7x above the reference range. The patient was commenced on DMSA 10 mg/kg/d three days "on" and 4 days "off" (cyclic dosing is used to avoid toxicity in general and bone marrow toxicity [neutropenia] in particular), selenium 800 mcg/d to promote excretion of toxic metals and to support renal and antioxidant protection, vegetable juices to provide potassium and citrate for urinary alkalinization and enhanced excretion of xenobiotics[325], and a proprietary phytochelatin (metal-binding peptides from plants) concentrate to bind toxic metals in the gut and thereby promote their fecal excretion by blocking enterohepatic recycling/recirculation. DMSA chelation is approved by the US Food and Drug Administration (FDA) for the treatment of lead toxicity in children.[326] The use of DMSA for children and adults is supported by peer-reviewed literature.[327,328,329,330,331] After approximately 8 months of treatment, the patient was completely free of pain, and the clinical improvement was associated with a reduction in both lead and mercury of approximately 50% as demonstrated by follow-up laboratory testing. This case was published in peer-reviewed literature for continuing medical education (CME) for physicians.[332]

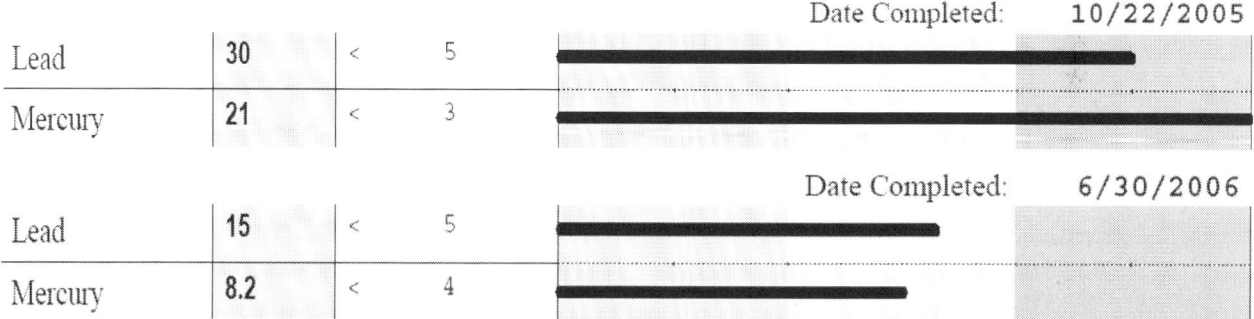

Marked accumulation of lead and mercury in a patient diagnosed with FM—complete elimination of pain and stiffness following identification and reduction in the body burden of lead and mercury: 54yo woman presents with nontraumatic widespread pain consistent with a diagnosis of fibromyalgia; the diagnosis of FM is confirmed by two clinicians. All laboratory test results were normal except for urine toxic metal testing which showed 6x elevations of lead and 7x elevations of mercury. Treatment (DMSA, citrate, selenium, and phytochelatins per above) for 8 months effected safe reductions in body burden of lead and mercury and alleviation of all pain.

---

[325] Crinnion WJ. Environmental medicine, part3:long-term effects of chronic low-dose mercury exposure. *Altern Med Rev* 2000 Jun;5:209-23
[326] "The Food and Drug Administration has recently licensed the drug DMSA (succimer) for reduction of blood lead levels >/= 45 micrograms/dl. This decision was based on the demonstrated ability of DMSA to reduce blood lead levels. An advantage of this drug is that it can be given orally." Goyer RA, Cherian MG, Jones MM, Reigart JR. Role of chelating agents for prevention, intervention, and treatment of exposures to toxic metals. *Environ Health Perspect*. 1995 Nov;103(11):1048-52
[327] Bradstreet et al. A case-control study of mercury burden in children with autistic spectrum disorders. *Journal of American Physicians and Surgeons* 2003; 8: 76-79
[328] Crinnion WJ. Environmental medicine, part three: long-term effects of chronic low-dose mercury exposure. *Altern Med Rev*. 2000 Jun;5(3):209-23
[329] Forman et al. A cluster of pediatric metallic mercury exposure cases treated with meso-2,3-dimercaptosuccinic acid (DMSA). *Environ Health Perspect*. 2000 Jun;108(6):575-7
[330] Miller AL. Dimercaptosuccinic acid (DMSA), a non-toxic, water-soluble treatment for heavy metal toxicity. *Altern Med Rev*. 1998 Jun;3(3):199-207
[331] DMSA. *Altern Med Rev*. 2000 Jun;5(3):264-7 thorne.com/altmedrev/.fulltext/5/3/264.pdf
[332] Vasquez A. *Musculoskeletal Pain: Expanded Clinical Strategies*. Institute for Functional Medicine. 2008

Small intestine bacterial overgrowth (SIBO) is the primary cause of and most logical explanation for FM; all other characteristics of the disease can be understood by understanding the protean effects of SIBO.

In the restructuring of this work in 2015 from its previous versions, I am—based on a more confident description of FM that moves from hypothesis to assertion—choosing to organize the pathophysiologic descriptions, upon which are based the therapeutic interventions, in the following four categories, each with their respective evidence:
1. The primary cause of fibromyalgia is SIBO—small intestinal bacterial overgrowth.
2. As a result of SIBO, fibromyalgia patients suffer somatic/body fatigue from mitochondrial dysfunction.
3. As a result of SIBO, fibromyalgia patients suffer increased sensitivity to pain due to heightened sensitivity of the brain and spinal cord—central sensitization—as well as from peripheral sensitization and impaired muscle function due to the previously established mitochondrial dysfunction.
4. All other biochemical and pathophysiologic abnormalities seen in fibromyalgia are explained from SIBO.

The primary focus of this section—establishing that SIBO is the cause of FM—is the provision of a cohesive and coherent explanation of FM; examples of treatments will be provided in this section on pathophysiology because the efficacy of treatments substantiates and helps prove the model that I have developed. Following these proofs substantiating this pathophysiologic model, a separate section detailing treatments will be provided. Allowing some repetition of citations used in the establishment of the pathophysiologic model, in the section on therapeutics I will review treatments using the format of my FINDSEX ® acronym to maintain consistency with my treatment protocols. Several diagrams will be provided, each providing novelty, emphasis, and repetition.

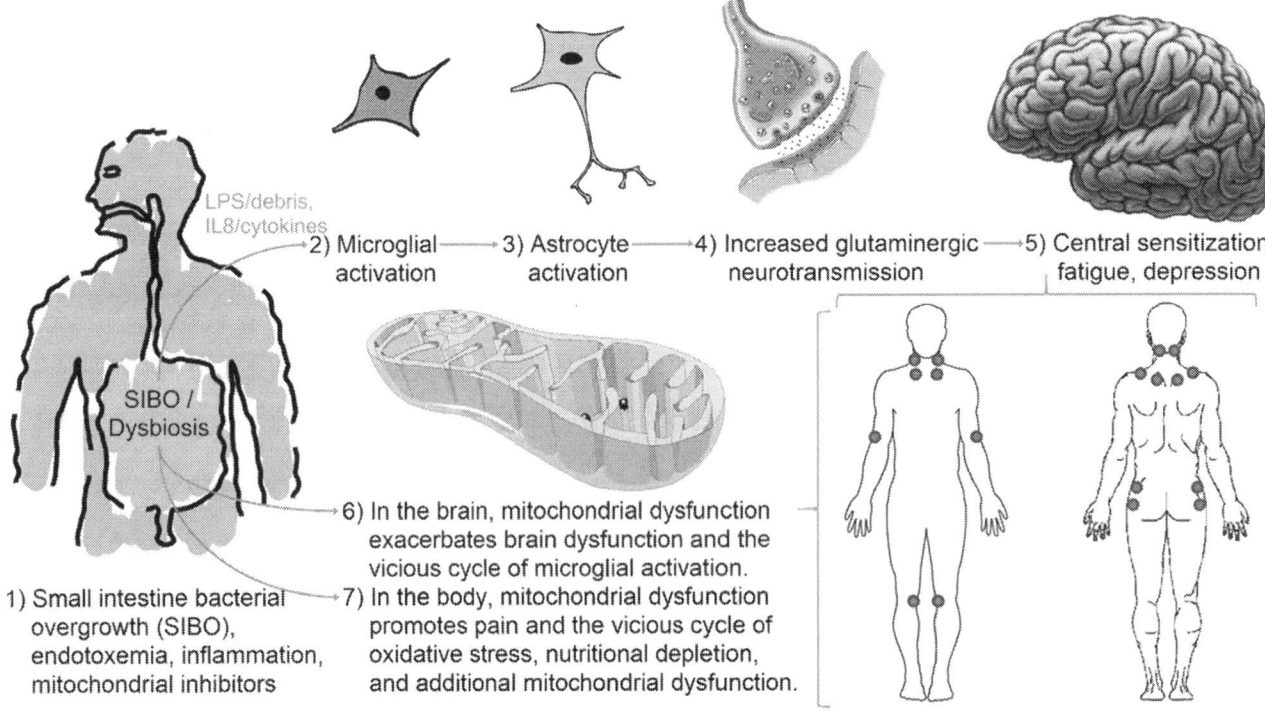

A simple integrated model of fibromyalgia, emphasizing dysbiosis-induced glial activation and mitochondrial dysfunction—first published in _Nutritional Perspectives_ 2015 Oct: Small intestine bacterial overgrowth (SIBO) elaborates endotoxin/lipopolysaccharide (LPS) with other inflammogens and mitochondrial inhibitors (including D-lactate and hydrogen sulfide [H2S]). Microglial activation can be triggered directly by LPS or indirectly by peripheral and central cytokines (especially IL-8), and it then triggers astrocyte activation and results in increased glutaminergic neurotransmission, which promotes central sensitization and the resulting depression, central fatigue, and pain sensitivity. In the brain, mitochondrial dysfunction exacerbates brain dysfunction and the vicious cycle of microglial activation. In the body, mitochondrial dysfunction promotes pain and the vicious cycles of oxidative stress, nutritional depletion, and additional mitochondrial dysfunction. Vitamin D deficiency, common in many conditions of persistent pain, exacerbates central pain by allowing increased microglial activation while also contributing to peripherally-sourced pain from muscle (myalgia) and bone (osteomalacia). Illustration by Vasquez; image of brain by IsaacMao per Flickr.com via creativecommons.org/licenses/by/2.0. See educational videos and updates at www.inflammationmastery.com/pain

❶ **Primary Cause of FM—SIBO: The primary cause of fibromyalgia is small intestinal bacterial overgrowth.** Small intestine bacterial overgrowth (SIBO)—also referred to as "intestinal bacterial overgrowth" or simply "bacterial overgrowth"—provides the single best model for explaining the clinical and pathophysiological manifestations of fibromyalgia. Although commonly underappreciated by many clinicians, SIBO is common in clinical practice, affecting for example approximately 40% of patients with rheumatoid arthritis, 84% of patients with IBS, and 90% to 100% of patients with fibromyalgia. **In a study of 42 fibromyalgia patients, all 42 FM patients showed laboratory evidence of SIBO, and the severity of the intestinal bacterial overgrowth correlated positively with the severity of the fibromyalgia**, thus indicating the plausibility of a causal relationship.[333] The links between fibromyalgia and IBS are also strong; **most IBS patients meet strict diagnostic criteria for fibromyalgia, and most fibromyalgia patients meet strict criteria for IBS**. Lubrano et al[334] showed that fibromyalgia severity correlated with IBS severity among patients who met strict diagnostic criteria for both conditions. The high degree of overlap between these two diagnostic labels suggests that these conditions are two variations of a common pathophysiological process—SIBO.[335] SIBO causes altered bowel function, immune activation, and visceral hypersensitivity, and it is the best causative explanation for the clinical and pathophysiological manifestations of IBS; for more details and citations, see the excellent review by Lin published in *Journal of the American Medical Association* in 2004.[336] IBS is characterized by *visceral* hyperalgesia (hypersensitivity to pain), just as fibromyalgia is characterized by *skeletal muscle* hyperalgesia. Given that strong evidence indicates that IBS is caused by SIBO and that IBS and fibromyalgia are variations of the same pathophysiological process, then fibromyalgia may therefore be caused by SIBO. However, these links and interconnections require substantiation, as provided throughout this section.

Bacterial LPS and other antigens absorbed from the intestine during SIBO contribute to a subclinical inflammatory state that results in pain hypersensitivity and increased cytokine release, both of which are characteristics of fibromyalgia. In animal models and in human research studies, exposure to bacterial endotoxin/LPS has been shown to increase the brain's sensitivity to and perception of pain. Immune-mediated and inflammation-mediated pathways that promote pain sensitivity and pain perception include ❶ increased production of nitric oxide with ❷ increased production of prostaglandins and cytokines, resulting in ❸ the sensitization of peripheral and/or central neurons to pain perception/transmission. In support of this concept, Lin[337] wrote in 2004, **"The immune response to bacterial antigen in SIBO provides a framework for understanding the hypersensitivity in both fibromyalgia and IBS"**. A later paper by Othmanm, Agüero, and Lin[338] in 2008 stated, "…a recent animal study demonstrated that exposure to endotoxin increased the production of prostaglandins and simultaneously decreased nitrous oxide production, resulting in inflammatory hyperalgesia" and "These observations suggest that SIBO is a common feature in both [IBS and FM] disorders and that altered gut microbiota in SIBO may play a role in the induction of somatic or visceral hypersensitivity, with affected patients meeting the diagnostic criteria for IBS, fibromyalgia or both disorders."

Gut bacteria also affect CNS/brain neurotransmission via vagal stimulation. According to a 2014 review by Galland, "Intrinsic primary afferent neurons (IPANs) are cellular targets of neuroactive bacteria and transmit microbial messages to the brain via the vagus nerve. Live bacteria may not be needed for these effects; in the case of *B. fragilis*, a lipid-free polysaccharide is both necessary and sufficient for IPAN activation. … Gut bacteria influence reactivity of the HPA axis and the induction and maintenance of nREM sleep. They may influence mood, pain sensitivity and normal brain development."[339]

- Fibromyalgia is tightly correlated with irritable bowel syndrome, a condition caused by small intestine bacterial overgrowth: Fibromyalgia and IBS are strongly convergent, and the evidence indicates that IBS is caused largely or completely by SIBO; again, for more details and citations, see the brilliant article by Lin, cited previously. Small intestine bacterial overgrowth is highly prevalent in fibromyalgia. Several studies have

---

[333] Pimentel et al. A link between irritable bowel syndrome and fibromyalgia may be related to findings on lactulose breath testing. *Ann Rheum Dis*. 2004 Apr;63(4):450-2
[334] Lubrano E, et al. Fibromyalgia in patients with irritable bowel syndrome. An association with the severity of the intestinal disorder. *Int J Colorectal Dis*. 2001 Aug;16(4):211-5
[335] Veale et al. Primary fibromyalgia and the irritable bowel syndrome: different expressions of a common pathogenetic process. *Br J Rheumatol*. 1991 Jun;30(3):220-2
[336] Lin HC. Small intestinal bacterial overgrowth: a framework for understanding irritable bowel syndrome. *JAMA*. 2004 Aug 18;292(7):852-8
[337] Lin HC. Small intestinal bacterial overgrowth: a framework for understanding irritable bowel syndrome. *JAMA*. 2004 Aug 18;292(7):852-8
[338] Othman M, Agüero R, Lin HC. Alterations in intestinal microbial flora and human disease. *Curr Opin Gastroenterol*. 2008 Jan;24(1):11-6
[339] Galland L. The gut microbiome and the brain. *J Med Food*. 2014 Dec;17(12):1261-72

shown that 90% to 100% of fibromyalgia patients have evidence of SIBO; such a strong correlation and the dose-response relationship imply causality and must be integrated into any science-based model of fibromyalgia.
- <u>Clinical study: Patients with FM have evidence of frequent and severe bacterial overgrowth in the intestines</u> (*Annals of the Rheumatic Diseases* 2004 Apr[340]): The **breath hydrogen test** is used for the detection of SIBO and involves orally administering a carbohydrate (such as lactulose, a source of sugar for bacteria) which is converted to hydrogen through bacterial fermentation; the exhaled hydrogen in the breath is measured as an indirect quantification of the amount of bacteria in the intestines. In this study, 20% of "healthy" control patients were found to have intestinal bacterial overgrowth via an abnormal hydrogen breath test compared with 93/111 (84%) subjects with IBS and **42/42 (100%) with fibromyalgia**. Subjects with fibromyalgia had higher hydrogen production (indicating more severe SIBO), peak hydrogen, and area under the curve than subjects with IBS. **The degree of somatic pain in fibromyalgia correlates significantly with the hydrogen level seen on the breath test.**

- <u>Small intestine bacterial overgrowth leads to systemic absorption of toxins that impair brain/nerve and muscle/mitochondrial function</u>: SIBO is associated with overproduction and absorption of bacterial cellular debris (e.g., lipopolysaccharide [LPS], bacterial DNA, peptidoglycans, teichoic acid, exotoxins) and antimetabolites—substances which are directly toxic to cellular energy/ATP production and muscle and nerve function—such as D-lactic acid, tyramine, tartaric acid, hydrogen sulfide. Intestinal gram-negative bacteria produce endotoxin (also known as lipopolysaccharide, LPS), which impairs skeletal muscle energy/ATP production (by stimulating skeletal muscle sodium-potassium-ATPase). Endotoxin also raises blood lactate (indicating impaired cellular energy production) under aerobic conditions in humans.[341] **Thus, via direct and indirect effects on cellular metabolism, chronic low-dose bacterial LPS/endotoxin exposure can result in impaired muscle metabolism and reduced ATP synthesis via impairment of mitochondrial function.** Intestinal bacteria also produce D-lactate, a well-known metabolic toxin in humans; SIBO often results in variable levels of D-lactate acidosis, severe cases of which can progress from fatigue and malaise to encephalopathy (e.g., confusion, ataxia, slurred speech, altered mental status) and death.[342] Supporting the proposal that bacterial overgrowth with D-lactate-producing bacteria is a contributor to the chronic fatigue syndromes including fibromyalgia is an excellent study published in 2009 showing that **patients with chronic fatigue syndrome have intestinal overgrowth of bacteria that produce the cellular toxin D-lactate**; specifically the research showed that these chronic fatigue patients have **a 7-fold increase in D-lactate producing *Enterococcus* and 1,100-fold increase in D-lactate producing *Streptococcus***. Energy/ATP underproduction and lactate overproduction cause muscle fatigue and muscle pain. An additional cellular toxin produced by intestinal bacteria is hydrogen sulfide (H2S), which causes DNA damage[343] (noted previously to be increased in fibromyalgia patients) and which impairs cellular energy production, a finding relevant to *but not necessarily limited to* the pathogenesis of ulcerative colitis.[344,345] Bacteria and yeast in the intestines produce H2S, which can bind to the mitochondrial enzyme cytochrome c oxidase (part of Complex IV of the electron transport chain), thereby impairing oxidative phosphorylation and ATP production; this may partly explain the association of gastrointestinal dysbiosis and small intestine bacterial overgrowth (SIBO) with conditions such as chronic fatigue syndrome (CFS) and fibromyalgia.[346] Given that sulfur-containing molecules such as sulfite and hydrogen sulfide bind to vitamin B12[347,348], we should reasonably expect that patients with excess exposure to H2S from the gastrointestinal tract would have an increased prevalence of vitamin B12 deficiency, and indeed this has been documented; vitamin B12 deficiency in CFS and FM patients promotes fatigue and brain dysfunction via the effects of vitamin B12 deficiency directly (ie, vitamin B12 deficiency is well known to cause

---

[340] Pimentel et al. A link between irritable bowel syndrome and fibromyalgia may be related to findings on lactulose breath testing. *Ann Rheum Dis*. 2004 Apr;63(4):450-2
[341] Bundgaard et al. Endotoxemia stimulates skeletal muscle Na+-K+-ATPase and raises blood lactate under aerobic conditions in humans. *Am J Physiol Heart Circ Physiol*. 2003 Mar;284(3):H1028-34
[342] Vella A, Farrugia G. D-lactic acidosis: pathologic consequence of saprophytism. *Mayo Clin Proc*. 1998 May;73(5):451-6
[343] Attene-Ramos MS, Wagner ED, Gaskins HR, Plewa MJ. Hydrogen sulfide induces direct radical-associated DNA damage. *Mol Cancer Res*. 2007 May;5(5):455-9
[344] Magee et al. Contribution of dietary protein to sulfide production in large intestine: in vitro and controlled feeding study in humans. *Am J Clin Nutr*. 2000 Dec;72(6):1488-94
[345] Babidge W, Millard S, Roediger W. Sulfides impair short chain fatty acid beta-oxidation at acyl-CoA dehydrogenase level in colonocytes. *Mol Cell Biochem*. 1998 Apr:117-24
[346] Lemle MD. Hypothesis: chronic fatigue syndrome is caused by dysregulation of hydrogen sulfide metabolism. *Med Hypotheses*. 2009 Jan;72(1):108-9
[347] Añíbarro et al. Asthma with sulfite intolerance in children: a blocking study with cyanocobalamin. *J Allergy Clin Immunol*. 1992 Jul;90(1):103-9
[348] Fujita et al. A fatal case of acute hydrogen sulfide poisoning caused by hydrogen sulfide: hydroxocobalamin therapy for acute hydrogen sulfide poisoning. *J Anal Toxicol*. 2011 Mar;35(2):119-23

Chapter 5.1—Functional Inflammology Protocol for Metabolic Inflammation: Migraine and Fibromyalgia

nerve damage and brain damage) and indirectly via impaired metabolism of homocysteine, which then triggers pain sensitivity and accelerated neurodegeneration via activation of NMDA receptors in the brain.[349]

- Experimental study: Effect of *E. coli* endotoxin on mitochondrial form and function (*Annals of Surgery* 1971 Dec[350]): Authors of this paper show that treatment of normal rat liver mitochondria with *E. coli* endotoxin results in mitochondrial impairment. They note previous research showing that animal exposure to *E. coli* endotoxin causes inhibition of mitochondrial respiration and uncoupling of oxidative phosphorylation. Near their conclusion, the authors write, "Thus we have evidence to show that topical *E. coli* endotoxin **has pathologic effects on both membrane integrity and internal mechanochemical systems of isolated mitochondria**." Readers should appreciate that *E. coli* is a common inhabitant of the gastrointestinal tract of humans and that its population is quantitatively increased during states of bacterial overgrowth of the small bowel, as is commonly seen in most patients with fibromyalgia. More recently, research has shown that impairment of mitochondrial function (noted in patients with fibromyalgia) can lead to destruction of mitochondria by a process termed "mitophagy" (noted in patients with fibromyalgia); over time, loss of mitochondria via mitophagy leads to reduced numbers of mitochondria in muscle and other tissues (noted in patients with fibromyalgia) and contributes to the fatigue and other symptoms which characterize FM.

- Clinical study: Increased D-lactic acid intestinal bacteria in patients with chronic fatigue syndrome (*In Vivo* 2009 Jul-Aug[351]): The authors of this 2009 study state in the summary of their research, "Patients with chronic fatigue syndrome (CFS) are affected by symptoms of cognitive dysfunction and neurological impairment, the cause of which has yet to be elucidated. However, these symptoms are strikingly similar to those of patients presented with D-lactic acidosis. A significant increase of Gram-positive facultative anaerobic fecal microorganisms in 108 CFS patients as compared to 177 control subjects is presented in this report. The viable count of D-lactic acid producing *Enterococcus* and *Streptococcus* spp. in the fecal samples from the CFS group (3.5 x 10(7) cfu [colony forming units]/L and 9.8 x 10(7) cfu/L respectively) were significantly higher than those for the control group (5.0 x 10(6) cfu/L and 8.9 x 10(4) cfu/L respectively). [**Note: This is approximately a 7x increase in D-lactate producing *Enterococcus* and 1,100x increase in D-lactate producing *Streptococcus*.**] Analysis of exometabolic profiles of *Enterococcus faecalis* and *Streptococcus sanguinis*, representatives of *Enterococcus* and *Streptococcus* spp. respectively, by NMR and HPLC showed that these organisms produced significantly more lactic acid from (13)C-labeled glucose, than the Gram negative *Escherichia coli*. Further, **both *E. faecalis* and *S. sanguinis* secrete more D-lactic acid than *E. coli*.** This study suggests a probable link between intestinal colonization of Gram-positive facultative anaerobic D-lactic acid bacteria and symptom expressions in a subgroup of patients with CFS. Given the fact that **this might explain not only neurocognitive dysfunction in CFS patients but

> **Patients with "chronic fatigue syndrome" and the associated neurologic dysfunction and muscle dysfunction have intestinal overgrowth of bacteria that produce D-lactic acid, a known neurotoxin and metabolic poison**
>
> In 2007 and 2008, the current author (AV) wrote and published *Musculoskeletal Pain: Expanded Clinical Strategies** with the Institute for Functional Medicine; this chapter on fibromyalgia is derived and updated from that work. In that publication, I reviewed evidence that fibromyalgia—at that time considered mysterious, idiopathic, chronic, relentless, and treatable only by pain-relieving drugs—was most likely caused by small intestine bacterial overgrowth (SIBO) and the resultant absorption of metabolic toxins and immunogenic debris. This perspective has been supported by numerous publications, particularly the article published by Sheedy et al** in 2009, which showed for the first time that patients with chronic fatigue syndrome—a condition tightly correlated with and which often overlaps with fibromyalgia—have SIBO with various bacteria that are high-output producers of D-lactic acid, a known neurotoxin and metabolic poison which potentially contributes to many of the main clinical, biochemical, and histologic manifestations of FM, namely mental fatigue and dyscognition (difficulty thinking), muscle fatigue and pain, biochemical evidence of mitochondrial impairment, and histologic evidence of mitochondrial myopathy.
>
> *Vasquez A. *Musculoskeletal Pain*. Institute for Functional Medicine, 2008. **Sheedy, et al. Increased d-lactic acid intestinal bacteria in patients with chronic fatigue syndrome. *In Vivo*. 2009 Jul

---

[349] Regland et al. Increased concentrations of homocysteine in the cerebrospinal fluid in patients with fibromyalgia and chronic fatigue syndrome. *Scand J Rheumatol*. 1997;26(4):301-7
[350] White et al. Effect of E. coli endotoxin on mitochondrial form and function. *Ann Surg*. 1971 Dec;174(6):983-90
[351] Sheedy JR, Wettenhall RE, Scanlon D, et al. Increased d-lactic acid intestinal bacteria in patients with chronic fatigue syndrome. *In Vivo*. 2009 Jul-Aug;23(4):621-8

also mitochondrial dysfunction, these findings may have important clinical implications." A note of personal experience: the current author (Dr Vasquez) had one event of severe headache and dyscognition following consumption of a prebiotic (FOS) supplement during my 6-year bout with a CFS-related condition; from this experience, I furthered my understanding of dysbiosis and continued to do so *via personal experience* for the following 10 years. Of further note, based on my personal experience and review of the biomedical literature, I proposed in the CME monograph *Musculoskeletal Pain: Expanded Clinical Strategies* (Institute for Functional Medicine, 2008) that D-lactic acidosis was one of the pathophysiologic mechanisms by which microbial overgrowth of the intestines causes fibromyalgia; I was therefore gratified to see this article—published in 2009, nearly 2 years after I had proposed this mechanism—verifying my hypothesis.

- <u>Commensal microbiota are necessary for the development of inflammatory pain</u> (*Proc Natl Acad Sci.* 2008 Feb[352]): In this remarkably insightful article, the authors introduce their work by stating, "The sensation of pain can be enhanced by acute or chronic inflammation", and that in their experimental model using germ-free and "conventional" (colonized) mice, they "show that inflammatory hypernociception induced by carrageenan, lipopolysaccharide, TNF-alpha, IL-1beta, and the chemokine CXCL1 was reduced in germ-free mice" while hypernociception induced by prostaglandins and dopamine was not altered by the presence/absence of bacteria. However, "reposition of the microbiota" or systemic administration of LPS essentially restored the pain and inflammation that was absent in germ-free mice, which also produced more IL-10, which—via experiments using anti-IL-10 antibody—proved to mediate both anti-inflammation and anti-nociception. The authors concluded that, "Therefore, these results show that contact with commensal microbiota is necessary for mice to develop inflammatory hypernociception. ... Therefore, these results show that contact with commensal microbiota [or LPS] is necessary for mice to develop inflammatory hypernociception possibly in a TLR-dependent manner. "

- <u>Low-dose LPS 'priming' of muscle provides an animal model of persistent elevated mechanical sensitivity for the study of chronic pain</u> (*Eur J Pain* 2011 Aug[353]): In this experiment using intramuscular hypertonic saline with either high or low doses of LPS, the authors found that low-dose LPS exacerbated long-term pain, while high-dose LPS caused the expected acute inflammatory response but did not promote development of chronic pain; the authors speculate that the low dose of LPS "primed" inflammatory and neurologic pathways for the persistence of pain perception while the higher dose may have invoked counter-inflammatory mechanisms ("larger-dose of LPS in this experiment may have provoked a protective effect such as invoking negative feedback loops") that failed to promote the development of central sensitization. By showing that low-level locally-administered systemically-circulated LPS could promote the development of chronic pain, these authors have supported a model consistent with an integrated model of fibromyalgia, in which SIBO/LPS can explain the entirety of this common condition.

---

[352] Amaral et al. Commensal microbiota is fundamental for the development of inflammatory pain. *Proc Natl Acad Sci U S A*. 2008 Feb 12;105(6):2193-7
[353] Yamaguchi et al. Low rather than high dose lipopolysaccharide 'priming' of muscle provides an animal model of persistent elevated mechanical sensitivity for the study of chronic pain. *Eur J Pain*. 2011 Aug;15(7):724-31

Chapter 5.1—Functional Inflammology Protocol for Metabolic Inflammation: Migraine and Fibromyalgia

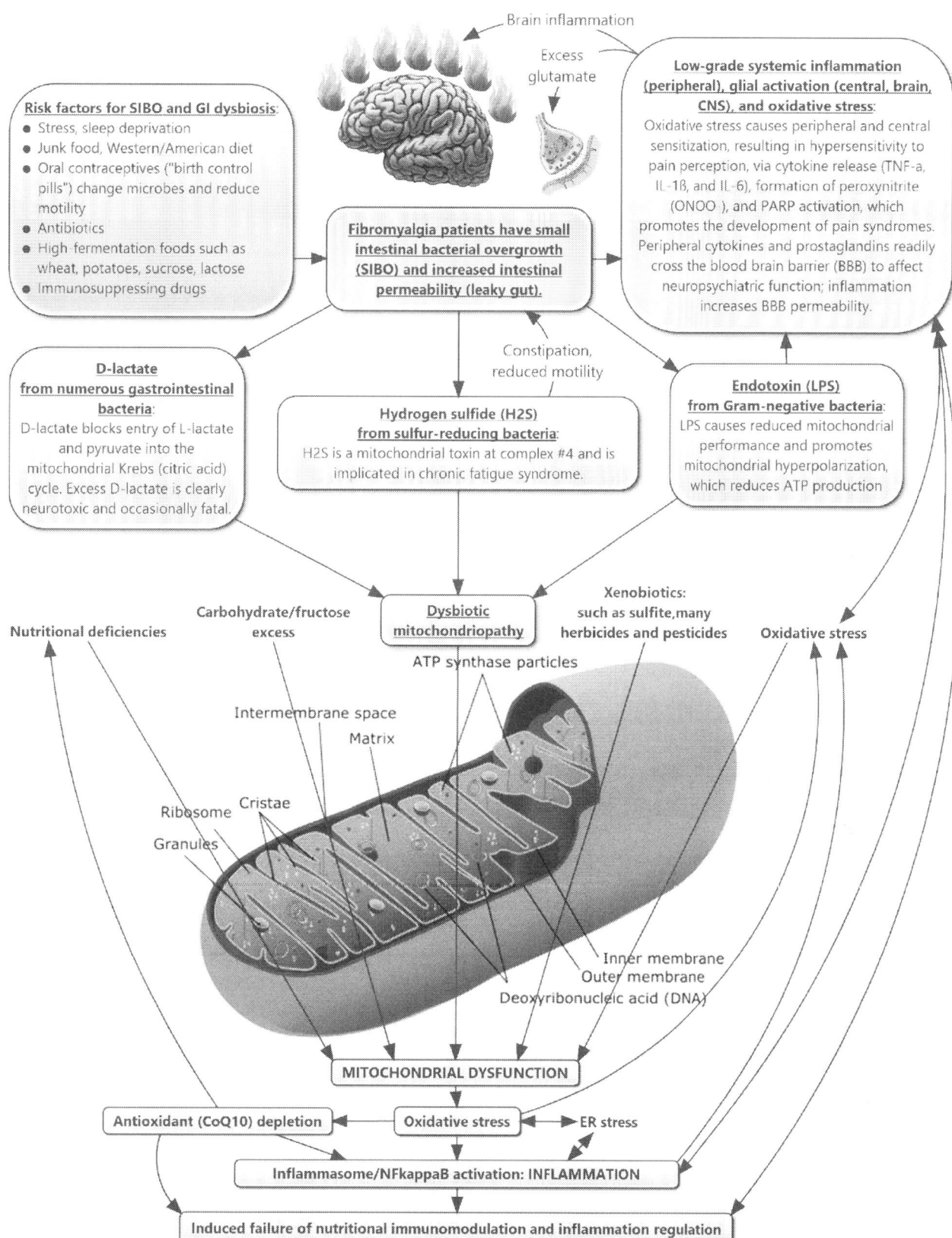

Fibromyalgia is a unique combination of SIBO-induced mitochondrial impairment with SIBO-induced glial activation: The synergistic combination of mitochondrial impairment (clearly evidenced in FM) with glial activation and the resulting central sensitization (clearly evidenced in FM) produces the clinical pattern of fibromyalgia. Image copyright © 2016 by Dr Alex Vasquez, all rights reserved and enforced. Image of brain by IsaacMao per Flickr.com via creativecommons.org/licenses/by/2.0.

❷ **Secondary Cause of FM (part 1) — Mitochondrial Dysfunction: Fibromyalgia patients suffer somatic/body and central/brain fatigue from mitochondrial dysfunction as a result of SIBO.**
The evidence is clear to the point of being irrefutable that FM patients have mitochondrial impairment; given the critical role of mitochondria in somatic and cerebral function, this mitochondrial impairment (dysmitochondriosis) alone would be sufficient to explain virtually all manifestations of FM. Dysmitochondriosis is evident biochemically and histologically, thus obviously accounting for — or at least contributing to — the muscle/somatic pain and fatigue; what makes the situation in FM unique is the coupling of mitochondrial impairment with central nervous system inflammation, ie, brain inflammation, glial activation, and central sensitization. Glial activation is exacerbated by mitochondrial dysfunction, because — at least in part — the hyperglutaminergic neurotransmission induced by microglial-astrocyte activation puts heavier demands on energy/ATP production to maintain neuronal homeostasis, for maintaining the increased metabolic activity in general and for maintaining intracellular calcium homeostasis in particular. The excess glutaminergic neurotransmission contributes to additional glial activation and mitochondrial impairment, thereby forming a reinforced vicious cycle. Mitochondrial dysfunction causes excess ROS production and resultant depletion of multifunctional nutrients/chemicals such as CoQ10, which again feeds back to promote additional mitochondrial impairment and reduction in antioxidant defenses, leading to altered and consequential ROS signaling and the perpetuation of inflammation (e.g., inflammasome activation), mitochondrial impairment, and glial activation. Clinicians and researchers need to appreciate that because of their pathophysiologic connections and consequences, microglial activation leads to astrocyte activation which leads to excessive glutaminergic (hyperglutaminergic) neurotransmission and the brain/neocortical hyperexcitation that promotes the manifestations of depression (and other components of sickness behavior) and pain via central sensitization; as such these terms become interconnected and largely interchangeable when discussed in their totality of effect.

Mitochondrial dysfunction promotes central sensitization via oxidative stress and cytokine release: Mitochondrial dysfunction increases free radical production, which promotes "neurologic hypersensitivity" to pain, i.e., the pain in fibromyalgia is not simply due to muscle fatigue due to mitochondrial dysfunction although that is clearly a major component. The mitochondrial dysfunction also promotes central sensitization, which should be treated directly via alleviating the mitochondrial dysfunction and the causative dysbiosis, in addition to patient-specific factors.

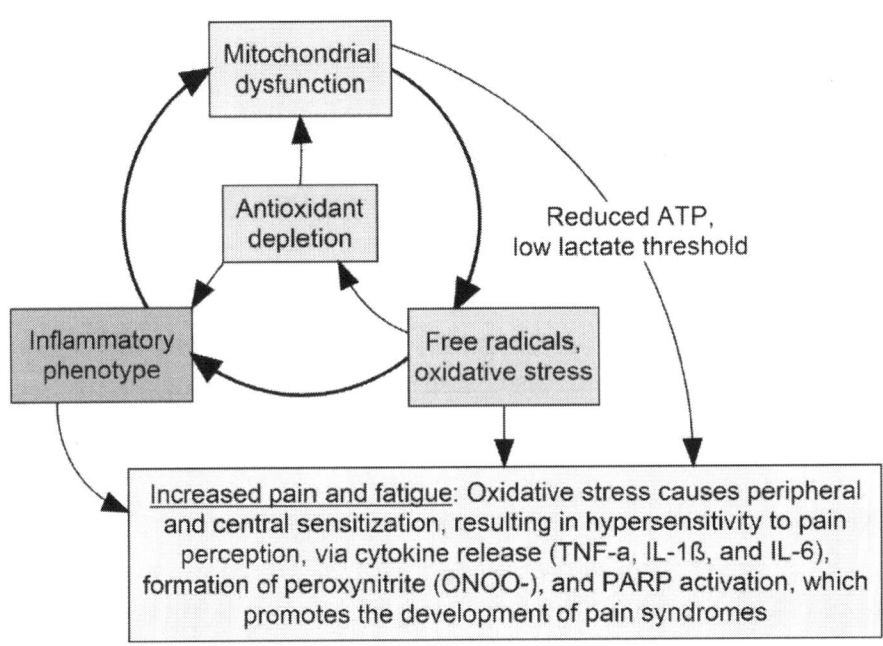

- Antimetabolites — microbial products that directly and indirectly impair mitochondrial performance: Yeast and bacteria can produce certain molecules which *jam up, monkey wrench*, or otherwise interfere with normal human cellular metabolism. The best example is **D-lactic acid**, which impairs human metabolic pathways that are designed to work with the "human" form of this metabolite — the levo isomer — L-lactic acid. Commonly resulting in headache, fatigue, depression, and sometimes death, D-lactic acidosis is extensively well documented in the medical research literature and commonly occurs in association with bacterial overgrowth of the intestine, particularly following intestinal bypass surgery.[354] Other antimetabolites produced from

---

[354] "D-Lactic acidosis is a potentially fatal clinical condition seen in patients with a short small intestine and an intact colon. Excessive production of D-lactate by abnormal bowel flora overwhelms normal metabolism of D-lactate and leads to an accumulation of this enantiomer in the blood." Vella A, Farrugia G. D-lactic acidosis: pathologic consequence of saprophytism. *Mayo Clin Proc*. 1998 May;73(5):451-6

(intestinal) microbes which are associated with human disease and dysfunction include **ammonia, tryptamine, tyramine, octopamine, mercaptates, aldehydes, alcohol (ethanol), tartaric acid, indolepropionic acid, indoleacetic acid, skatole, indole, putrescine,** and **cadaverine**. Many of these metabolites are seen in higher amounts in patients with migraine, depression, weakness, confusion, schizophrenia, agitation, hepatic encephalopathy, chronic arthritis and rheumatoid arthritis. **Gut-derived neurotoxins** from bacteria and yeast may contribute to autistic symptomatology[355,356], and case reports have consistently demonstrated that excess absorption of bacterial metabolites can alter behavior in humans and result in acute neurocognitive decline and behavioral abnormalities in children.[357] **Hydrogen sulfide (H2S)**, produced by intestinal bacteria such as *Citrobacter freundii*[358], is a mitochondrial poison[359] and is strongly associated with disease activity in ulcerative colitis.[360] Degradation of tryptophan by bacterial tryptophanase predisposes the host to a "functional tryptophan deficiency" and may result in insufficiency of serotonin which would contribute to hyperalgesia, depression, impaired adrenal responsiveness[361] ("hypoadrenalism"), and insomnia; **indole** and **skatole**, which are gut-derived bacterial degradation products of tryptophan, produce an inflammatory arthritis identical to rheumatoid arthritis in animal models.[362,363]

> **D-lactate triad: SIBO, CHO, and IP—bacterial overgrowth, carbohydrate, and increased permeability; the solutions are to reduce SIBO, reduce/modify CHO, heal gut, promote motility**
> "D-lactate is usually produced in excess when small bowel resection allows delivery of a high carbohydrate load to the colon. Elevation of D-lactate in plasma may also occur after other types of abdominal surgery, as a result of increased intestinal permeability and bacterial translocation across the intestinal mucosal barrier. Nonsurgical causes of intestinal hyperpermeability also increase absorption of D-lactate from the intestinal lumen."
>
> Galland L. *J Med Food*. 2014 Dec

- Increased D-lactic acid intestinal bacteria in chronic fatigue syndrome (*In Vivo* 2009 Jul[364]): The authors report finding a **significant increase of Gram-positive facultative anaerobic fecal microorganisms** in 108 CFS patients as compared to 177 control subjects. Specifically, they report, "The viable count of D-lactic acid producing *Enterococcus* and *Streptococcus* spp. in the fecal samples from the CFS group (3.5 x 10(7) cfu/L and 9.8 x 10(7) cfu/L respectively) were significantly higher than those for the control group (5.0 x 10(6) cfu/L and 8.9 x 10(4) cfu/L respectively). **Readers should note that this is approximately a 10-fold increase in *Enterococcus* and a >1,000-fold increase in *Streptococcus* in the CFS group compared with the control group. These Gram-positive bacteria produced not only more lactic acid in general but specifically they produced more of the dextro isomer D-lactic acid** than the Gram negative *Escherichia coli*. The authors correctly conclude that these findings "might explain not only neurocognitive dysfunction in CFS patients but also mitochondrial dysfunction, these findings may have important clinical implications."
- Short bowel syndrome and D-lactic acidosis (*Arch Dis Child* 1980 Oct[365]): The three-sentence abstract of this case report reads, "Metabolic acidosis in a 3-year-old child with short bowel syndrome led to the discovery of massive D-lactic aciduria. After normalization of the intestinal bacterial flora, D-lactate disappeared together with the acidosis. Dysbacteriosis with excessive production of D-lactate by intestinal bacteria (unidentified) and subsequent absorption explains this unusual cause of metabolic acidosis." Note the use

---

[355] Sandler RH, Finegold SM, Bolte ER, et al. Short-term benefit from oral vancomycin treatment of regressive-onset autism. *J Child Neurol*. 2000 Jul;15(7):429-35
[356] Shaw et al. Increased urinary excretion of analogs of Krebs cycle metabolites and arabinose in two brothers with autistic features. *Clin Chem* 1995;41(8Pt1):1094-104
[357] "The neurological features consisted of a depressed conscious state, confusion, aggressive behaviour, slurred speech and ataxia. The organic acid profile of urine demonstrated increased amounts of lactic, 3-hydroxypropionic, 3-hydroxyisobutyric, 2-hydroxyisocaproic, phenyllactic, 4-hydroxyphenylacetic and 4-hydroxyphenyllactic acids. Of the lactic acid 99% was D-lactic acid." Haan et al. Severe illness caused by products of bacterial metabolism in child with short gut. *Eur J Pediatr*. 1985;144:63-5
[358] Lennette EH (editor in chief). *Manual of Clinical Microbiology. Fourth Edition*. Washington DC; American Society for Microbiology: 1985, page 269. See also web.indstate.edu/thcme/micro/GI/general/sld038.htm Accessed 10/27/2005
[359] "Treatment of H2S poisoning may benefit from interventions aimed at minimizing ROS-induced damage and reducing mitochondrial damage." Eghbal MA, Pennefather PS, O'Brien PJ. H2S cytotoxicity mechanism involves reactive oxygen species formation and mitochondrial depolarisation. *Toxicology*. 2004 Oct 15;203(1-3):69-76
[360] "CONCLUSIONS: Metabolic effects of sodium hydrogen sulfide on butyrate oxidation along the length of the colon closely mirror metabolic abnormalities observed in active ulcerative colitis, and increased production of sulfide in ulcerative colitis suggests that the action of mercaptides may be involved in the genesis of ulcerative colitis." Roediger WE, et al. Reducing sulfur compounds of the colon impair colonocyte nutrition: implications for ulcerative colitis. *Gastroenterology* 1993 Mar;104:802-9
[361] "This hypothesis is supported by the findings in chronic MS patients of a significantly diminished adrenal cortisol reactivity to insulin-induced hypoglycemia which is considered a stress response mediated through the 5-HT system. Consequently, since patients with MS exhibit an abnormal response to stress it follows that increased tryptophan availability through dietary supplementation would diminish their vulnerability to psychological stress." Sandyk R. Tryptophan availability and the susceptibility to stress in multiple sclerosis: a hypothesis. *Int J Neurosci*. 1996 Jul;86(1-2):47-53
[362] Nakoneczna I, Forbes JC, Rogers KS. The arthritogenic effect of indole, skatole and other tryptophan metabolites in rabbits. *Am J Pathol*. 1969 Dec;57(3):523-38
[363] Rogers KS, Forbes JC, Nakoneczna I. Arthritogenic properties of lipophilic, aryl molecules. *Proc Soc Exp Biol Med*. 1969 Jun;131(2):670-2
[364] Sheedy JR, et al. Increased d-lactic acid intestinal bacteria in patients with chronic fatigue syndrome. *In Vivo*. 2009 Jul-Aug;23(4):621-8
[365] Schoorel EP, Giesberts MA, Blom W, van Gelderen HH. D-Lactic acidosis in a boy with short bowel syndrome. *Arch Dis Child*. 1980 Oct;55(10):810-2

of "dysbacteriosis" which is more commonly abbreviated to "dysbiosis", with the latter being more accurate in its nonspecificity since the disease-producing microbes may be from several different kingdoms and thus not exclusive to bacteria. The young boy in this case report underwent bowel resection secondary to multiple congenital malformations and complications from infection, i.e., mesenteric thrombosis tertiary to dehydration secondary to infectious diarrhea. After several episodes of neurocognitive dysfunction including weakness and dyspnea, the child was eventually diagnosed with multiple nutritional deficiencies, malabsorption, and lactic acidosis secondary to intestinal overgrowth of Gram-positive D-lactate-producing bacteria and an insufficiency of Gram-negative bacteria. The child was treated only with probiotic supplementation—no antibiotics were given—and D-lactate levels fell quickly within 4 days and were virtually undetectable by day 11. Thus, probiotic therapy alone may be sufficient treatment for some cases of D-lactic acidosis, particularly when an "insufficiency dysbiosis" of beneficial bacteria and a relative or absolute "overgrowth dysbiosis" of harmful bacteria coexist.

❸ **Secondary Cause of FM (part 2)—Central Sensitization: As a result of SIBO, fibromyalgia patients suffer increased sensitivity to pain due to heightened sensitivity of the brain and spinal cord—as well as from peripheral sensitization and impaired muscle function due to the previously established mitochondrial dysfunction. In other words, inflammation induced by microbial debris promotes sensitization to pain.**
- The *existence* of central sensitization in fibromyalgia—real and generally accepted: Central sensitization—the increased perception of "pain" from otherwise nonpainful stimuli—is a well-accepted component of fibromyalgia; in fact, some authors and medical societies have claimed that sensitization of the brain and spinal cord is indeed the sole cause of fibromyalgia. The emphasis on *central* is to specify that the sensitization is localized in the *central* nervous system (comprised only of the brain and spinal cord) and not in the *peripheral* nervous system—the peripheral nerves (e.g., in the arms and legs) nor in their receptors (e.g., in skin, muscles, and other tissues). Again, the consensus is that central sensitization is present in fibromyalgia, that these patients have heightened sensitivity to pain and the perception of pain from stimuli that would not otherwise be painful to "normal" persons whom do not have fibromyalgia.
- The *origin* of central sensitization—an issue of the highest importance in the treatment of fibromyalgia: What is needed in the conversation on central sensitization is not debating is *presence* but rather an understanding of its *cause*, or purported lack of cause. In medicine, when we say that a condition is "primary" we are saying that conceptually the condition has *no known cause*, that it is *idiopathic*, that the primary origin is *inherent* within the disease condition itself; to say that something is *primary* is to say that it itself is the origin of the disease process. In contrast, when we describe a condition or aspect of disease as *secondary*, we are saying that it is due to a preceding primary problem, that it follows some other event, that it is second in line in the disease process. Likewise, we can say that a problem that occurs *causally* (not simply *chronologically*) after a secondary problem is a tertiary problem, and so on. If two events happen at the same time but one event does not cause the other, then the events are *associated* (somehow related) or *concomitant* (occurring at the same time).

> **Dysbiosis—SIBO and LPS—alter brain function, neurotransmission, to promote pain, fatigue, depression**
>
> "Structural bacterial components such as **lipopolysaccharides** provide low-grade tonic stimulation of the innate immune system. Excessive stimulation due to bacterial **dysbiosis, small intestinal bacterial overgrowth**, or increased intestinal permeability may produce systemic and/or central nervous system inflammation."
>
> Galland L. *J Med Food.* 2014 Dec

  o Example—distinguishing *primary*, from *secondary*, from *tertiary*, from *associated*: If a person has an automobile or sporting accident (the *primary* event) and suffers a painful injury, we would say that the injury is *secondary* to trauma, and that any resulting psychological distress would be *secondary* to the pain and impairment from the injury, or *tertiary* to the primary trauma. If the psychological distress were due to the accident itself (e.g., distressing memories of what occurred), then the psychological distress would be *secondary* to experiencing the event of the accident itself, independent from or complicated by any pain or impairment. If the patient also had a skin disease that existed previously, we would say that the patient has a *concomitant* skin disease but that it is unrelated to the accident or the injury, unless perhaps the patient intentionally injured himself/herself as a result of the skin disease (for example in a suicide attempt or some other form of self-harm secondary to the primary illness).

In the conversation on central sensitization in fibromyalgia, the distinction between *primary* and *secondary* is of the highest importance. If we say that central sensitization in fibromyalgia is *primary*, we are saying that the problem originates in the brain and spinal cord, and that the appropriate treatments are therefore those that target the brain and spinal cord, such as pain-relieving drugs; other treatments might be useful, but they are of secondary importance to directly influencing the *primary* problem in the brain and spinal cord. For ethical and professional reasons, doctorate-level physicians are obligated to address the primary cause of disease whenever possible; this is a matter of acumen and beneficence because failing to address the primary cause of the problem when possible firstly allows the primary disease to fester and develop and progress while secondly leaving the patient dependent upon—enslaved to—symptom alleviation.

- Example—distinguishing *professional* and *ethical* behavior (e.g., treating the primary problem) from *unprofessional/irresponsible/unethical* behavior (e.g., failing to treat the primary problem, promoting dependency and exposing the patient to excess risk and expense, etc) among physicians: If a patient sits on a nail but cannot see the nail and then reports to the physician's office with a complaint of pain, the physician is professionally obligated to assess the situation by providing treatment of the primary problem—in this case, by removing the nail. The cause of the pain is the embedded nail, removing the nail will shortly alleviate the pain (assuming no infection or other complication). But let's say that an unethical physician is paid by a drug company to sell analgesic drugs, and the physician neglects his/her professional responsibility and his/her ethical responsibility to the patient; this physician fails to address the primary problem (the painful nail) and instead prescribes a dangerous drug to partly alleviate the pain at a cost of $200 and the doctor also receives a cash "gift" of $50 from the drug company for having promoted the sale of the drug. Because the doctor failed to address the *primary* problem (the painful nail) and only addressed the *secondary* problem (the pain), the doctor has failed to provide professional and ethical care, and this is further complicated by the physician's nondisclosed conflict of interest and receipt of a cash reward for having ordered the patient to spend $200 on a drug that he/she did not need. What is worse, the patient might be injured from the drug, or might eventually develop and infection because of the ongoing nature of the embedded nail. The doctor has created dependency (e.g., return visits and fees for more prescriptions) and income (e.g., from office fees and from payments by the drug company) while failing to treat the primary cause of the patient's problem.

If we say that pain and central sensitization in fibromyalgia are *primary*—without identifiable cause—then we have to treat symptomatically by suppressing/targeting the pain itself; this is reasonable, but increasingly unlikely as science advances and we better understand the nature and causes of diseases. If we say that pain and central sensitization in fibromyalgia are *secondary*—caused by a primary problem—then we are professionally and ethically obligated to address the primary cause of the pain. If physicians and medical groups say that pain and central sensitization in fibromyalgia are *primary* when in fact the research has made clear the cause of and treatment for the pain, then these physicians and medical groups—possibly to advance their own importance in society and income from drug companies—are behaving unprofessionally and unethically and unscientifically by cheating their patients of the opportunity for cure and putting these patients at risk for complications from the primary disease, at risk for adverse drug effects including injury and death, and burdening patients and the healthcare system with the costs of thousands of dollars individually and billions of dollars collectively/systematically by frauding the nature of the illness and its treatment. There it is; that is the defining line between *ethical* and *professional* behavior and *unethical* and *unprofessional* behavior, whether by individual physicians affecting the lives of hundreds of patients or by medical organizations affecting millions of patients for billions of dollars. If fibromyalgia has a legitimate cause, then doctors have an obligation to treat the cause of the problem rather than profit by unnecessary clinical patronage (coerced by pain, made necessary by the "necessary" need for repeated assessments and prescription refills). Relatedly, medical organizations that fraud the practice of medicine by distributing false information and promoting inefficacious treatment protocols that culminate in nontreatment of the primary disorder, the unnecessary use of dangerous treatments that harm patients, and unnecessary expenses to healthcare systems in the billions of dollars would be culpable for same.

- <u>Describing central sensitization in fibromyalgia as "primary" is unscientific, magical thinking that promotes drug dependency, drug sales, and physician dependency</u>: This promulgation, based on the selective ignoring of a vast and readily integrated body of scientific body of information, harms patients and costs healthcare systems billions of dollars.
- <u>Describing central sensitization in fibromyalgia as "secondary" is scientific, logical thinking</u>: This affirmation, based on the integration of a large body of scientific information, benefits patients, empowers doctors, and saves healthcare systems billions of dollars.

Given that the existence of central sensitization in animal and human models of pain, and that its presence is reasonably well established in human patients with fibromyalgia, the main question(s) to address are ❶ What is the cause of the central sensitization?, ❷ How can the cause be effectively treated?, ❸ Only if no cause can be found are we then allowed to ask, "What treatments—natural or pharmaceutical—are appropriate for the direct treatment of central sensitization?" Anytime that doctors and policymakers are discussing treatments, we have to consider 1) cost, availability, distributive justice, 2) effectiveness, 3) cost-effectiveness ratio, 4) safety, contraindications, adverse effects, and 5) drug-drug and drug-nutrient interactions; in shorthand, we can think of the risk:benefit ratio as a summation of these considerations, wherein "risk" is everything negative and "benefit" is everything positive. Except perhaps in Emergency Medicine and Urgent Care, doctors for the most part are trained to ignore the cause of problems and to just prescribe the "appropriate" drug for each diagnosis[366]; this is why medical care for nonacute/chronic/persistent diseases is so abysmal—because doctors have been trained to turn off their investigative brains, to recite the *idiopathic dogma*[367] that "chronic diseases are a complex interplay genes and environment, and while we don't yet know the exact cause, we can give you medicines that will help", to focus on molecules rather than context and cause, and to basically just practice as MDs—medicine dispensers. Exacerbating the problem are the medical/science writers and the medical organizations (e.g., specialists organizations in pain, rheumatology, and general medicine) that are paid hundreds of thousands and millions of dollars, respectively, to selectively ignore *causal* data and emphasize the *drug-selling* data.

In the following sections, I will establish that ❶ central sensitization in fibromyalgia is caused by microbial debris and secondary metabolic and inflammatory effects, that ❷ rational treatment of microbe-induced pain and inflammation must focus on 1) the eradication/modulation of the inflammatory microbial load, 2) the restoration of barrier defenses (e.g., the defensive lining of the skin and the gut wall) to reduce absorption of and exposure to microbial molecules (enhancing elimination of already-absorbed microbial toxins is a possibility, but is of lesser importance and is mentioned in some of my other writings), and ❸ treatments to directly address the inflammatory pain and central sensitization.

---

[366] Ely JW, Osheroff JA, Ebell MH, et al. Analysis of questions asked by family doctors regarding patient care. *BMJ*. 1999 Aug 7;319(7206):358-61
[367] Vasquez A. Twilight of the Idiopathic Era and the Dawn of New Possibilities in Health and Healthcare. *Naturopathy Digest* 2006 Mar naturopathydigest.com/archives/2006/mar/idiopathic.php. See also: Ely et al. Analysis of questions asked by family doctors regarding patient care. *BMJ*. 1999 Aug:358-61

Central sensitization in fibromyalgia is caused by microbial debris and secondary metabolic and inflammatory effects: My main thesis is stated immediately above, that central sensitization in fibromyalgia is caused by microbial debris and the secondary metabolic and inflammatory effects. I think this is easily demonstrated in graphic form, and so I will use the following image and its caption do most of the explaining, then I will follow the caption with a bit more discussion before concluding this section with a few notes and comments on research.

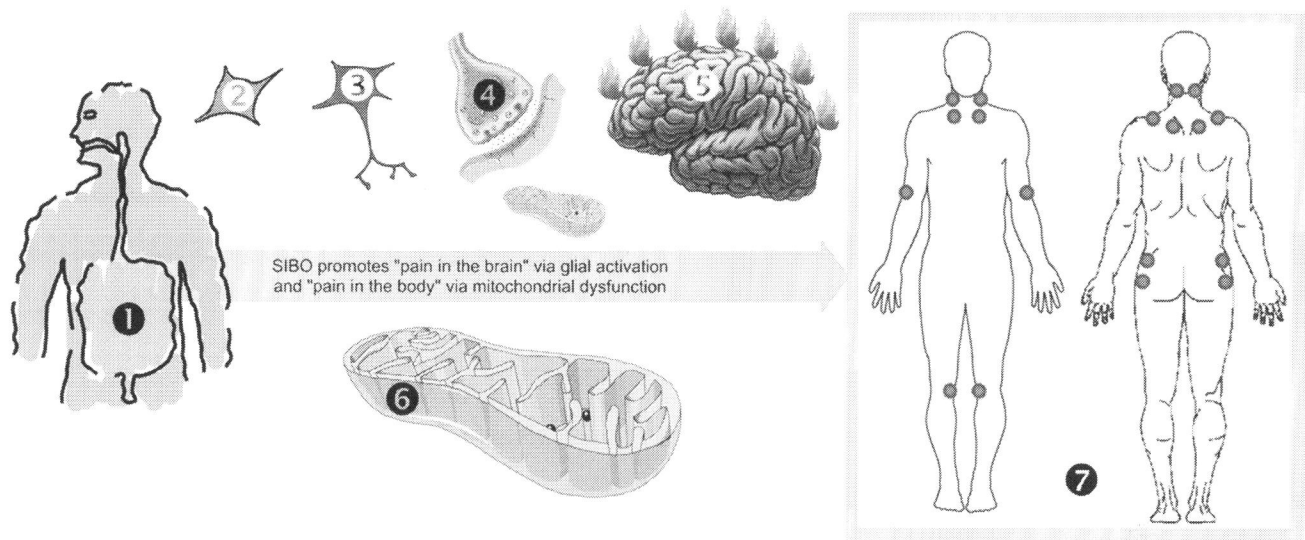

SIBO promotes glial activation and mitochondrial impairment which are the major causes of fibromyalgia syndrome: *From gut to "brain pain"*: ❶ Small intestine bacterial overgrowth (SIBO) leads to intestinal absorption and systemic distribution of low levels of bacterial endotoxin (endotoxemia) and other **inflamm**ation-**gen**erating molecules (inflammogens); this results in low-grade inflammation (including release of cytokines, prostaglandins and other inflammatory mediators and oxidants).[368] Microbial inflammogens cause systemic inflammation, and cytokines and prostaglandins produced peripherally (ie, outside of the central nervous system [CNS], which is the brain and spinal cord) can readily traverse the blood-brain barrier (BBB) and enter the CNS to promote glial activation—brain inflammation.[369] Some of these microbial inflammogens may be able to bypass the BBB directly, when the BBB becomes permeable/leaky following induction of systemic inflammation. SIBO can also elaborate mitochondrial inhibitors such as endotoxin (lipopolysaccharide, LPS)[370], hydrogen sulfide (H2S)[371], and D-lactate.[372] In the brain, mitochondrial dysfunction exacerbates brain dysfunction and the vicious cycle of microglial activation.[373] ❷ Microglia are immune cells in the brain that respond to cytokines, prostaglandins, and microbial inflammogens; when microglia become stimulated or "activated" by inflammatory triggers/signals, the microglia signal/activate/irritate the nearby ❸ astrocytes, which are cells in the brain that respond by causing an increase in neuron-to-neuron communication (neurotransmission) via the neurotransmitter glutamate, which is stimulatory to neurons.[374] ❹ While glutamate is necessary in small and regulated amounts, higher levels of glutamate promote central sensitization, pain amplification, "brain fatigue", depression and anxiety; when very elevated, glutamate can promote migraine headaches, seizures and epilepsy.[375] ❺ High levels of glutamate cause excitation of brain neurons, and this increased activity leads to increased production of free radicals, which cause additional local inflammation and mitochondrial dysfunction within the brain, leading back to microglial activation for a vicious cycle. The brain is now in a "positive feedback loop" which promotes additional pain/fatigue/depression independently from ongoing stimulation from the original trigger. As you can see from this description, microglial activation causes astrocyte activation in a close relationship; since microglia—the brain's immune cells—and astrocytes—the brain's supportive or helper or "nurse" cells—are both categorized as glial cells (glia = glue = the mass of cells that creates the supporting structure for the neurons of the brain), you can see that *microglial activation* and *astrocyte activation* and *glial activation* are extensions of each other and can be used somewhat interchangeably. Excess or prolonged microglial activation promotes neurodegeneration via hyperexcitation of neurons, basically causing them to "burn out" in a process

---

[368] Patel et al. Human experimental endotoxemia in modeling pathophysiology, genomics, and therapeutics of innate immunity in complex cardiometabolic diseases. *Arterioscler Thromb Vasc Biol* 2015 Mar:525-34. Ferguson et al. Omega-3 PUFA supplementation response to endotoxemia in healthy volunteers. *Mol Nutr Food Res* 2014 Mar;601-13
[369] Wilson CJ, Finch CE, Cohen HJ. Cytokines and cognition--the case for a head-to-toe inflammatory paradigm. *J Am Geriatr Soc*. 2002 Dec;50(12):2041-56
[370] Scirocco et al. Exposure of Toll-like receptors 4 to bacterial lipopolysaccharide (LPS) impairs human colonic smooth muscle cell function. *J Cell Physiol*. 2010 May; 442-50
[371] Lemle MD. Hypothesis: chronic fatigue syndrome is caused by dysregulation of hydrogen sulfide metabolism. *Med Hypotheses*. 2009 Jan;72(1):108-9
[372] Sheedy et al. Increased d-lactic Acid intestinal bacteria in patients with chronic fatigue syndrome. *In Vivo*. 2009 Jul-Aug;23(4):621-8
[373] Nguyen et al. A new vicious cycle involving glutamate excitotoxicity, oxidative stress and mitochondrial dynamics. *Cell Death Dis*. 2011 Dec 8;2:e240
[374] Béchade C, Cantaut-Belarif Y, Bessis A. Microglial control of neuronal activity. *Front Cell Neurosci*. 2013 Mar 28;7:32
[375] Devinsky O, Vezzani A, Najjar S, De Lanerolle NC, Rogawski MA. Glia and epilepsy: excitability and inflammation. *Trends Neurosci*. 2013 Mar;36(3):174-84

that has been described as "brain on fire."[376] The exception to this occurs after a period of particularly protracted microglial activation, which can cause damage or "burn out" of the astrocytes, too; this "astrocyte degeneration" leads to neurodegeneration when the astrocytes become impaired and cannot perform their supportive and "nursing" functions to the neurons. ❻ *From gut to "body pain and fatigue"*: SIBO can also elaborate mitochondrial inhibitors such as endotoxin (lipopolysaccharide, LPS), hydrogen sulfide (H2S), and D-lactate, as previously stated and cited. In the body, mitochondrial dysfunction promotes pain and the vicious cycle of oxidative stress, nutritional depletion, and additional mitochondrial dysfunction. Mitochondrial dysfunction in muscle leads to the cellular/cytologic and histologic/tissue changes that are typical and well-documented in cell and muscle samples of patients with fibromyalgia.[377] These peripheral (e.g., non-brain) changes in muscle also prove beyond any doubt that fibromyalgia is not a "brain disease" or solely a "disorder of pain processing." ❼ **Thus, fibromyalgia can be easily explained/understood as SIBO-induced central sensitization and mitochondrial dysfunction, resulting in pain and fatigue**; all other abnormalities in FM can be traced back to these key problems. Image of brain by IsaacMao per Flickr.com via creativecommons.org/licenses/by/2.0. See educational videos and updates at www.inflammationmastery.com/pain

Central sensitization (enhanced and autonomous pain hypersensitivity) seen in FM can be caused by bacterial LPS: Somewhat independent from the immune/inflammation-mediated hyperalgesia induced by LPS is the hyperalgesia mediated by central nervous system responses. The central sensitization seen with fibromyalgia[378] might be explained as being caused by intestinally-derived bacterial toxins. **Bacterial LPS/endotoxin promotes central sensitization via (microglia-driven astrocyte-induced glutamate-mediated) activation of NMDA receptors and by inducing hyperalgesia (elevated pain perception) and anti-analgesia (reduced response to pain inhibition).**[379] Accumulated evidence suggests that fibromyalgia may be a disorder of somatic hypersensitivity induced by bacterial toxins derived from quantitative excess or qualitative abnormalities in gut bacteria.[380]

> **Exposure to the bacterial endotoxin lipopolysaccharide (LPS) causes increased sensitivity to painful stimuli (hyperalgesia) and a reduction in opioid analgesia (anti-analgesia)**
>
> "Intraperitoneal injection of toxins, such as the bacterial endotoxin lipopolysaccharide (LPS), is associated with a well-characterized increase in sensitivity to painful stimuli (hyperalgesia) and a longer-lasting reduction in opioid analgesia (anti-analgesia) when pain sensitivity returns to basal levels."
>
> Johnston et al. Inhibition of morphine analgesia by LPS. *Behav Brain Res* 2005 Jan

Gut dysbiosis—generally speaking—can promote syndromes of pain, fatigue and depression via systemic inflammation (triggered directly by microbial molecules), immune activation via increased intestinal permeability, and increased glutaminergic neurotransmission via glial activation—microglial activation followed by astrocyte activation. As the name implies, central sensitization describes a condition of heightened perception of / responsiveness to sensory stimuli, particularly pain. In an authoritative review by Woolf[381] in 2011, the following introduction provides both definition and description, "Nociceptor inputs can trigger a prolonged but reversible increase in the excitability and synaptic efficacy of neurons in central nociceptive pathways, the phenomenon of central sensitization. Central sensitization manifests as pain hypersensitivity, particularly dynamic tactile allodynia, secondary punctate or pressure hyperalgesia, aftersensations, and enhanced temporal summation." Woolf goes on to note that central sensitization (CS) can be elicited in humans by diverse experimental noxious stimuli to skin, muscles or viscera, and that CS "results in secondary changes in brain activity that can be detected by electrophysiological or imaging techniques." Clinical conditions in which CS plays a role include fibromyalgia, osteoarthritis, headache, temporomandibular joint disorders, chronic musculoskeletal pain, dental pain, neuropathic pain, visceral pain hypersensitivity disorders (such as irritable bowel syndrome, IBS), and post-surgical/traumatic pain. In essence, anything that causes pain (e.g., injury) or contributes to increased pain perception (e.g., sleep deprivation, magnesium deficiency, vitamin D deficiency) can contribute to central sensitization simply by virtue of neuronal plasticity, which makes repeatedly activated pathways—in this case the perception of and response to pain—become more permanent via synaptogenesis, e.g., pain chronification via activity/repetition-dependent synaptic plasticity. In experimental models, bacterial LPS has been shown to promote central sensitization via activation of microglial release of extracellular ATP which in turn triggers astrocytes to

---

[376] Cohen G. The brain on fire? *Ann Neurol.* 1994 Sep;36(3):333-4
[377] Cordero et al. Mitochondrial dysfunction and mitophagy activation in blood mononuclear cells of fibromyalgia patients: implications in the pathogenesis of the disease. *Arthritis Res Ther.* 2010;12(1):R17. Olsen NJ, Park JH. Skeletal muscle abnormalities in patients with fibromyalgia. *Am J Med Sci.* 1998 Jun;315(6):351-8
[378] Meeus et al. Central sensitization: biopsychosocial explanation for chronic widespread pain in patients with fibromyalgia and chronic fatigue syndrome. *Clin Rheumatol.* 2007 Apr;26(4):465-73
[379] Johnston IN, Westbrook RF. Inhibition of morphine analgesia by LPS: role of opioid and NMDA receptors and spinal glia. *Behav Brain Res.* 2005;156(1):75-83
[380] Othman M, Agüero R, Lin HC. Alterations in intestinal microbial flora and human disease. *Curr Opin Gastroenterol.* 2008 Jan;24(1):11-6
[381] Woolf CJ. Central sensitization: implications for the diagnosis and treatment of pain. *Pain.* 2011 Mar;152(3 Suppl):S2-15

promote glutaminergic neurotransmission, thereby increasing neurocortical hyperexcitability and promoting pain, depression, central fatigue, and migraine/seizure. As paraphrased here and noted in the diagram from Béchade et al[382], LPS stimulates microglia to externalize the DAMP extracellular ATP, recruiting astrocytes to release glutamate leading to increased excitatory transmission via a glutamate, the major excitatory neurotransmitter. Clinicians should appreciate that LPS is a prototype agonist but certainly not the only "environmental factor" capable of triggering hyperglutaminergic neurotransmission (excess glutamate activating the glutamate-sensitive NMDA receptors) or what might be called glutamate-mediated hypertransmission (normal levels of glutamate triggering a hypersensitive receptor, or a hypersensitive or "loaded" or "primed" intracellular cascade). The clinical correlates of this model are relevant for chronic pain, depression, central fatigue, neurodegeneration, post-traumatic pain, post-traumatic stress disorder (PTSD), epilepsy and migraine. LPS is one of many "noxious stimuli" that can trigger microglia activation for eventual hyperglutaminergic neurocortical hyperexcitability and excitotoxicity; other factors include trauma and inflammatory cytokines and prostaglandins. LPS, other microbial immunogens, and other inflammatory triggers need not originate in the gut, as many long-term/chronic "dysbiotic-like" infections are localized within the brain and central nervous system, such as HSV[383], *Chlamydia pneumoniae*[384], and "algal" chlorovirus ATCV-1[385]; neuropathologic infections, antigens and neurotoxins such as the aluminum used in vaccines/immunizations as an adjuvant can be trafficked from the periphery into the brain by the immune system.[386] This author's perspective is that gut-derived microbial signals do indeed evoke central sensitization, most likely via direct CNS entry (via a "leaky" blood-brain barrier; perhaps trafficked by immunocytes), but at the very least indirectly by peripheral immune activation and the systemic proinflammatory state characterized by increased production of cytokines and prostaglandins which readily cross the blood-brain barrier to produce microglial activation and the subsequent hyperexcitation, excitotoxicity, and sickness behavior—including fatigue, depression, sensitivity. Microglial activation promotes neuronal hyperexcitation/excitotoxicity, and the reverse is also true: neuronal (hyper)activity promotes microglial activation; thereby forming a vicious cycle. Additionally, factors such as heightened glutamate/NMDA receptor sensitivity, depleted antioxidant defenses and specific nutrient deficiencies (e.g., pyridoxine, magnesium, zinc, vitamin D, n3 fatty acids), mitochondrial impairment, and an pro-inflammatory microenvironment combine additively/synergistically to promote the progressively accelerated pace of reciprocal microglial activation and neuronal hyperexcitation/excitotoxicity noted in several neurodegenerative states. This could be complicated by gut *Clostridia* production of 3-3-hydroxyphenyl-3-hydroxypropionic acid (HPHPA) and p-cresol causing neurotransmitter imbalance—elevated dopamine and reduced norepinephrine.

Defining and describing what is meant by "brain inflammation" relative to central sensitization in pain syndromes: Encephalitis (*enceph*—brain, *itis*—inflammation) classically refers to acute brain inflammation resultant from brain infection, such as a viral infection, e.g., herpes encephalitis. However, we now appreciate more subtle forms of chronic/persistent brain inflammation—metabolic-immunologic inflammation—in neurologic and psychiatric conditions such as chronic pain, autism, Parkinson's and Alzheimer's DZs, as well as chronic depression and schizophrenia. As such, we can reasonably expand the use of the term *encephalitis* beyond the acute and infectious to include the chronic and metabolic, ie

> **Glial activation = brain inflammation = promotion of central sensitization = pain hypersensitivity and other manifestations of sickness behavior**
>
> Typical manifestations of brain inflammation and sickness behavior include:
> - Reduced physical activity, inertia
> - Reduced food intake
> - Reduced social interaction
> - Reduced sexual behavior
> - Reduced mood, depression;
> - Impaired memory
> - Heightened sensitivity to pain
>
> Maier SF, Watkins LR. Consequences of the Inflamed Brain. *Report on Progress 2012*, University of Colorado at Boulder. Dana Alliance 2012 Aug

---

[382] Béchade C, Cantaut-Belarif Y, Bessis A. Microglial control of neuronal activity. *Front Cell Neurosci.* 2013 Mar 28;7:32
[383] Ball et al. Intracerebral propagation of Alzheimer's disease: strengthening evidence of a herpes simplex virus etiology. *Alzheimers Dement.* 2013 Mar;9(2):169-75
[384] Hammond et al. Immunohistological detection of Chlamydia pneumoniae in the Alzheimer's disease brain. *BMC Neurosci.* 2010 Sep 23;11:121
[385] Yolken et al. Chlorovirus ATCV-1 is part human oropharyngeal virome associated changes in cognitive functions in humans and mice. *Proc Natl Acad Sci* 2014 Nov:16106-11
[386] "We previously showed that poorly biodegradable aluminum-coated particles injected into muscle are promptly phagocytosed in muscle and the draining lymph nodes, and can disseminate within phagocytic cells throughout the body and slowly accumulate in brain." Gherardi et al. Biopersistence and brain translocation of aluminum adjuvants of vaccines. *Front Neurol* 2015 Feb;6:4. See also: "Detection of Al(III) in tissues indicated presence of aluminum in the nervous tissue of experimental animals." Luján et al. Autoimmune/autoinflammatory syndrome induced by adjuvants (ASIA syndrome) in commercial sheep. *Immunol Res* 2013 Jul:317-24

chronic/persistent metabolic-immunologic brain inflammation. Not all brain inflammation leads to central sensitization, but all central sensitization has a (neuro)inflammatory component. If we say "central sensitization", then we are a bit lost, because we are describing too many things at once (e.g., pain, emotions, MRI changes, early experiences, neurotransmitters...); as such, overuse of the term central sensitization can lead us astray or leave us unclear and therefore easily manipulated. If we say "brain inflammation" then we can reasonably appreciate a process that is: 1) causal—only so many things cause brain inflammation, so we can organize a plan of discovery and treatment, and 2) limited—we appreciate that most inflammatory disorders and responses should resolve. Any inflammatory trigger will be expected to trigger brain inflammation; while insults to the brain are obvious, peripheral cytokines and immunocytes can cross the blood brain barrier (BBB) to trigger inflammation in the brain, even when the problem started in the periphery. If you know that "the brain has an immune system", and that when the brain's immune system triggers inflammation—whether via central/brain insult or peripheral/body inflammation—that the result is neuroinflammation, pain, depression, and expedited neurodegeneration, then you can appreciate the intermixing of systemic inflammation with brain inflammation, with microglial activation and the resultant changes in neurotransmission that lead to pain sensitivity and changes in mood/affect called "sickness behavior." In this work, I show that fibromyalgia is a unique combination of dysbiosis-induced glial activation and mitochondrial impairment; both of these components need to be treated effectively and at the same time in order to alleviate the brain inflammation—unique—in fibromyalgia. "Glial activation with mitochondrial impairment" is different from and worse than glial activation or mitochondrial impairment by itself; the combination of the two together exacerbates both while also causing a vicious cycle that promotes ongoing pain, fatigue, and emotional/psychiatric changes.

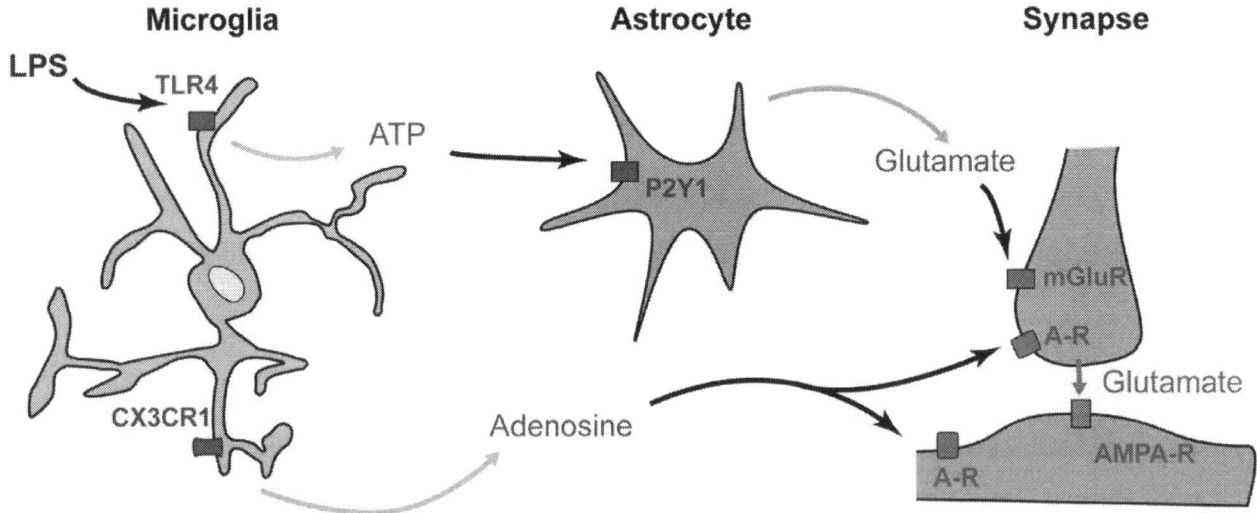

Microglial LPS reception leads to (extracellular) ATP elaboration, which stimulates astrocytes to produce glutamate, which increases glutaminergic neurotransmission: "Upon LPS stimulation, microglia rapidly produce ATP, which recruits astrocytes. Astrocytes subsequently release glutamate, and this leads to increased excitatory transmission via a metabotropic glutamate receptor-dependent mechanism." Illustration and quote from: Béchade C, Cantaut-Belarif Y, Bessis A. Microglial control of neuronal activity. *Front Cell Neurosci*. 2013 Mar 28;7:32[387]

Verified in animal models and likely contributory to clinical pain syndromes such as fibromyalgia is the observation that bacterial endotoxin/LPS can also contribute to central sensitization. This concept is introduced here and will be further substantiated in a following section on clinical pain syndromes and more so in separate writings, whether specific to fibromyalgia[388] or general to rheumatology.[389] The basic pathophysiology is quite simple and linear, as depicted below and itemized with citations thereafter.

Bacterial LPS → microglial activation → astrocyte hyperglutaminogenesis → neurocortical hyperexcitation → pain

---

[387] Béchade C, Cantaut-Belarif Y, Bessis A. Microglial control of neuronal activity. *Front Cell Neurosci*. 2013 Mar 28;7:32 doi: 10.3389/fncel.2013.00032 journal.frontiersin.org/article/10.3389/fncel.2013.00032/abstract. Copyright © 2013 Béchade, Cantaut-Belarif and Bessis. *Front Cell Neurosci* is an open-access journal distributed under terms of Creative Commons Attribution License, which permits use/distribution/reproduction in other forums, provided original authors/source are credited.
[388] Vasquez A. *Fibromyalgia in a Nutshell: A Safe and Effective Functional Medicine Strategy*. 2012, and later versions.
[389] Vasquez A. *Functional Medicine Rheumatology v3.5*. 2014, and later versions.

However, the correlation of serum LPS (or its indirect marker, anti-LPS antibodies) with increased intestinal permeability provides another interpretation, wherein serum LPS activity is simply a correlate with increased IP which promotes pain via nonspecific and non-LPS mechanisms, as shown in the following sequence:

SIBO/LPS-induced mucosal damage → absorption of gut-derived immunogens and toxins → microglial activation → astrocyte-mediated hyperglutaminogenesis → neurocortical hyperexcitation → pain/depression/fatigue and associated low-grade immune activation

- Endotoxemia induces visceral hypersensitivity and altered pain evaluation in healthy humans (*Pain*. 2012 Apr[390]): "…transient systemic immune activation results in decreased visceral sensory and pain thresholds and altered subjective pain ratings."
- LPS-induced hyperalgesia in healthy humans (*Brain Behav Immun* 2014 Oct[391]): "Our results revealed widespread increases in musculoskeletal pain sensitivity in response to a moderate dose of LPS (0.8 ng/kg), which correlate both with changes in IL-6 and negative mood."
- Widespread hyperalgesia in adolescents with symptoms of irritable bowel syndrome (*J Pain*. 2014 Sep[392]): "We examined pain sensitivity in 961 adolescents from the general population (mean age 16.1 years), including pain threshold and tolerance measurements of heat (forearm) and pressure pain (fingernail and shoulder) and cold pressor tolerance (hand). … Our results indicate that adolescents in the general population with IBS symptoms, like adults, have widespread hyperalgesia. … Our results suggest that central pain sensitization mechanisms in IBS may contribute to triggering and maintaining chronic pain symptoms."

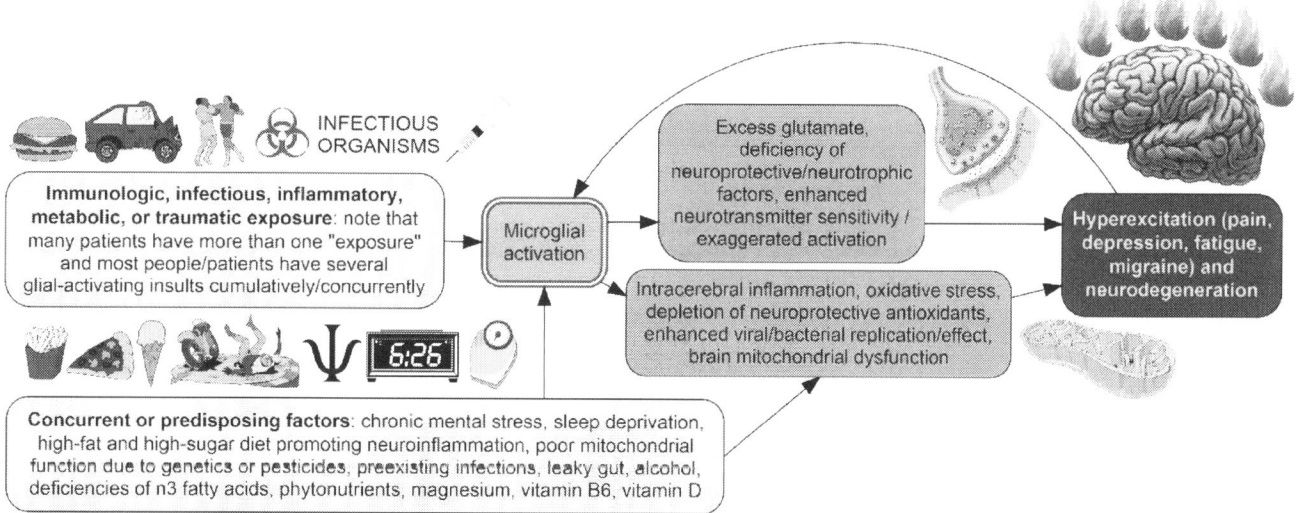

Numerous insults—immunologic, inflammatory, infectious, metabolic, traumatic, toxins—can "team up" or act individually to incite glial activation: Very importantly, glial activation can remain in an activated state promoting neurotransmitter imbalance, hyperexcitation, and neurodegeneration for several years after a single biochemical, infectious, or physical assault. High-fat foods, physical trauma from sports and accidents, psychological and mental stress, sleep deprivation, and the metabolic stress of hyperglycemia, insulin resistance, and obesity all contribute to brain inflammation—also called "the brain on fire" per Cohen, *Annals of Neurology* 1994—to promote pain, depression, migraine/seizure, and neurodegeneration. Images and descriptions above © 2015-2015 by Dr Vasquez except for image of brain by IsaacMao per Flickr.com via creativecommons.org/licenses/by/2.0. See educational videos and updates at www.inflammationmastery.com/pain

Proof of accelerated neurodegeneration in fibromyalgia: The model I have presented—dysbiosis-induced glial activation and mitochondrial dysfunction leading to both pain sensitivity and accelerated neurodegeneration—is supported by research showing accelerated brain aging and loss of gray matter (neuron cell bodies) in patients with fibromyalgia. As such, I consider my model of fibromyalgia to be very complete and consistent.

---

[390] Benson et al. Acute experimental endotoxemia induces visceral hypersensitivity and altered pain evaluation in healthy humans. *Pain*. 2012 Apr;153(4):794-9
[391] Wegner et al. Inflammation-induced hyperalgesia: timing, dosage, and negative affect on somatic pain in human endotoxemia. *Brain Behav Immun*. 2014 Oct;41:46-54
[392] Stabell et al. Widespread hyperalgesia in adolescents with symptoms of irritable bowel syndrome: results from a large population-based study. *J Pain*. 2014 Sep;15(9):898-906

- Accelerated brain gray matter loss and premature aging of the brain in fibromyalgia patients (*J Neurosci* 2007 Apr[393]): "In this study, we investigate anatomical changes in the brain associated with fibromyalgia. Using voxel-based morphometric analysis of magnetic resonance brain images, we examined the brains of 10 female fibromyalgia patients and 10 healthy controls. We found that fibromyalgia patients had significantly less total gray matter volume and showed a 3.3 times greater age-associated decrease in gray matter than healthy controls. The longer the individuals had had fibromyalgia, the greater the gray matter loss, with each year of fibromyalgia being equivalent to 9.5 times the loss in normal aging. In addition, fibromyalgia patients demonstrated significantly less gray matter density than healthy controls in several brain regions, including the cingulate, insular and medial frontal cortices, and parahippocampal gyri. The neuroanatomical changes that we see in fibromyalgia patients contribute additional evidence of CNS involvement in fibromyalgia. In particular, fibromyalgia appears to be associated with an acceleration of age-related changes in the very substance of the brain."

Additional comments and perspectives on central sensitization—integrating the model of SIBO-induced central sensitization via glial activation and other mechanisms

We appreciate that FM contains the aspect of central sensitization and the associated pain amplification; our intellectual task is to understand the most likely origin or—alternatively—to ascribe the central sensitization

> **Follow the pathophysiologic trail from bacterial exposures to IL-8 to brain inflammation and central sensitization**
> Various microbial exposures are known to cause elevations in IL-8, elevated CSF levels of which are characteristic of and perhaps unique to fibromyalgia, a condition causatively associated with intestinal bacterial overgrowth. The best summation of data at this point reads that the brain inflammation of fibromyalgia pain is triggered by microbial exposure. The alternate hypothesis that all of these pathophysiologic characteristics are unrelated is illogical and subintellectually immeritous.

to a magical/idiopathic origination. My contention is that microbial exposure leads to pain amplification via central sensitization in fibromyalgia; the means by which this can occur are both direct/primary causality and indirect/secondary, etc. Most of these were summarized in the above image and caption; in the sections that follow I will review some additional and also new information and show how it further supports this model.

- Systemic inflammation causes suppression of brain noradrenergic signaling (experimental study; *Science* 1983 Aug[394]) and is likely to contribute to fatigue, depression, pain sensitivity, and a heightened stress response: This experimental study shows that systemic inflammation leads to reduction in norepinephrine signaling in the (rat) brain. This would be expected to lead to fatigue, depression, and pain sensitivity. Secondarily, lack of adrenergic stimulation of the central alpha-2 receptor would be expected to stimulate the sympathetic nervous system; we note that bacterial overgrowth of the intestines (SIBO) causes enhanced sympathetic activity that is noted in patients with IBS which is closely related to fibromyalgia.[395] Low-grade systemic inflammation is known to be a component of FM, and generally speaking the molecules in or near the human body that are most likely to trigger inflammation are of microbial origin. Patients with IBS and FM are both known to have small intestine bacterial overgrowth. **In sum, we can reasonably opine that SIBO in FM and IBS leads to low-grade inflammation which leads to suppression of brain noradrenergic signaling, leading directly to fatigue and depression while likely also contributing to a heightened stress/sympathetic response; all of these components are noted in patients with IBS/FM.**
- FM patients have elevated CSF levels of IL-8 (*J Neuroimmunol* 2012 Jan[396] and *J Neuroimmunol* 2015 Mar[397]): Kadetoff et al did good work in 2012 when they were the first to find that patients with FM showed elevated levels of the inflammatory cytokine IL-8 in the cerebrospinal fluid (CSF) of patients with FM; research suggests that IL-8 promotes central sensitization via brain inflammation, and again this is consistent with FM. Although these authors note that "The release of pro-inflammatory cytokines/chemokines (including IL-8), by glia cells

---

[393] Kuchinad et a. Accelerated brain gray matter loss in fibromyalgia patients: premature aging of the brain? *J Neurosci*. (2007 Apr 11;27(15):4004-7
[394] "We report here that the immune response elicits a decrease in NA synthesis in the hypothalamus and that soluble products of activated immunological cells induce a decrease in NA content in the hypothalamus." Besedovsky et al. The immune response evokes changes in brain noradrenergic neurons. *Science*. 1983 Aug 5;221(4610):564-6
[395] Lin HC. Small intestinal bacterial overgrowth: a framework for understanding irritable bowel syndrome. *JAMA*. 2004 Aug 18;292(7):852-8
[396] "To our knowledge, this is the first study assessing intrathecal concentrations of pro-inflammatory substances in fibromyalgia. We report elevated cerebrospinal fluid and serum concentrations of interleukin-8, but not interleukin-1beta, in FM patients." Kadetoff D, Lampa J, Westman M, Andersson M, Kosek E. Evidence of central inflammation in fibromyalgia-increased cerebrospinal fluid interleukin-8 levels. *J Neuroimmunol*. 2012 Jan;242:33-8
[397] Kosek E, Altawil R, Kadetoff D, et al. Evidence of different mediators of central inflammation in dysfunctional and inflammatory pain–interleukin-8 in fibromyalgia and interleukin-1 β in rheumatoid arthritis. *J Neuroimmunol*. 2015 Mar 15;280:49-55

can be triggered by stress, immune activation and afferent nociceptive input [pain]", they ultimately pin the blame on pain by concluding in their abstract that their findings are "in accordance with FM symptoms being mediated by sympathetic activity...and supports the hypothesis of glia cell activation in response to pain mechanisms." The error these authors make is when they prematurely and without citation ascribe the elevation of IL-8 to pain and sympathetic activation in FM. The authors are thereby stating in essence that pain causes stress which causes more pain; they completely failed to both 1) substantiate this claim, and/or to 2) consider alternate hypotheses. Their attribution of the sympathetic activation to pain appears premature at best; they provide no substantiation for this claim, and it appears to have been promulgated and accepted simply because it is convenient and in accord with the prevailing drug-funded model of fibromyalgia which states that, in fibromyalgia, pain/stress causes pain/stress via a vicious cycle of glial activation (central sensitization, brain inflammation). In 2015, this same team of researchers replicated their 2012 finding of elevated IL-8 in FM; however they exacerbated their error of attribution by stating that their findings are—citing the work of Clauw—"in line with the proposal to regard FM as a sympathetically mediated pain syndrome." I believe that these authors are inappropriately connecting their finding of elevated IL-8 in FM with elevated sympathetic activity and pain in FM, and then concluding conveniently and hastily that these are the only variables worth considering and that therefore these must cause each other; they give zero consideration to the fact that these FM patients have SIBO[398] and that microbial molecules and intestinal inflammation can cause elevations in IL-8. So while we can reasonably accept that IL-8 is elevated in FM and may contribute to pain sensitization via brain inflammation and central sensitization, ascribing the elevation of IL-8 to the same pain that it promotes is circular thinking which in this case is also faulty thinking without consideration and exclusion of the obvious microbial overload noted in these patients with FM. Their citation to Clauw is particularly questionable since Clauw has been heavily funded by drug companies and consistently cheerleads the idea that FM is a primary disorder of the brain that needs—in accord with his sources of funding—perpetual drug treatment.[399] While stress and pain may lead to elevations in IL-8, an inflammatory cytokine that appears to promote central pain sensitization, stress and pain alone are less likely to be the primary contributors to the central sensitization in FM than are microbial and mitochondrial contributors, especially in combination. If stress and pain were sufficient to cause central sensitization, then virtually all medical students/residents (and working single mothers, persons living in war zones, persons living in prison and in conditions of gross social inequality, etc) and all competitive athletes would have central sensitization and FM; thus the general lack of population-wide correlation argues against the stress/pain model of *IL-8 induced central sensitization* in FM. Furthermore and even more clearly, the stress/pain model of pain sensitization in FM absolutely fails to account for the mitochondrial and other molecular abnormalities. In contrast, SIBO and microbial exposure is the perfect explanation of/for central sensitization in FM.

- Elevations in IL-8 can be triggered by various bacterial exposures: In an experimental cell culture model, exposure to the Gram-negative bacterium *Pseudomonas aeruginosa* was shown to increase production of IL-8.[400] Similarly, exposure to protein from the Gram-negative bacterium *Streptococcus pyogenes* also triggered expression of IL-8.[401]
- Elevations in IL-8 correlate with gastrointestinal infection with *Blastocystis* in patients with IBS, known to be correlated with FM (*PLoS One* 2015 Sep[402]): "Among 109 (IBS n = 35 and non-IBS n = 74) adults, direct stool

---

[398] Pimentel et al. A link between irritable bowel syndrome and fibromyalgia may be related to findings on lactulose breath testing. *Ann Rheum Dis*. 2004 Apr;63(4):450-2

[399] American Pain Society. Fibromyalgia Has Central Nervous System Origins. May 16, 2015. americanpainsociety.org/about-us/press-room/fibromyalgia-clauw and "Dr Clauw has received grants/research support from Pfizer and Forest Laboratories. He is a consultant and a member of the advisory boards for Pfizer, Eli Lilly and Company, Forest Laboratories, Cypress Biosciences, Pierre Fabre Pharmaceuticals, UCB, and AstraZeneca. Dr Arnold has received grants/research support from Eli Lilly and Company, Pfizer, Cypress Biosciences, Wyeth Pharmaceuticals, Boehringer Ingelheim, Allergan, and Forest Laboratories. She is a consultant for Eli Lilly and Company, Pfizer, Cypress Biosciences, Wyeth Pharmaceuticals, Boehringer Ingelheim, Forest Laboratories, Allergan, Takeda, UCB, Theravance, AstraZeneca, and sanofi-aventis. Dr McCarberg has received honoraria from Cephalon, Eli Lilly and Company, Endo Pharmaceuticals, Forest Laboratories, Merck & Co, Pfizer, and Purdue Pharma. The FibroCollaborative group was sponsored by Pfizer." "Editorial support was provided by Gayle Scott, PharmD, of UBC Scientific Solutions and funded by Pfizer." Clauw DJ, Arnold LM, McCarberg BH; FibroCollaborative. The science of fibromyalgia. *Mayo Clin Proc*. 2011 Sep;86(9):907-11

[400] "Thus, IL-8 mRNA expression was prolonged after P. aeruginosa stimulation in CF epithelial cells, and this sustained IL-8 expression may contribute to the excessive inflammatory response in CF." Joseph T, Look D, Ferkol T. NF-kappaB activation and sustained IL-8 gene expression in primary cultures of cystic fibrosis airway epithelial cells stimulated with Pseudomonas aeruginosa. *Am J Physiol Lung Cell Mol Physiol*. 2005 Mar;288(3):L471-9

[401] "Our results showed that following exposure to SPE B or G308S, the levels of IL-8 protein and mRNA were increased and the increase was inhibited by the addition of anti-Fas antibody, suggesting that the increased production of IL-8 by SPE B is mediated through Fas receptor." Chang CW, Wu SY, Chuang WJ, Lin YS, Wu JJ, Liu CC, Tsai PJ, Lin MT. The IL-8 production by Streptococcal pyrogenic exotoxin B. *Exp Biol Med* (Maywood). 2009 Nov;234(11):1316-26

[402] Ragavan et al. Blastocystis sp. in Irritable Bowel Syndrome (IBS) - Detection in Stool Aspirates during Colonoscopy. *PLoS One*. 2015 Sep 16;10(9):e0121173

examination and culture of colonic aspirates were initially negative for *Blastocystis*. However, PCR analysis detected *Blastocystis* in 6 (17%) IBS and 4 (5.5%) non-IBS patients. In the six positive IBS patients by PCR method, subtype 3 was shown to be the most predominant (3/6: 50%) followed by subtype 4 (2/6; 33.3%) and subtype 5 (1/6; 16.6%). IL-8 levels were significantly elevated in the IBS Blasto group and IBS group (p<0.05) compared to non-IBS and non-IBS Blasto group. ... Meanwhile, the IL-5 levels were significantly higher in IBS Blasto group (p<0.05) compared to non-IBS and non-IBS Blasto group. This study implicates that detecting Blastocystis by PCR method using colonic aspirate samples during colonoscopy, suggests that this may be a better method for sample collection due to the parasite's irregular shedding in Blastocystis-infected stools. Patients with IBS infected with parasite showed an increase in the interleukin levels demonstrate that *Blastocystis* does have an effect in the immune system." Additionally, *Blastocystis* is known to cause intestinal damage and to thereby increase intestinal permeability[403], and fibromyalgia patients are also noted to have increased intestinal permeability[404]; the possibility exists that *Blastocystis* induces increased IL-8 via intestinal mucosal damage which leads to increased absorption of microbial debris/inflammogens rather than directly, although an additive and synergistic mechanism/effect is more likely.

- Elevations in IL-8 are seen in patients with IBS, known to be triggered by SIBO and causatively dose-dependently correlated with FM (*Am J Gastroenterol. 2010 Oct*[405]): If my model is correct that the elevated IL-8 in FM is caused by the SIBO and not by the FM itself, then we should see elevated in IL-8 in patients with IBS, which is caused by SIBO. In a study with more than 120 human subjects, results showed that patients with IBS/SIBO have increased plasma levels of IL-6 and IL-8. Note that this study used blood/plasma/serum testing rather than cerebrospinal fluid (CSF) as in the previously reviewed studies; IL-8 levels in blood and CSF tend to correlate, as demonstrated in FM patients who typically have elevations in both locations (ie, peripherally and centrally).

- Bacterial overgrowth in ulcerative colitis patients correlates with increased inflammatory mediator production, including IL-8 (*J Crohns Colitis. 2014 Aug*[406]): Among patients with ulcerative colitis, SIBO is common and correlates with reduced intestinal motility and increased production of inflammatory mediators: "observed that there was a significant correlation between SIBO with IL-6, IL-8, TNF-α, and IL-10, LPO and GSH."

- Small intestinal bacterial overgrowth association with TLR4 expression and IL-8 (*Dig Dis Sci. 2011 May*[407]): SIBO was more common in nonalcoholic steatohepatitis (NASH) patients than control subjects (77% v 31.25%) and only IL-8 levels were significantly higher in patients than control and correlated positively with TLR-4 expression. "NASH patients have a higher prevalence of small intestinal bacterial overgrowth which is associated with enhanced expression of TLR-4 and release of IL-8. SIBO may have an important role in NASH through interactions with TLR-4 and induction of the pro-inflammatory cytokine, IL-8."

Microbe-induced brain inflammation without brain infection and without high fever, vasculitis or autoimmunity

Viral infections
- Measles
- Mumps
- Rubella
- Influenza A and B
- Herpes simplex
- Epstein-Barr virus
- Varicella
- Vaccinia

Bacterial infections
- Mycoplasma pneumoniae
- Chlamydia
- Legionella
- Streptococcus
- Campylobacter
- Shigella

Immunizations*
- Rabies
- DPT—Diphtheria, pertussis tetanus
- Smallpox
- Measles
- Japanese B encephalitis
- Influenza

*Of important note, vaccinations commonly contain numerous adjuvants, toxic metals such as mercury and aluminum, highly allergenic antibiotics and cell culture components including egg, aborted human cells, and monkey kidney cells; as such, the inflammatory encephalitis that results cannot be ascribed solely to the microbial component of the vaccination.

cdc.gov/vaccines/pubs/pinkbook/downloads/appendices/B/excipient-table-2.pdf Accessed 2015 Nov

---

[403] "The IP was found to have increased in patients with protozoan infections compared with control patients (7.20+/-5.52 vs. 4.47+/-0.65%, P=0.0017). The IP values were 9.91+/-10.05% in Giardia intestinalis group, 6.81+/-2.25% in Blastocystis hominis group, 5.78+/-2.84% in Entamoeba coli group. In comparison with the control group, the IP was significantly higher in G. intestinalis and B. hominis patients (P=0.0025, P=0.00037, respectively), but not in E. coli patients. In conclusion, the IP increases in patients with G. intestinalis and B. hominis but not with E. coli infection. This finding supports the view that IP increases during the course of protozoan infections which cause damage to the intestinal wall while non-pathogenic protozoan infections have no effect on IP. The increase in IP in patients with B. hominis brings forth the idea that B. hominis can be a pathogenic protozoan." Dagci H, Ustun S, Taner MS, Ersoz G, Karacasu F, Budak S. Protozoon infections and intestinal permeability. *Acta Trop*. 2002 Jan;81(1):1-5
[404] Goebel et al. Altered intestinal permeability in patients with primary fibromyalgia and in patients with complex regional pain syndrome. *Rheumatology* 2008 Aug;47(8):1223-7
[405] Scully et al. Plasma cytokine profiles in females with irritable bowel syndrome and extra-intestinal co-morbidity. *Am J Gastroenterol*. 2010 Oct;105(10):2235-43
[406] Rana et al. Relationship of cytokines, oxidative stress and GI motility with bacterial overgrowth in ulcerative colitis patients. *J Crohns Colitis*. 2014 Aug;8(8):859-65
[407] Shanab et al. Small intestinal bacterial overgrowth in nonalcoholic steatohepatitis: toll-like receptor 4 and plasma levels of interleukin 8. *Dig Dis Sci*. 2011 May;1524-34

Stress/pain is an insufficient explanation for the origination of FM: While stress and pain may lead to elevations in IL-8, an inflammatory cytokine that appears to promote central pain sensitization, stress and pain alone are less likely to be the primary contributors to the central sensitization in FM than are microbial and mitochondrial contributors, especially in combination. If stress and pain were sufficient to cause central sensitization, then virtually all medical students/residents (and working single mothers, persons living in war zones, etc) and all competitive athletes would have central sensitization and FM; thus the general lack of population-wide correlation argues against the stress/pain model of IL8-induction of central sensitization in FM. Furthermore and even more clearly, the stress/pain model of pain sensitization in FM absolutely fails to account for the mitochondrial and other molecular abnormalities. In contrast, **SIBO and microbial exposure is the perfect explanation of/for FM.**

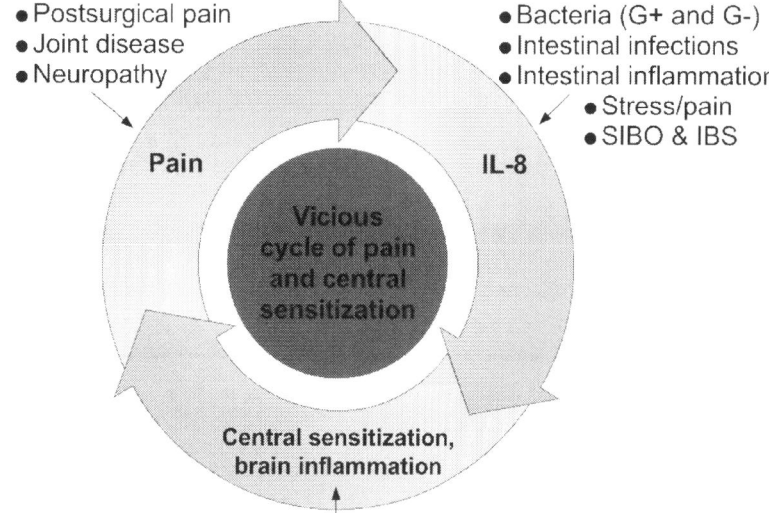

In sum, the data clearly show that microbial exposure in general and bacterial overgrowth of the small intestine in particular trigger increased IL-8 production. While elevated IL-8 is not specific to a particular disease per se, it is consistently elevated in patients with small intestine bacterial overgrowth, regardless of disease association. Many conditions causatively related to SIBO are noted to have elevated IL-8 levels, and we should reasonably attribute the elevated IL-8 levels in fibromyalgia to bacterial exposure, especially given the exceptionally high and dose-dependent correlation of SIBO with FM. IL-8 promotes pain sensitization, and the hyperalgesic effects of IL-8 are mediated in part by activation of beta-adrenergic receptors[408], chronic/sustained activation of which would be expected to promote depression, anxiety, and fatigue in addition to pain. IL-8 promotes activation of the sympathetic nervous system via beta-adrenergic receptors, and activation of the sympathetic nervous system via beta-adrenergic receptors feeds back to cause additional IL-8 production, thereby promoting a vicious cycle; my contention is that microbial debris is the initiating factor in this cycle, and that exposure to microbial inflammogens from SIBO is both more powerful (microbial debris is clearly pro-inflammatory, and LPS in particular is the most potent inflammogen known to science) and more durable (24/7 for months and years) than any "stress response." As such, the microbial inflammatory load is the more reasonable recipient of the burden of physiologic guilt than is the "stress level" of these patients. Stress is transient, and highly stressed people (e.g., medical students and residents, air traffic controllers, etc) do not show an undue burden of pain and mitochondrial impairment as do patients with an excessive total microbial load, specifically—as previously shown per literature review—SIBO. Thus having established that SIBO is sufficient to induce brain inflammation, I will now review more severe cases of what might be considered true gastrointestinal infection as a cause of severe brain inflammation; I am moving from the *mild and functional* to the truly *infectious and pathologic* to further exemplify this model.

- Gastrointestinal infection/dysbiosis as a cause of severe brain inflammation: criteria and review: We might reasonably expect that if *comparatively/relatively mild* cases of dysbiosis and SIBO can cause *mild* brain inflammation, then cases may have been reported wherein *more severe* gastrointestinal infections have led to *more severe* brain inflammation (ie, encephalitis and encephalomyelitis), and indeed such cases are impressively

---

[408] Kosek E, Altawil R, Kadetoff D, et al. Evidence of different mediators of central inflammation in dysfunctional and inflammatory pain--interleukin-8 in fibromyalgia and interleukin-1 β in rheumatoid arthritis. *J Neuroimmunol*. 2015 Mar 15;280:49-55. Kadetoff D, Lampa J, Westman M, Andersson M, Kosek E. Evidence of central inflammation in fibromyalgia-increased cerebrospinal fluid interleukin-8 levels. *J Neuroimmunol*. 2012 Jan;242:33-8

abundant. The most important distinction is that of what is being discussed here, specifically brain inflammation induced by gastrointestinal-specific microbes, their inflammatory debris, and the resulting inflammatory response within the brain, *versus* an infection that originated in the gut and then spread to the brain, or an infection that induced sepsis and systemic complications that resulted in altered cognition. Accordingly, viral infections which can easily penetrate the brain are selected against (e.g., essentially omitted) in this discussion; emphasis here is on *bacterial infections* with *no evidence of direct brain/meningeal infection* in *otherwise healthy people* resulting in severe brain inflammation via "indirect" inflammatory effects. Inflammatory bowel disease (IBD) is known to be causatively/contributively associated with gut dysbiosis; IBD patients can develop brain inflammation and brain/CNS antibody-mediated autoimmunity[409], apparently triggered by the gut dysbiosis, but such cases are excluded from the discussion below for the sake of clarity in establishing a more pure causative connection between gut dysbiosis and brain inflammation *mediated via cytokines* and *microglial activation* in *otherwise healthy people*. Likewise, cases of infection-induced encephalitis wherein high fever or vasculitis were present are likewise excluded. As I have discussed and presented recently in 2015[410], dysbiosis-induced disease is generally effected via multiple mechanisms, one of which is the inflammatory-cytokine response, which in this conversation would "spill from the periphery into the brain" to cause microglial activation and the resulting brain inflammation—encephalitis.

- Acute disseminated encephalomyelitis (ADEM) secondary to transient gastroenteritis in an otherwise healthy adult (*Case Rep Neurol Med.* 2015 Jun[411]): "A 62-year-old man presented with encephalopathy and rapid neurological decline following a gastrointestinal illness. A brain MRI revealed extensive supratentorial white matter hyperintensities consistent with ADEM and thus he was started on high dose intravenous methylprednisolone. He underwent a brain biopsy showing widespread white matter inflammation secondary to demyelination. At discharge, his neurological exam had significantly improved with continued steroid treatment and four months later, he was able to perform his ADLs."
- Acute disseminated encephalomyelitis with Campylobacter gastroenteritis (*J Neurol Neurosurg Psychiatry* 2004 May[412]): "We report a case of acute disseminated encephalomyelitis (ADEM) temporally associated with Campylobacter gastroenteritis in a previously fit man. ... Two days after admission (day 16 of illness), his family reported a change in his personality and he complained of slurring of speech, intermittent diplopia, and difficulty in walking. Examination revealed mild dysarthria, left sided facial weakness, mild left pyramidal limb weakness, and decreased sensation in the left leg. Tendon reflexes were brisk but plantar responses were flexor. His gait was ataxic. ... In the majority of cases, the condition develops after systemic viral infections most commonly measles, mumps, rubella, influenza A and B, herpes simplex, Epstein-Barr virus, varicella, and vaccinia. It has also been reported following bacterial infections with Mycoplasma pneumoniae, Chlamydia, Legionella, and Streptococcus, or following immunizations for rabies, diphtheria/tetanus/pertussis, smallpox, measles and Japanese B encephalitis."
- Bickerstaff's brainstem encephalitis related to Campylobacter jejuni gastroenteritis (*J Clin Pathol.* 2007 Oct[413]): "Here we report a case of BBE following a gastrointestinal infection with Campylobacter jejuni. The patient presented with acute onset of confusion and ophthalmoplegia. The cerebrospinal fluid (CSF) showed lymphocytic pleocytosis and raised protein. This acute presentation was preceded by an episode of Campylobacter-related diarrhea as confirmed by high titers of Campylobacter-specific IgM antibodies."
- Acute encephalopathy preceding Shigella infection (*Isr Med Assoc J.* 2001 May[414]): This is a case of a 3yo girl with encephalitis which preceded mild mucus diarrhea which lead to the discovery of *Shigella sonnei*. Thus, this case is doubly unique for 1) the fact that the encephalopathy *preceded* any gastrointestinal manifestations, and 2) the gastrointestinal manifestations were impressively *mild*.

---

[409] Yamamoto et al. Bickerstaff's brainstem encephalitis associated with ulcerative colitis. *BMJ Case Rep*. 2012 Sep 21;2012
[410] The presentations to which I make reference here are specifically video presentations 1-3 wherein I describe the molecular mechanisms of dysbiosis-induced disease in our CE/CME program hosted at http://www.nutritionandfunctionalmedicine.org/lms/ and derived from the printed monograph *Human Microbiome and Dysbiosis in Clinical Disease*. International College of Human Nutrition and Functional Medicine 2015
[411] Mahdi et al. A Case of Acute Disseminated Encephalomyelitis in a Middle-Aged Adult. *Case Rep Neurol Med*. 2015;2015:601706
[412] Orr et al. Acute disseminated encephalomyelitis temporally associated with Campylobacter gastroenteritis. *J Neurol Neurosurg Psychiatry*. 2004 May;75(5):792-3
[413] Hussain et al. Bickerstaff's brainstem encephalitis related to Campylobacter jejuni gastroenteritis. *J Clin Pathol*. 2007 Oct;60(10):1161-2
[414] Somech R, Leitner Y, Spirer Z. Acute encephalopathy preceding Shigella infection. *Isr Med Assoc J*. 2001 May;3(5):384-5

Chapter 5.1—Functional Inflammology Protocol for Metabolic Inflammation: Migraine and Fibromyalgia

Means by which inflammatory cytokines in the periphery (body) can result in inflammation in the brain
1. Cytokines enter the brain where the blood–brain barrier (BBB) is weak or non-existent (i.e. circumventricular organs).
2. Cytokines are transported into the brain by selective uptake systems (transporters), thus bypassing BBB.
3. Cytokines may act directly or indirectly on peripheral nerves that can send afferent signals to the brain.
4. Cytokines can act on peripheral tissues, inducing the secretion of molecules whose ability to penetrate the brain is not limited by the barrier. A major target appears to be endothelial cells, which bear receptors for IL-1 and endotoxin.
5. Cytokines can be synthesized by immune cells that infiltrate the brain.
6. Cytokines and the resultant increased ROS production can also impair mitochondrial function; impaired mitochondrial function in the periphery/body drains antioxidants (such as glutathione) and nutritive substances (such as CoQ10, tocopherols, and lipoic acid) that are important for the maintenance of cellular function and homeodynamics. Thereby, peripheral inflammation and mitochondrial impairment becomes a "sink" for draining nutrients and protectants while also becoming a "faucet" for inflammatory mediators, molecular debris and alarm signals. Increased production of glutamate, QUIN, and NO- fuel NMDAr activation and mitochondrial dysfunction. (DrV)
7. Cytokines may be transported in a retrograde manner from the periphery, via axonal mechanisms.
8. Cytokines may be transported in a retrograde manner from the periphery, via axonal mechanisms. (Zhang, An)
Contents 1-5 of this table are fully credited to Dunn AJ. Effects of cytokines and infections on brain neurochemistry. *Clin Neurosci Res*. 2006 Aug;6(1-2):52-68. Zhang JM, An J. Cytokines, inflammation, and pain. *Int Anesthesiol Clin*. 2007 Spring;45(2):27-37

Beyond glutamate in microbe-triggered brain inflammation: alterations in tryptophan, serotonin and additional triggering of NMDA receptors for more brain/pain activation and less inhibition

- Introduction to altered intracerebral tryptophan metabolism: Microbial, psychological, and inflammatory stressors cause elevations in IFN-g and TNF-a; "IFN-g induces the enzyme indoleamine 2,3-dioxygenase (IDO, found in immune cells such as macrophages and dendritic cells), which causes reduction in tryptophan availability, leading to a reduction in serotonin synthesis in the brain."[415] Paraphrasing the brilliantly excellent 2015 review by Jo, Zhang, Emrich, and Dietrich[416]: Following induction, IDO catabolizes/converts tryptophan into kynurenine (KYN), is converted to kynurenic acid (KA) or quinolinic acid (QUIN); "KA and QUIN have contrasting roles influencing the glutamatergic system, the first acting as antagonist and the latter as agonist of the glutamate N-methyl-D-aspartate receptor (NMDAr). Microglia are the main producers of QUIN in the brain, whereas astrocytes are the CNS-key cells involved in KA synthesis. This is explained by the fact that microglia express kynurenine 3-monooxygenase (KMO), the rate-limiting enzyme in the production of QUIN. Conversely, astrocytes exclusively express kynurenine aminotransferases, which are essential in the conversion of KYN to KA." Thus, in relevant summary, glial activation in the brain leads to cerebral tryptophan depletion, thereby depleting this important precursor of serotonin and melatonin while also leading to additional NMDAr agonism via QUIN. As such we are able to advance our understanding of this utmost-important pathophysiology that is occurring in the brain following/during inflammatory events/exposures, as illustrated and additionally detailed in the caption.

---

[415] Hurley LL, Tizabi Y. Neuroinflammation, neurodegeneration, and depression. *Neurotox Res*. 2013 Feb;23(2):131-44
[416] Jo WK, Zhang Y, Emrich HM, Dietrich DE. Glia in the cytokine-mediated onset of depression: fine tuning the immune response. *Front Cell Neurosci*. 2015 Jul 10;9:268

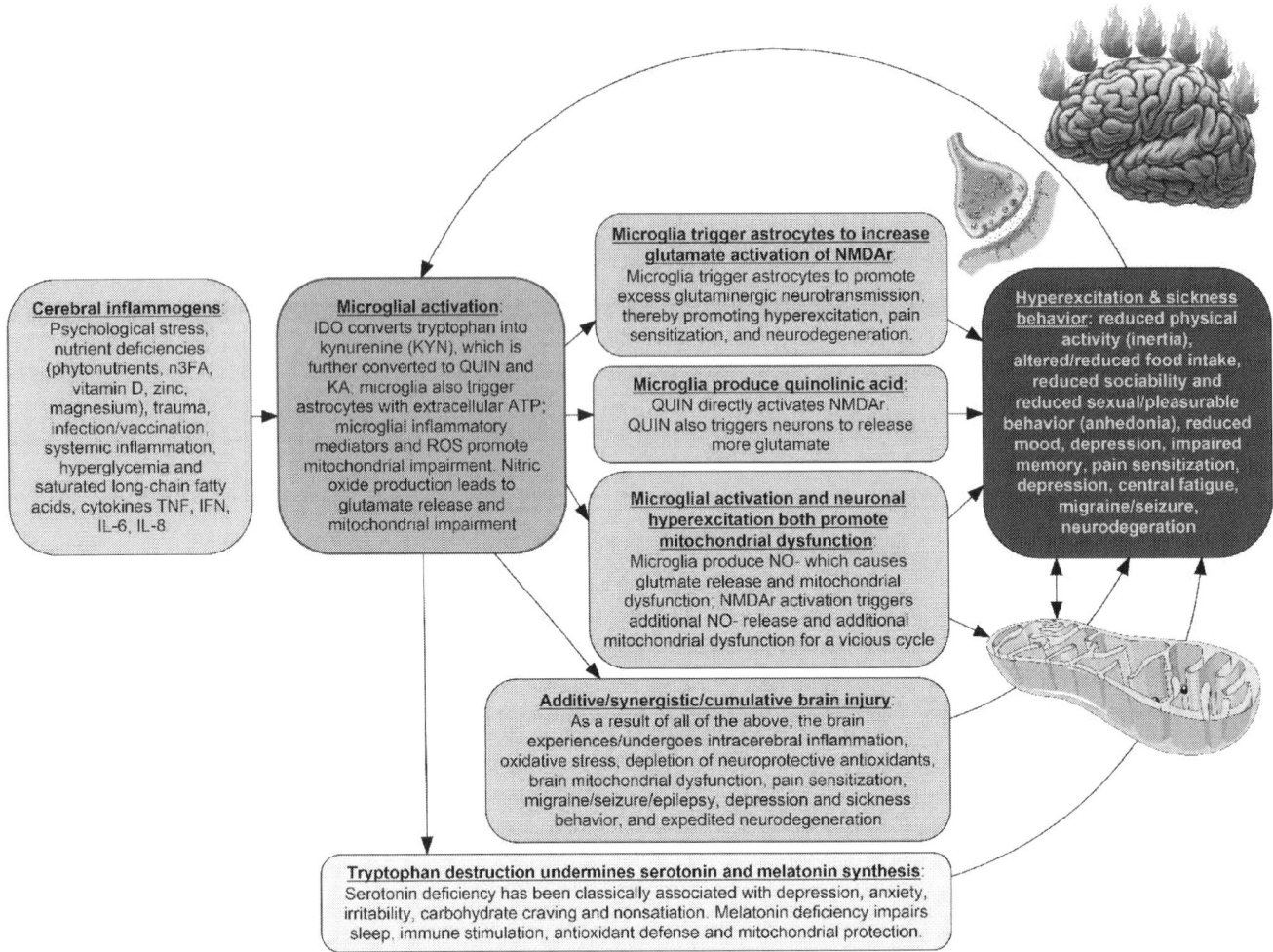

Brain inflammation leads directly to triple enhancement of NMDAr activation (via NO-, astrocytes, and QUIN) and triple impairment of mitochondrial function (via NO-, ROS, and inflammatory barrage): Details and citations provided in the following caption. Images above by DrV. Image of brain by IsaacMao per Flickr.com via creativecommons.org/licenses/by/2.0. See educational videos and updates at www.inflammationmastery.com/pain.

- **Cerebral inflammation can be triggered by any indirect or direct insult**, including psychological stress, nutrient deficiencies (phytonutrients, n3FA, vitamin D, zinc, magnesium), trauma, infection/vaccination, systemic inflammation, hyperglycemia and saturated long-chain fatty acids, cytokines TNF, IFN, IL-6, IL-8. We need to change the way that we appreciate inflammation: we have traditionally/historically/conveniently/simplistically thought of inflammation as needing a trigger and occurring in response to that trigger; while that remains largely but not absolutely true, we need to appreciate that inflammation—especially metabolic inflammation—occurs as a manifestation of cellular dysfunction in general and mitochondrial dysfunction in particular and as such can be initiated by nutritional deficiency, including phytonutrient deficiency. For example, the fact that nutritional supplementation in physiologic subpharmacologic doses provides an antiinflammatory benefit (e.g., multivitamin/mineral[417] and vitamin D3[418]) is proof that the preexisting nutritional deficiency—even though slight and subclinical—was itself the cause of the inflammatory response. We have crystalline clarity now that peripheral inflammation leads to central/brain inflammation, whether from trauma, infection, or vaccination[419,420,421] and as such we can reasonably state—and should accept as true—that any significant inflammatory/immunologic response in the periphery is going to cause/promote central/brain inflammation and the resulting hyperexcitation, pain sensitization, and sickness behavior. Important in this conversation is that while the model of "microglial activation = astrocyte activation = neuroinflammation / central sensitization / sickness behavior" is generally accurate, this same model is actually quite nuanced and contains different variations, leading to different clinical phenotypes. Data is giving shape to the idea that different cytokines have different effects on brain cellular components, leading to different phenotypic manifestations. Taking a single example, we note the recent 2015 report showing that beta amyloid induced IL-8 mediated microglial activation, and that the resulting astrocytic activation is mediated by a different (nonIL-8) pathway; also noted is the 2015 report showing

---

[417] Church TS, Earnest CP, Wood KA, Kampert JB. Reduction of C-reactive protein levels through use of a multivitamin. *Am J Med*. 2003 Dec 15;115(9):702-7
[418] Timms et al. Circulating MMP9, vitamin D and variation in the TIMP-1 response with VDR genotype. *QJM*. 2002 Dec;95(12):787-96
[419] Wright CE et al. Acute inflammation and negative mood: mediation by cytokine activation. *Brain Behav Immun*. 2005 Jul;19(4):345-50
[420] Harrison NA et al. Inflammation causes mood changes through alterations in subgenual cingulate activity and mesolimbic connectivity. *Biol Psychiatry*. 2009 Sep:407-14
[421] Wilson CJ, Finch CE, Cohen HJ. Cytokines and cognition—the case for a head-to-toe inflammatory paradigm. *J Am Geriatr Soc*. 2002 Dec;50(12):2041-56

that central/brain neuroinflammation in RA is mediated by IL-1 while that in FM is mediated by IL-8. As such, we should speak of neuroinflammations and glial activations in the plural rather than singular forms.

- **Microglial activation drains tryptophan** by triggering IDO conversion of tryptophan into kynurenine (KYN), which is further converted to QUIN and KA.[422] Tryptophan destruction undermines serotonin and melatonin synthesis: Serotonin deficiency has been classically associated with depression, anxiety, irritability, carbohydrate craving and nonsatiation. Melatonin deficiency impairs sleep, immune stimulation, antioxidant defense and mitochondrial protection.
- **Microglia trigger astrocytes to increase glutamate activation of NMDAr**: Microglia trigger astrocytes with extracellular ATP. Microglial activation triggers astrocytes to promote excess glutaminergic neurotransmission[423], thereby promoting hyperexcitation, pain sensitization, and neurodegeneration.
- **Microglial activation and increased nitric oxide (NO-) production leads to glutamate release and mitochondrial impairment**. Microglial inflammatory mediators and ROS promote mitochondrial impairment. Readers should appreciate the implications of this vicious cycle, where in "fire in the brain" can readily burn out of control; in context with other information in this section, this understanding helps explain, for example, how immunologic triggers in the periphery such as vaccination/immunization can cause devastating brain injury, basically leading to "metabolic-inflammatory meltdown of the brain" as seen in the most horrific of vaccination responses—vaccination encephalopathy/encephalitis.[424,425] Encephalitis can also result from gastrointestinal infection.[426]
- **Microglia produce quinolinic acid (QUIN) to activate NMDAr**: Microglia produce QUIN which activates NMDAr while QUIN also triggers neurons to release more glutamate, thereby leading to additional contribution to NMDAr activation.[427] Astrocytes can produce the NMDAr agonist kynurenic acid (KA); however, in settings of glial activation, NMDAr activation clearly predominates.
- **Microglial activation and neuronal hyperexcitation both promote mitochondrial dysfunction**.[428] Microglia produce (or induce astrocytic production of*) NO- which causes glutamate release and mitochondrial dysfunction; NMDAr activation triggers additional NO- release and additional mitochondrial dysfunction for a vicious cycle. Neuronal excitation and microglial activation both promote mitochondrial dysfunction; this is physiologically disastrous because neurons need optimized mitochondrial performance generally and especially when faced with increased/dysregulated demand. Very important here is the appreciation that mitochondrial dysfunction is itself pro-inflammatory (via ROS and DAMP/PAMP receptor activation by mitochondrial fragments), thereby adding to the vicious cycle. Further, mitochondrial dysfunction triggers excess release of inflammatory cytokines from a wide range of cell types; the range of cell types showing increased pro-inflammatory cytokine release following induction of mitochondrial impairment is so wide that possibly all cells (in this conversation, including microglia and neurons) are involved/affected. *Kim and Nagai[429] noted that human microglia do not produce NO- and that NO- production in the human brain glia originates chiefly from astrocytes; still, in the context of an intact brain, we might summarize that the microglia induce NO- production by triggering astrocytes, even if human microglia—distinguished from other animal models—cannot directly produce NO-.
- **Additive/synergistic/cumulative brain injury manifesting generally and chronically as increased brain fragility**: As a result of all of the above, the brain experiences/undergoes intracerebral inflammation, oxidative stress, depletion of neuroprotective antioxidants, enhanced viral/bacterial replication/effect, brain mitochondrial dysfunction, pain sensitization, migraine/seizure/epilepsy, depression and sickness behavior, and expedited neurodegeneration. Since many of these metabolic impairments are both common and silent, these lead to the promotion of neurodegeneration and "brain fragility" (or per Morley and Seneff: "diminished brain resilience syndrome"[430]) wherein the brain is supranormally vulnerable to other insults, such as trauma, dietary insult, glyphosate/pesticide exposure, infections and vaccinations.

As clinicians, pragmatists, and intellectuals, we are obligated to employ the available data: We already have sufficient evidence for this model, and clinicians should move forward and implement this model in clinical practice; use of this model is considerably more attractive than ascribing the associated conditions to spontaneous generation and condemning the associated patients to eternal medicalization. For ethical and logistical reasons, we will probably never have "perfect proof" of this model, because such would require induction of disease in healthy people, and this would be clearly risky and unethical and would therefore never be approved by any competent IRB (institutional review board, tasked with approving research investigations). What we might call *perfect proof* of this would have to be established by *prospective* studies showing that ❶ administration of a sufficient total microbial load (TML)—specifically SIBO in the case of FM—causes central sensitization as assessed by some reliable

---

[422] Jo WK, Zhang Y, Emrich HM, Dietrich DE. Glia in the cytokine-mediated onset of depression: fine tuning the immune response. *Front Cell Neurosci.* 2015 Jul 10;9:268
[423] Béchade C, Cantaut-Belarif Y, Bessis A. Microglial control of neuronal activity. *Front Cell Neurosci.* 2013 Mar 28;7:32
[424] Alicino et al. Acute disseminated encephalomyelitis with severe neurological outcomes following virosomal seasonal influenza vaccine. *Hum Vaccin Immunother.* 2014;10(7):1969-73
[425] Lee et al. An adverse event following 2009 H1N1 influenza vaccination: a case of acute disseminated encephalomyelitis. *Korean J Pediatr.* 2011 Oct;54(10):422-4
[426] Mahdi N, Abdelmalik PA, Curtis M, Bar B. A Case of Acute Disseminated Encephalomyelitis in a Middle-Aged Adult. *Case Rep Neurol Med.* 2015;2015:601706
[427] Jo WK, Zhang Y, Emrich HM, Dietrich DE. Glia in the cytokine-mediated onset of depression: fine tuning the immune response. *Front Cell Neurosci.* 2015 Jul 10;9:268
[428] Brown GC, Bal-Price A. Inflammatory neurodegeneration mediated by nitric oxide, glutamate, and mitochondria. *Mol Neurobiol.* 2003 Jun;27(3):325-55
[429] Kim SU Nagai A. Microglia as immune effectors of the central nervous system: Expression of cytokines and chemokines. *Clin Experiment Neuroimmunol.* 2010 May; 1: 61–69
[430] Morley WA, Seneff S. Diminished brain resilience syndrome: A modern day neurological pathology of increased susceptibility to mild brain trauma, concussion, and downstream neurodegeneration. *Surg Neurol Int.* 2014 Jun ; 5: 97

technology such as functional magnetic resonance imaging (fMRI) of the brain or evidence of inflammatory changes within the cerebrospinal fluid (CSF). A prospective trial of this nature would almost certainly be considered unethical and would be almost impossible to perform, either technically (how would the microbial load be delivered safely, effectively, and *temporarily* without risk of long-term harm?) and the related near-impossibility of being passed/approved by a responsible research IRB. **Natural life provides this trial via the observation that people who naturally develop small intestine bacterial overgrowth and other forms of gastrointestinal dysbiosis frequently develop syndromes of central sensitization such as irritable bowel syndrome, fibromyalgia, and complex regional pain syndrome[431]; prospective experimental studies have shown that animals exposed to microbial inflammogens indeed develop central sensitization.** Secondarily and of lesser importance, we would also want prospective evidence that ❷ the above-mentioned microbe-induced central sensitization can be alleviated by removal of the inflammatory microbial load. Ideally, we would look for *direct* evidence of alleviation of central sensitization, again by looking at the brain and its surrounding fluid for functional/molecular evidence of normalization, but without the ideal situation, we could look for *indirect* evidence of alleviation of central sensitization by looking for alleviation of pain. **Clinical studies with humans have already shown that removing the excessive microbial load—specifically SIBO in the case of FM—alleviates the clinical manifestations of central sensitization, namely excessive pain, fatigue, and other physical and mental manifestations of the illness.** Supportively, we can appreciate data showing that ❸ mitochondrial dysfunction (mitodysfunction) and systemic inflammation in general promote central sensitization, and thereby the alleviation of either mitodysfunction or inflammation by alleviating the TML/SIBO would be expected to alleviate central sensitization. In a world of imperfect research, incomplete data, and the vast majority of doctors/researchers having had zero training in nutrition, mitochondrial optimization and treatments for dysbiosis, let us look at the support for this thesis while emphasizing the importance of incorporating the available data while recalling that absence of evidence (ie, the studies have not been performed) is neither evidence of refutation (ie, no evidence refutes the thesis) nor evidence of absence (ie, we have an abundance of supporting data, even if we are currently and perhaps always will be lacking a perfect prospective study of microbe-induced central sensitization followed by relief of same via antimicrobial/antidysbiotic interventions).

Basic science research and the practice of medicine are shaped by powerful financial interests

"A clinician who is unaware of the political forces that shape healthcare policy and research is analogous to a captain of an oceangoing ship not knowing how to use a compass, sextant, or coastline map. Medical science and healthcare policy are influenced by a myriad of powerful private interests motivated by their own goals, at times different from the stated goal of medicine, which purports to hold paramount the patient's welfare. Scientific objectivity and the guiding ethical principles of informed consent, beneficence, autonomy, and non-malfeasance are subject to different interpretations depending on the lens through which a dilemma is viewed. This gives rise to a disarrayed tug-of-war between factions and private interests, with paradigmatic victory often being awarded to those with the best marketing campaigns and political influence while less importance is given to safety, efficacy, and the economic burden to consumers. To be ignorant of such considerations is to be blind to the nature of research, policy, and our own biased inclinations for and against particular paradigms, assessments, and interventions. Research articles and sources of authority must be approached with an artist's delicacy and with a willingness to consider new information that may contradict deeply rooted beliefs."

Vasquez A. *Musculoskeletal Pain: Expanded Clinical Strategies*. Institute for Functional Medicine, 2008

The medicomonetary reality: The goal of most research is not cure of disease, but translation of biology into drug sales: In order to further understand the nature, goals, and constraints of "biomedical research", one first needs to have some *insights into the obvious*: namely, that the general goal of *biomedical* research is to understand enough of nature and biology so that a drug can be developed to address the condition being studied. This is the *translation* of biology into the practice of medicine, and generally this culminates in the development and sales of drugs. Of note, these drugs generally do not cure the disease nor solve the problems, but rather alleviate select and occasionally irrelevant biochemical or clinical indexes of the disease to show "improvement" and therefore justify regulatory "approval" and therefore empower drug sales. Consciously or unconsciously, this is how the system operates, and this is how people within the system generally think; in medical research centers—ie, medical research centers are almost always affiliated with hospitals or medical schools—generally everyone within these systems is a devotee of the medical paradigm, which generally holds that diseases are *idiopathic* and *need to be treated with drugs*, generally multiple drugs (polypharmacy) for

---

[431] Reichenberger et al. Establishing a relationship between bacteria in the human gut and complex regional pain syndrome. *Brain Behav Immun*. 2013 Mar;29:62-9

indefinite periods of time. The lifeblood of biomedical research centers is funding, and this is generally tied to the expectation of "advancing the practice of medicine" which by definition means using—and therefore developing for sale—more drugs. (Drugs are tangible, and drugs are profitable; drugs keep doctors employed, and they keep so-called healthcare centers [e.g., clinics and hospitals] open and profitable because patients are obligated to seek consultations and constantly renew their prescriptions—the profitability and repeat business are guaranteed when patients are told, "You have a disease that is not understood, and this disease will kill or harm you unless you take and continue to take these drugs." Drugs also maintain our power-over paradigm[432]

> **Paradigms—goals, components, needs and fears—are created to support the prevailing power structures**
>
> 248   PUBLIC OPINION
>
> consciously what facts, in what setting, in what guise he shall permit the public to know.
>
> 4
>
> That the manufacture of consent is capable of great refinements no one, I think, denies. The process by which public opinions arise is certainly no less intricate than it has appeared in these pages, and the opportunities for manipulation open to anyone who understands the process are plain enough.
>
> Lippmann W. *Public Opinion*. Harcourt & Brace, 1922

which holds that our worldview is centered on the feeling that we have *power over our problems* [rather than accepting them as they are and working with them as they are], and again we tend to seek this through concrete and obvious and simple-minded means because these are the means that have immediate appeal.) Any basic science or medical researcher who finds that a disease can be cured and the problem solved by a simple treatment will at best be the *Hero of the Day* before being forgotten (and—quite often—later fired); the doctor or researcher who finds that a drug *might* help a problem and who thereafter receives speaking opportunities at national and international medical society meetings (always directly or indirectly funded by drug companies) and who receives several hundred thousand dollars in research grants (always directly or indirectly funded by drug companies and/or the private interests tied to them) will be remembered and championed.

Limited thought—systematic addictions to searching for parts without appreciating the whole, and to disregarding the obvious as simple and therefore without merit: Within the medical paradigm is the belief that diseases are complex and serious and that therefore the treatments must be complex and serious; ironically, drug therapy is actually based on simplistic thinking evidenced by the fact that drugs almost always work on only *one single* pathway/process while complex chronic/persistent diseases are always *multifaceted*. When diseases are indeed simple or at least understandable, they must be made to appear complex and enigmatic by the confusion of research and the perpetual reenactment of ignoring previous research and starting from zero. In this manner, researchers can apply for grants using the words and phrases "new" and "innovative" and "translational" and "advancing medical science to advance patient care" and thus fall into the line of *common thought*—drug production, drug sales, medical dependency to which we are all indoctrinated and accustomed—which makes these ideas and phrases easily acceptable. A doctor or researcher who advances the idea that a disease can be treated by simple means runs the risk of appearing himself/herself as *simple* rather than *insightful*. Collectively as a society, we have created illusions of complexity on many facets of life, ranging from poverty to diabetes to perpetual war; each of these—like the majority of illnesses that bind patients and busy/occupy healthcare systems—can be understood and managed effectively, but not if we allow ourselves to be convinced that they are complex and not if we allow the profiteers to hold the reigns of conversation and intervention. I state that biomedical researchers by virtue of the society in which they live and the environments in which they work are inclined to state and perhaps believe (ie, honestly accept as their own perception) that fibromyalgia is complex, not understood, and ultimately only treatable with drugs—that is the "prescribed" line of medical thinking; we get that paradigm of thought hammered into us via reading, exams, and sleep deprivation in medical school, and medical school courses are the initial training ground for both doctors and their biomedical research colleagues (e.g., many PhD students have to take medical school courses if the PhD program is housed within a university system that contains a medical school). Thus, virtually everyone in "medicine" whether as a physician or researcher is indoctrinated with the *idiopathic-*

---

[432] Largent C, Breton D. *The Paradigm Conspiracy: Why Our Social Systems Violate Human Potential*. Hazelden, 1998

polypharmacy model. Finally, we must also appreciate the profiteering bias manifested by medical journals—who profit from sales of drug-friendly articles when drug companies pay for associated advertising and for article reprints[433]—and magazines/newspapers/websites/television that also receive more advertising money when they publish drug-friendly articles and news. In this way, research and news is inherently biased toward publishing drug-friendly articles and news.

> **Western culture is built upon the denial of nature, the denigration of what is natural, and a denial of the here and now**
>
> "Anti-naturalness [the denial of nature, naturalness, and our human nature] assumed the throne [dominating paradigm]. With relentless logic, one arrived at the demand to deny nature."
>
> Nietzsche FW. *Will to Power*. 1901, essay #245

Private control over public paradigms: Notably in the United States—the country that most strongly influences international healthcare, science, and policy—the largest media outlets for news, science, and television are owned by a small handful of multicorporation conglomerates that are interconnected—either directly, paradigmatically, or financially—with drug companies; indeed, each facet of health and healthcare in American society—including medical education itself[434] and the lack of labeling of GMO foods and the horrific failure to regulate the pesticide industry's contamination of food, air, and water[435,436]—is strongly influenced by lobbying and money from drug companies and private business interests.[437]

Denial of the obvious, especially that which is natural and immediate: Western culture has been largely built on the denial of the present moment and denial of what is natural, hence our fascination with anything new, "modern", "sophisticated", and laboratory-clean. As such, our treatment paradigms tend to avoid the *actionable* and *clear* and *natural* in favor of the passive (e.g., drugs), squeaky-clean (e.g., drugs), and the future (e.g., idiopathic now, but we are supposed to wait for future developments while postponing action and thought and in the meanwhile resign our consciousness to "the powers that be" which are supposed to make the decisions for us by their "virtue" of divine insight). Given that Western culture in its entirety is based on denial of now, natural, and actionable (except when stirred to war)—and any look at our generally passive, listless, and deferent society will confirm this—we should not be surprised that our medical paradigm is largely the same, characterized by passive drugs and passive medical care, apathy in personal thought and action (including medical professionals themselves who are taught

---

[433] Smith R. Medical Journals Are an Extension of the Marketing Arm of Pharmaceutical Companies. *PLoS Med.* 2005 May; 2(5): e138
[434] Drug-company influence on medical education in USA. *Lancet.* 2000 Sep 2;356(9232):781
[435] Bøhn T, Cuhra M, Traavik T, Sanden M, Fagan J, Primicerio R. Compositional differences in soybeans on the market: glyphosate accumulates in Roundup Ready GM soybeans. *Food Chem.* 2014 Jun 15;153:207-1 sciencedirect.com/science/article/pii/S0308814613019201
[436] Majewski MS, Coupe RH, Foreman WT, Capel PD. Pesticides in Mississippi air and rain: a comparison between 1995 and 2007. *Environ Toxicol Chem.* 2014 Jun;33:1283-93
[437] For anyone paying attention to America's political scene, especially since 1980, this statement will be self-evident and abundantly buttressed by the observance of national events that favor small numbers of rich and therefore powerful private and business interests over the interests of the American people and the overall welfare of the nation.

While some would reasonably argue that American society has always been a populist façade run amuck by a puppeteering financial elite, the population-wide data is very clear that the financial balance of the nation was essentially "inverted" in the 1980s during a time when—a fact that nobody can refute—the stated policy of the government was specifically to give more money, more power, and less responsibility to the rich at the expense of the rest of the nation; this was called "trickle-down economics" and was sold to the public via the illusion that by making the rich more rich the wealth of the nation would eventually "trickle down" to the financially lower segments of society and everyone would bathe in the wealth of the nation via the generosity and altruism of the financial elite. Obviously, this did not occur; along with the government's deregulation of industry and the demolition of public power generally and unions in particular, deceptively termed "free trade agreements" shipped American jobs overseas to nations with comparatively zero worker and environmental protections to the benefit of multinational corporations. As shown in the image from President Obama's 2015 State of the Union address, the largest section of the American population—the bottom 90%—now have access to but a small piece of the American national pie, while the top 1% own nearly 70% of the country's resources. As a result, rich individuals and companies gained even more power while the citizens saw their interests ignored; finally in 2014, the obvious was published: America no longer functioned as a democracy by and for the people but rather as an oligarchy, with the government controlled by private interests controlling policies ranging from public health to education and public transportation. Studies and press are listed below:

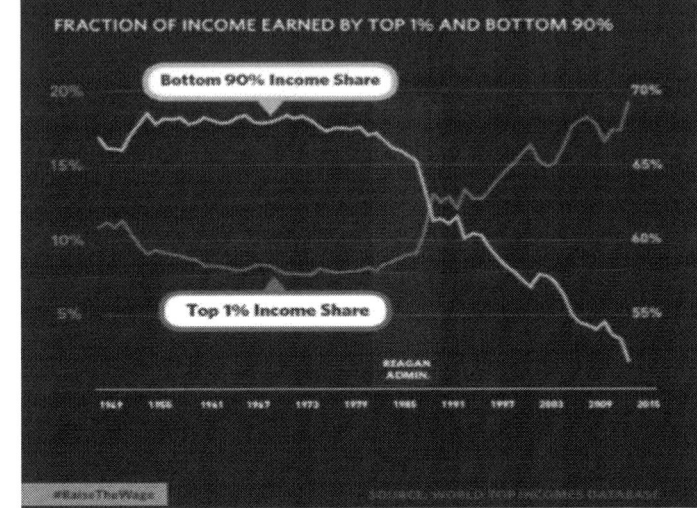

1. US is an oligarchy, not a democracy. bbc.com/news/blogs-echochambers-27074746
2. America is an oligarchy, not a democracy or republic, university study finds. washingtontimes.com/news/2014/apr/21/americas-oligarchy-not-democracy-or-republic-unive/
3. The US is an oligarchy, study concludes: "Report by researchers from Princeton and Northwestern universities suggests that US political system serves special interest organisations, instead of voters." telegraph.co.uk/news/worldnews/northamerica/usa/10769041/The-US-is-an-oligarchy-study-concludes.html
4. Is the USDA Silencing Scientists? theatlantic.com/science/archive/2015/11/is-the-usda-silencing-scientists/413803/
5. One Nation, Under Monsanto. counterpunch.org/2013/02/26/one-nation-under-monsanto Of special note, this article is written by Paul Craig Roberts PhD (Economics), a former Assistant Secretary of the US Treasury and Associate Editor of the *Wall Street Journal*.

not to think but rather to defer to authority and guidelines while they focus on diagnoses and drugs[438]). In Western medicine, natural treatments whether diets or vitamins are constantly attacked and obfuscated, despite their self-evident value and importance; meanwhile, drugs in general and vaccines in particular are sacrosanct, despite their generally artificial importance, indebting/bankrupting expenses, and adverse effects that often exceed their benefit.

❹ All other biochemical and pathophysiologic abnormalities seen in fibromyalgia can likewise be explained from the primary SIBO.

- <u>Restless leg syndrome and fibromyalgia commonly co-exist, and restless leg syndrome can be alleviated by eradication of SIBO</u>: Restless leg syndrome (RLS) occurs in approximately 30% of FM patients and can be effectively treated by addressing SIBO with a combination of antibiotics (drugs or botanical medicines that eradicate bacteria) and probiotics (products containing beneficial bacteria, which help restore "microbial balance" in the gastrointestinal tract).[439]

- <u>SIBO commonly causes nutrient malabsorption and thus predisposes to subclinical selective malnutrition, specifically micronutrient deficiency</u>: SIBO causes nutrient malabsorption[440] and can thereby contribute to the vitamin D and magnesium deficiencies that promote pain and mitochondrial dysfunction, respectively, and which are common in fibromyalgia. Intestinal bacterial overgrowth causes nutrient malabsorption via intestinal inflammation and villus atrophy (anatomic impairment) and impairment of digestion, specifically the enzymatic degradation of mucosal peptidases and disaccharidases by bacterial proteases (biochemical impairment). As reported by McEvoy and colleagues[441], bacterial contamination [overgrowth] of the small intestine is an important cause of occult malabsorption and malnutrition, especially in the elderly.

- <u>SIBO can be triggered or exacerbated by emotional stress</u>: SIBO can be triggered in humans by reduced mucosal immunity following stressful life events, and this helps explain the link between psychoemotional stress and the SIBO-related conditions IBS and FM. Chronic mental-emotional stress causes reduced production of the antibody secretory IgA (sIgA) which is the primary line of defense against bacteria and other microorganisms in the gastrointestinal tract; thus, mental-emotional stress can reduce intestinal immunity and thereby promote SIBO. Further, stress in humans triggers enhanced microbial pathogenicity via microbial endocrinology.

- <u>Oxidative stress triggers exaggerated pain perception—hyperalgesia (hypersensitivity to pain) and allodynia (perception of pain from normal stimuli)</u>: Patients with fibromyalgia show evidence of increased free radical (oxidant) production and reduced antioxidant defenses. Increased oxidative stress can be caused by immune activation and mitochondrial dysfunction; immune activation and mitochondrial dysfunction also promote oxidative stress and depletion of antioxidants, resulting in a vicious cycle, as illustrated. In the excellent review by Cordero et al[442], the authors note that recent studies have shown that oxidative stress causes peripheral and central sensitization and alters nerve sensitivity to pain (nociception), resulting in hyperalgesia—hypersensitivity to normal stimuli. The free radical (oxidant) superoxide promotes the development of pain through direct peripheral sensitization and the release of various cytokines (such as TNF-α, IL-1β, and IL-6), the formation of peroxynitrite (ONOO-), and PARP activation. PARP—poly-ADP-ribose-polymerase—is a nuclear enzyme activated by superoxide/peroxynitrite radicals; activation of PARP promotes the development of pain syndromes, including the components of small sensory fiber neuropathy, thermal and mechanical hyperalgesia, tactile allodynia, and exaggerated pain behavior in animal models of diabetic neuropathy.[443,444]

- <u>Low plasma levels of L-tryptophan seen in fibromyalgia patients can be caused by degradation of dietary tryptophan by the bacterial enzyme tryptophanase</u>: Patients with fibromyalgia have low blood levels of the amino acid L-tryptophan[445], which is used in the body to make serotonin (important for mood maintenance,

---

[438] Ely et al. Analysis of questions asked by family doctors regarding patient care. *BMJ*. 1999 Aug 7;319(7206):358-61 ncbi.nlm.nih.gov/pmc/articles/PMC28191/
[439] Weinstock et al. Restless Legs Syndrome in Patients with Irritable Bowel Syndrome: Response to Small Intestinal Bacterial Overgrowth Therapy. *Dig Dis Sci* 2007 May:1252-6
[440] Elphick HL, Elphick DA, Sanders DS. Small bowel bacterial overgrowth. An underrecognized cause of malnutrition in older adults. Geriatrics. 2006 Sep;61(9):21-6
[441] McEvoy et al. Bacterial contamination of the small intestine is an important cause of occult malabsorption in the elderly. *Br Med J* 1983 Sep 17;287(6395):789-93
[442] Cordero MD, et al. Mitochondrial dysfunction and mitophagy activation in blood mononuclear cells of fibromyalgia patients. *Arthritis Res Ther*. 2010;12(1):R17
[443] Wang ZQ, Porreca F, Cuzzocrea S et al. A newly identified role for superoxide in inflammatory pain. *J Pharmacol Exp Ther*. 2004 Jun;309(3):869-78
[444] Ilnytska O, et al. Poly(ADP-ribose) polymerase inhibition alleviates experimental diabetic sensory neuropathy. *Diabetes* 2006 Jun;55:1686-94
[445] "Plasma-free tryptophan is inversely related to the severity of subjective pain in 8 patients who fulfilled criteria for a variety of non-articular rheumatism, the "fibrositis syndrome". The observation is consistent with animal and human studies suggesting a relationship between reduced brain serotonin metabolism and pain reactivity." Moldofsky H, Warsh JJ. Plasma tryptophan and musculoskeletal pain in non-articular rheumatism ("fibrositis syndrome"). *Pain*. 1978 Jun;5(1):65-71

pain alleviation, and appetite control) and melatonin (important for normal sleep, support of mitochondrial function [stimulant of ETC complexes #1 and #4], and antioxidant protection, given that melatonin is one of the most powerful antioxidants). Bacteria such as *Escherichia coli, Proteus vulgaris,* and *Bacteroides* produce the enzyme tryptophanase[446], which destroys L-tryptophan in the gut before it is absorbed from ingested foods; thus, generalized bacterial overgrowth of the small intestine could reasonably be expected to exacerbate this phenomenon. In patients with fibromyalgia, higher tryptophan levels correlate positively with serotonin levels and with less pain and better sleep, while lower tryptophan levels are associated with sleep impairment, reduced serotonin levels, and higher levels of substance P, a neurotransmitter that promotes inflammation and pain perception.[447] Fibromyalgia patients produce 31% less melatonin than do healthy controls, and "this may contribute to impaired sleep at night, fatigue during the day, and changed pain perception."[448] Thus, a likely sequence of events is that, for example, a period of stressful life events can cause impair gastrointestinal immunity leading to intestinal bacterial overgrowth, which itself causes tryptophan degradation via elaboration of bacterial tryptophanase, causing tryptophan deficiency and resultant deficiencies of serotonin (leading to pain, depression, anxiety, and food/carbohydrate craving) and melatonin (leading to sleep disturbance, impaired antioxidant defense and mitochondrial function, and impaired immune responsiveness). Enhanced degradation of tryptophan can also be effected by the liver via enhanced activation of the enzyme tryptophan pyrrolase (tryptophan 2,3-dioxygenase, TDO) in response to "stress" and increased cortisol and by immune cells such as macrophages and dendritic cells via the enzyme indoleamine 2,3-dioxygenase (IDO) which is upregulated via the Th1-type cytokine interferon-gamma (IFNg).[449] Pregnancy induces accelerated tryptophan degradation; however, for the focus of this conversation we note that most FM patients are not pregnant and thus the likely causes of tryptophan deficiency in FM patients, who show evidence of SIBO, stress, and inflammation are ❶ intraluminal bacterial tryptophanase, ❷ hepatic tryptophan pyrrolase, and ❸ immune/inflammatory 2,3-dioxygenase.

- The therapies that help fibromyalgia share mechanisms of action consistent with the model presented here: As will be reviewed below under *Therapeutic Interventions,* essentially all of the most successful therapies for fibromyalgia have effects on intestinal flora, muscle perfusion/contractility, or mitochondrial bioenergetics (biological production of cellular energy/ATP). This is true for vegetarian diets (which favorably alter gut flora and improve antioxidant defenses), supplementation with tryptophan/melatonin (which preserve mitochondrial function during bacterial LPS/endotoxin exposure), physical treatments such as acupuncture (which improves tissue perfusion), and the use of nutrients such as magnesium, acetyl-L-carnitine, D-ribose, creatine, and coenzyme Q-10—all of which support or improve mitochondrial function.

Overall and when integrated together, the research literature provides compelling evidence linking intestinal bacterial overgrowth with the genesis and perpetuation of fibromyalgia. Chronic low-dose exposure to bacterial debris such as lipopolysaccharide/endotoxin and metabolic toxins such as hydrogen sulfide and D-lactic acid from SIBO is a plausible cause of impaired cellular energy production that results in chronic, widespread muscle fatigue and soreness and which may culminate in the clinical presentation of fibromyalgia. The individual components of this model have been substantiated by mechanistic studies in animals and/or research studies in humans. CFS also shares many epidemiological and clinical similarities with FM, and a similar pathophysiology is therefore highly probable.

A consistent report from many CFS and fibromyalgia patients is that of environmental intolerance (EI) and multiple chemical sensitivity (MCS), often grouped together as EI-MCS; these are complex disorders that the medical profession has failed to appreciate and which are characterized by adverse physiological responses to ambient levels of toxic chemicals and other environmental exposures. EI-MCS can be plausibly explained by SIBO because bacterial LPS/endotoxin impairs hepatic cytochrome P450 detoxification enzymes, resulting in reduced

---

[446] Demoss RD, Moser K. Tryptophanase in Diverse Bacterial Species. *Journal of Bacteriology* 1969; 98: 167-171
[447] "A strong negative correlation between SP and 5-HIAA as well as between SP and TRP could be demonstrated. High serum concentrations of 5-HIAA and TRP showed a significant relation to low pain scores. Moreover, 5-HIAA was strongly related to good quality of sleep, while SP was related to sleep disturbance." Schwarz et al. Relationship of substance P, 5-hydroxyindole acetic acid and tryptophan in serum of fibromyalgia patients. *Neurosci Lett.* 1999 Jan 15;259(3):196-8
[448] "The FMS patients had a 31% lower MT secretion than healthy subjects during the hours of darkness... Patients with fibromyalgic syndrome have a lower melatonin secretion during the hours of darkness than healthy subjects. This may contribute to impaired sleep at night, fatigue during the day, and changed pain perception." Wikner J, Hirsch U, Wetterberg L, Röjdmark S. Fibromyalgia syndrome associated with decreased nocturnal melatonin secretion. *Clin Endocrinol* (Oxf). 1998 Aug;49:179-83
[449] Schröcksnadel K, Wirleitner B, Winkler C, Fuchs D. Monitoring tryptophan metabolism in chronic immune activation. *Clin Chim Acta.* 2006 Feb;364(1-2):82-90

drug metabolism and impaired clearance of xenobiotics/toxins.[450] Accumulation of xenobiotics in CFS patients[451] might therefore be explained in part by LPS-induced inhibition of xenobiotic clearance secondary to SIBO. Further, the metabolic and immunologic effects of LPS can also account for the immune activation, neurological dysfunction, and musculoskeletal complaints noted in patients with CFS, IBS, and FM. A simplified yet accurate model of fibromyalgia which accounts for the major clinical and objective abnormalities seen with this condition is presented in the diagram that follows. Following the exclusion of diagnosable and treatable conditions that can contribute to or mimic fibromyalgia, and by using an integrated model, clinicians can design treatment plans based on the previously reviewed pathogenesis and on the therapeutic considerations detailed in the following section.

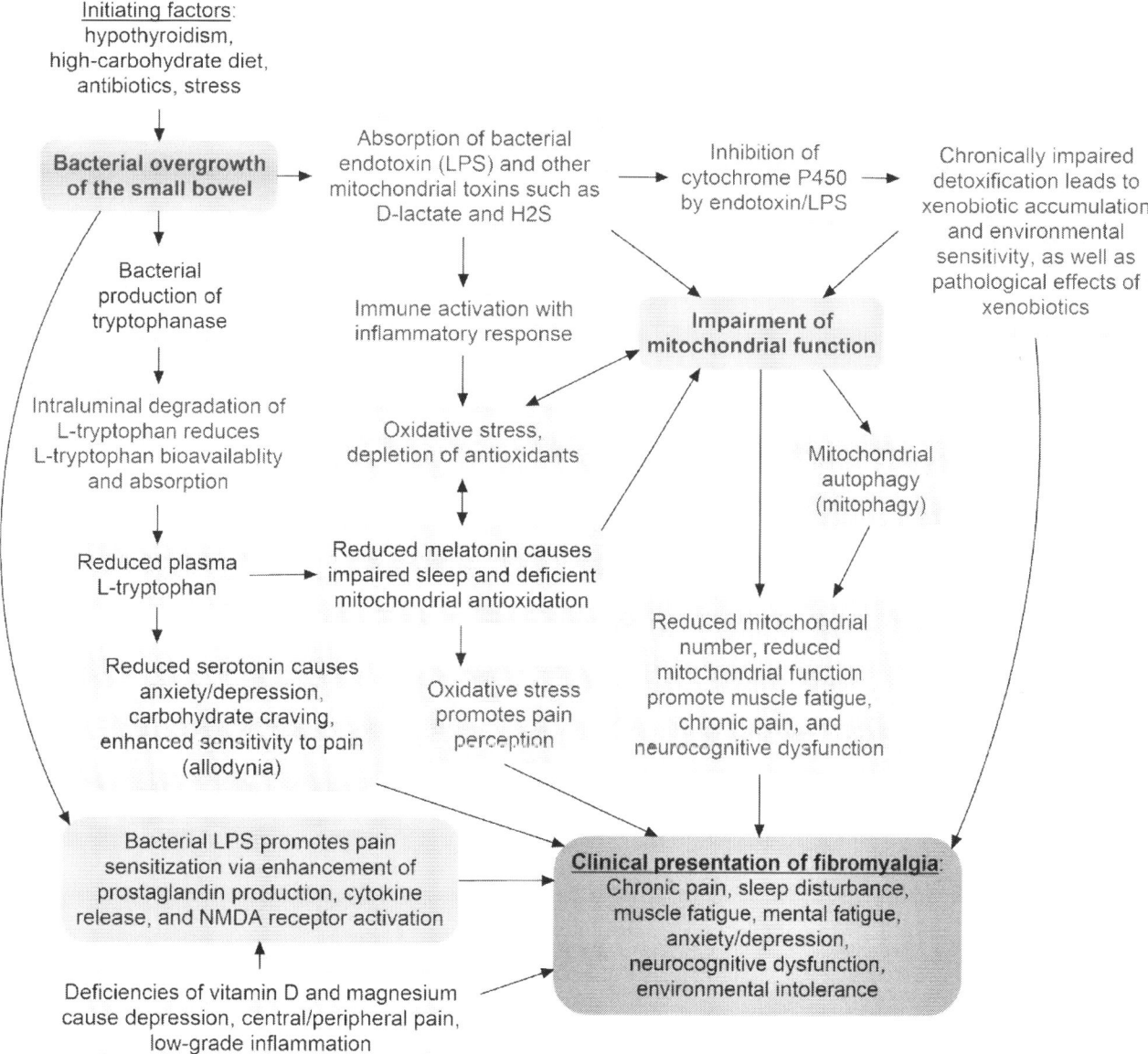

How small intestine bacterial overgrowth (SIBO) causes fibromyalgia: Bacterial overgrowth of the small bowel leads to chronic low-grade tryptophan insufficiency resulting in reduced endogenous production of serotonin (important for positive mood and relief from anxiety pain) and of melatonin (important for restful sleep and protection of mitochondria from oxidative stress). Bacterial mitochondrial toxins such as endotoxin, H2S, D-lactate cause impaired mitochondrial energy production, which leads to mitophagy, muscle fatigue, pain, and cognitive impairment. See educational videos and updates at www.inflammationmastery.com/pain

---

[450] Shedlofsky et al. Endotoxin administration to humans inhibits hepatic cytochrome P450-mediated drug metabolism. *J Clin Invest*. 1994 Dec;94(6):2209-14
[451] Dunstan et al. A preliminary investigation of chlorinated hydrocarbons and chronic fatigue syndrome. *Med J Aust*. 1995 Sep 18;163(6):294-7

Introduction to Therapeutic Approach: Rational treatment of microbe-induced pain and inflammation as seen in FM must focus on the eradication/modulation of the inflammatory microbial load as well as on the restoration of barrier defenses (e.g., the defensive lining of the skin and the gut wall) to reduce absorption of and exposure to microbial molecules (enhancing elimination of already-absorbed microbial toxins is a possibility, but is of lesser importance and is mentioned in some of my other writings). Enhancement of mitochondrial dysfunction and glial activation (brain inflammation) should also be pursued.

This section begins with a review of evidence showing important gut involvement in FM, CFS/SEID, and CRPS—all of which are variants on the same themes of microbe-induced microglial activation and mitochondrial dysfunction, as depicted in the following image, showing the relative contribution and combination of dysbiosis-induced mitochondrial impairment and brain inflammation (e.g., glial activation and central sensitization).

Complex regional pain syndrome: severe sensitization, often localized

Chronic fatigue syndrome (SEID): mitochondrial dysfunction predominates over sensitization

Fibromyalgia: moderate-severe sensitization combined with mitochondrial dysfunction

Irritable bowel syndrome: mild general sensitization, gastrointestinal symptoms predominate

Dysbiotic neurotoxicity / mitochondriopathy ———— Dysbiotic encephalitis / pain

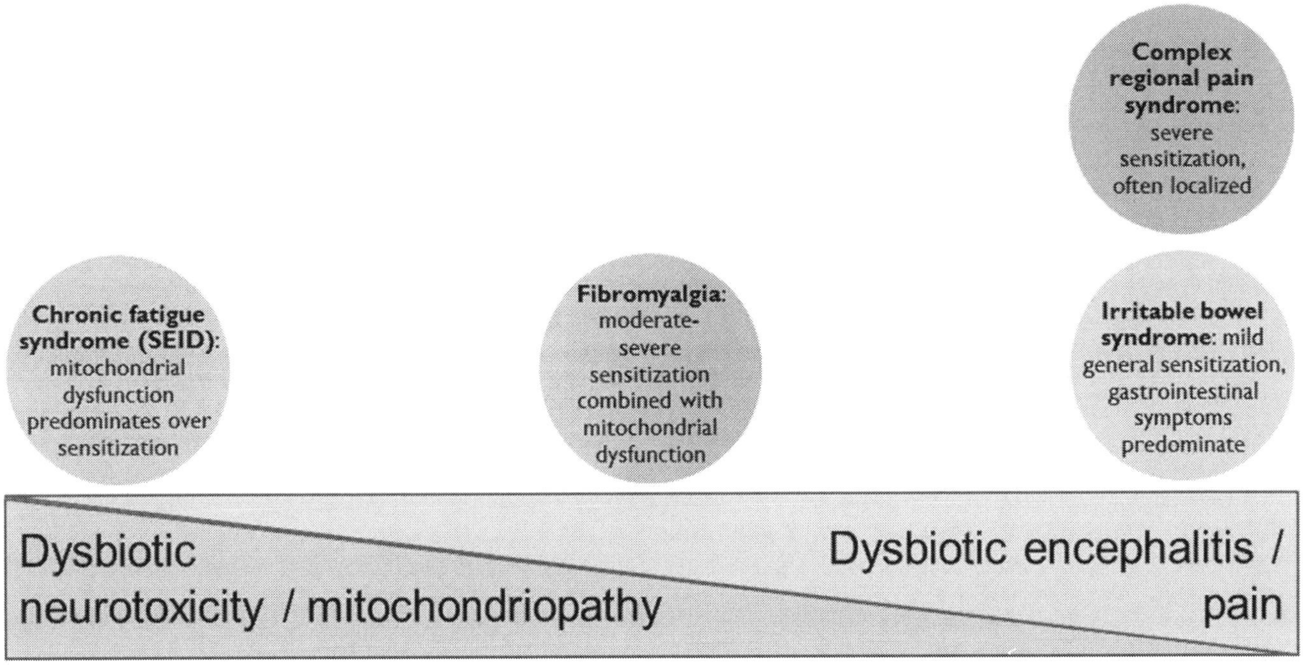

Microbe-induced mitochondriopathy/neurotoxicity and brain inflammation—a schematic diagram of relative contributions and combinations: In CFS/SEID, microbial induction of mitochondrial dysfunction and (D-lactate) neurotoxicity predominate, whereas fibromyalgia is a unique combination of both microbial mitochondriopathy with central sensitization. In IBS, central sensitization is potent, along with a more disruptive gut microbiome leading to bloating and constipation/diarrhea; meanwhile in CRPS, a neuroinflammatory dysbiosis clearly dominates the clinical picture.

- Patients with chronic fatigue syndrome (SEID), fibromyalgia (FM), and complex regional pain syndrome (CRPS) show evidence of increased intestinal permeability, leading to increased absorption of antigens, immunogens, inflammogens, and mitochondrial inhibitors (etc): Finally and most importantly, **antimicrobial therapy alleviates FM (and IBS) symptoms in direct proportion to the success of bacterial overgrowth eradication**, thus adding strong direct evidence in support of SIBO as a main cause of FM.[452,453] Recent clinical trials have shown that treatment of the fibromyalgia-related conditions IBS and SIBO by use of the nonabsorbed oral antibiotic rifaximin results in significant diminution of IBS-SIBO symptomatology with benefits lasting after the discontinuation of therapy.[454,455]
    - Increased intestinal permeability (IP) in patients with primary fibromyalgia (FM) and in patients with complex regional pain syndrome (CRPS) (*Rheumatology* 2008 Aug[456]): Authors of this study used well-established tests of mucosal permeability for gastroduodenal permeability (sucrose) and small intestinal permeability (lactulose and mannitol, as discussed in Chapter 1) to find that both FM and CRPS patients

---

[452] Wallace DJ, Hallegua DS. Fibromyalgia: the gastrointestinal link. *Curr Pain Headache Rep*.2004 Oct;8(5):364-8
[453] Pimentel et al. Improvement of symptoms by eradication of small intestinal overgrowth in FM: a double-blind study. [Abstract] *Arthritis Rheum* 1999, 42:S343
[454] Pimentel et al. The effect of a nonabsorbed oral antibiotic (rifaximin) on the symptoms of the irritable bowel syndrome. *Ann Intern Med*. 2006 Oct 17;145(8):557-63
[455] Sharara et al. A randomized double-blind placebo-controlled trial of rifaximin in patients with abdominal bloating and flatulence. *Am J Gastroenterol*. 2006 Feb;101(2):326-33
[456] Goebel et al. Altered intestinal permeability in patients with primary fibromyalgia and in patients with complex regional pain syndrome. *Rheumatology* 2008 Aug;47(8):1223-7

show increased burden of intestinal involvement; they write, "Patients with FM had a significantly higher IP than healthy volunteers. Indeed, both FM and CRPS groups had significantly higher IP values than healthy volunteers for both gastroduodenal permeability and small bowel permeability. The difference in gastroduodenal permeability between the two patient groups did not reach significance. Gastroduodenal permeability, as measured with the sucrose test, was increased in 13 patients with FM (32.5%) and in six patients with CRPS (35.3%) and in one healthy volunteer. The small bowel permeability index [using lactulose and mannitol] was increased in 15 patients with FM (37.5%), three patients with CRPS (17.6%) and in none of the volunteers." A wonderful addition to / extension of this work would have been the measurement of serum LPS levels, but this was not performed in the current study and has not been performed as of early 2015—noted and summarized immediately below is one study (Maes et al, *J Affect Disord* 2007 Apr) that measured antibodies to LPS (not LPS directly) in patients with CFS—chronic fatigue syndrome. Notably, increased IP in FM did not correlate with GI symptoms; however increased IP did correlate with serologic positivity against *Helicobacter pylori*, *Yersinia enterocolitica* and *Campylobacter jejuni*.

- **Increased serum IgA and IgM against LPS of enterobacteria in chronic fatigue syndrome (CFS) shows involvement of gram-negative enterobacteria and increased gut-intestinal permeability in the etiology of CFS** (*J Affect Disord* 2007 Apr[457]): These investigators measured IgA and IgM to LPS of the enterobacteria *Hafnia alvei*, *Pseudomonas aeruginosa*, *Morganella morganii*, *Proteus mirabilis*, *Pseudomonas putida*, *Citrobacter koseri*, and *Klebsiella pneumoniae* and "found that the prevalences and median values for serum IgA against the LPS of enterobacteria are significantly greater in patients with CFS than in normal volunteers and patients with partial CFS. Serum IgA levels were significantly correlated to the severity of illness, as measured by the FibroFatigue scale and to symptoms, such as irritable bowel, muscular tension, fatigue, concentration difficulties, and failing memory. The results show that enterobacteria are involved in the etiology of CFS and that an increased gut-intestinal permeability has caused an immune response to the LPS of gram-negative enterobacteria."
- Gut inflammation in chronic fatigue syndrome. (*Nutr Metab* 2010 Oct[458]): "Many CFS patients complain of gut dysfunction. In fact, patients with CFS are more likely to report a previous diagnosis of irritable bowel syndrome (IBS), a common functional disorder of the gut, and experience IBS-related symptoms. Recently, evidence for interactions between the intestinal microbiota, mucosal barrier function, and the immune system have been shown to play a role in the disorder's pathogenesis. Studies examining the microecology of the gastrointestinal (GI) tract have identified specific microorganisms whose presence appears related to disease; in CFS, a role for altered intestinal microbiota in the pathogenesis of the disease has recently been suggested. **Mucosal barrier dysfunction promoting bacterial translocation has also been observed.** ...For example, the administration of probiotics could alter the gut microbiota, improve mucosal barrier function, decrease pro-inflammatory cytokines, and have the potential to positively influence mood in patients where both emotional symptoms and inflammatory immune signals are elevated. Probiotics also have the potential to improve gut motility, which is dysfunctional in many CFS patients."
- Altered intestinal microbiota and organic acids may be the origin of symptoms in irritable bowel syndrome. (*Neurogastroenterol Motil.* 2010 May[459]): "Irritable bowel syndrome patients showed significantly higher counts of Veillonella and Lactobacillus than controls. They also expressed significantly higher levels of acetic acid, propionic acid, and total organic acids than controls. The quantity of bowel gas was not significantly different between controls and IBS patients. Finally, IBS patients with high acetic acid or propionic acid levels presented with significantly worse GI symptoms, QOL and negative emotions than those with low acetic acid or propionic acid levels or controls."
- Intestinal permeability and hypersensitivity in the irritable bowel syndrome. (*Pain* 2009 Nov[460]): "Here we demonstrate that diarrhea-predominant IBS (D-IBS) patients display increased intestinal permeability. We

---

[457] Maes et al. Increased serum IgA and IgM against LPS of enterobacteria in chronic fatigue syndrome (CFS): indication for the involvement of gram-negative enterobacteria in the etiology of CFS and for the presence of an increased gut-intestinal permeability. *J Affect Disord* 2007 Apr, 99:237–240
[458] Lakhan SE, Kirchgessner A. Gut inflammation in chronic fatigue syndrome. *Nutr Metab* (Lond). 2010 Oct 12;7:79
[459] Tana et al. Altered profiles of intestinal microbiota and organic acids may be the origin of symptoms in irritable bowel syndrome. *Neurogastroenterol Motil.* 2010 May;22(5):512-9, e114-5
[460] Zhou Q, Zhang B, Verne GN. Intestinal membrane permeability and hypersensitivity in the irritable bowel syndrome. *Pain.* 2009 Nov;146(1-2):41-6

have also found that increased intestinal membrane permeability is associated with visceral and thermal hypersensitivity in this subset of D-IBS patients. We evaluated 54 D-IBS patients and 22 controls for intestinal membrane permeability using the lactulose/mannitol method. ... We also evaluated the mean mechanical visual analogue scale (M-VAS) pain rating to nociceptive thermal and visceral stimulation in all subjects. ... Approximately 39% of diarrhea-predominant IBS patients had increased intestinal membrane permeability as measured by the lactulose/mannitol ratio. These IBS patients also demonstrated higher M-VAS pain intensity reading scale. Interestingly, the IBS patients with hypersensitivity and increased intestinal permeability had a higher Functional Bowel Disorder Severity Index (FBDSI) score (100.8 + or - 5.4) than IBS patients with normal membrane permeability and sensitivity (51.6 + or - 12.7) and controls (6.1 + or - 5.6) (p<0.001). A subset of D-IBS patients had increased intestinal membrane permeability that was associated with an increased FBDSI score and increased hypersensitivity to visceral and thermal nociceptive pain stimuli. Thus, increased intestinal membrane permeability in D-IBS patients may lead to more severe IBS symptoms and hypersensitivity to somatic and visceral stimuli."

- Antimicrobial/antibiotic treatment alleviates fibromyalgia in most FM patients, just as it also alleviates gastrointestinal symptoms in patients with irritable bowel syndrome (IBS): Finally and most importantly, **antimicrobial therapy alleviates FM (and IBS) symptoms in direct proportion to the success of bacterial overgrowth eradication**, thus adding strong direct evidence in support of SIBO as a main cause of FM.[461,462] Recent clinical trials have shown that treatment of the fibromyalgia-related conditions IBS and SIBO by use of the nonabsorbed oral antibiotic rifaximin results in significant diminution of IBS-SIBO symptomatology with benefits lasting after the discontinuation of therapy.[463,464]
    - Clinical trial: Rifaximin therapy for patients with irritable bowel syndrome without constipation (*New England Journal of Medicine* 2011 Jan[465]): Authors of this study evaluated rifaximin, a minimally absorbed antibiotic, as treatment for IBS. Subjects were given rifaximin at a dose of 550 mg or placebo, three times daily for 2 weeks and were followed for 10 weeks thereafter. "Significantly more patients in the rifaximin group than in the placebo group had adequate relief of global IBS symptoms during the first 4 weeks after treatment (40.8% vs. 31.2%). Similarly, more patients in the rifaximin group than in the placebo group had adequate relief of bloating (39.5% vs. 28.7%). In addition, significantly more patients in the rifaximin group had a response to treatment as assessed by daily ratings of IBS symptoms, bloating, abdominal pain, and stool consistency. The incidence of adverse events was similar in the two groups." Thus, among patients who had IBS without constipation, treatment with rifaximin for 2 weeks provided significant relief of IBS symptoms, bloating, abdominal pain, and loose or watery stools. *Comments by Dr Vasquez: Shortcomings of the intervention used in this IBS-rifaximin study include* ❶ *failure to use long-term treatment, which is often necessary in the treatment of chronic SIBO,* ❷ *failure to co-administer an antifungal agent to avert fungal growth in the intestines which commonly occurs as a result of antimicrobial/antibacterial drug treatment,* ❸ *failure to administer probiotics to re-establish beneficial flora, and* ❹ *failure to implement dietary modification to sustain the beneficial eradication of excess bacteria—allowing patients to continue their unhealthy diets and lifestyles is the most assured way to ensure that the condition (SIBO-IBS) will return.*
    - Review of clinical trials: Rifaximin as treatment for SIBO and IBS (*Expert Opinion on Investigational Drugs* 2009 Mar[466]): A recognized expert in the treatment of SIBO-related conditions, Dr Pimentel writes, "**Rifaximin is a broad-range, gastrointestinal-specific antibiotic that demonstrates no clinically relevant bacterial resistance**. Therefore, rifaximin may be useful in the treatment of gastrointestinal disorders associated with altered bacterial flora, including irritable bowel syndrome (IBS) and small intestinal bacterial overgrowth (SIBO)." He also notes regarding the use of rifaximin in the treatment of IBS, "Rifaximin improved global symptoms in 33 - 92% of patients and eradicated SIBO in up to 84% of patients with IBS, with results sustained up to 10 weeks post-treatment. Rifaximin caused a lower number of

---

[461] Wallace DJ, Hallegua DS. Fibromyalgia: the gastrointestinal link. *Curr Pain Headache Rep.* 2004 Oct;8(5):364-8
[462] Pimentel et al. Improvement of symptoms by eradication of small intestinal overgrowth in FM: a double-blind study. [Abstract] *Arthritis Rheum* 1999, 42:S343
[463] Pimentel et al. The effect of a nonabsorbed oral antibiotic (rifaximin) on the symptoms of the irritable bowel syndrome. *Ann Intern Med.* 2006 Oct 17;145(8):557-63
[464] Sharara et al. A randomized double-blind placebo-controlled trial of rifaximin in patients with abdominal bloating and flatulence. *Am J Gastroenterol.* 2006 Feb;101(2):326-33
[465] Pimentel M, Lembo A, Chey WD, et al. Rifaximin therapy for patients with irritable bowel syndrome without constipation. *N Engl J Med.* 2011 Jan 6;364(1):22-32
[466] Pimentel M. Review of rifaximin as treatment for SIBO and IBS. *Expert Opin Investig Drugs.* 2009 Mar;18(3):349-58

adverse events compared with metronidazole or levofloxacin and may have a more favorable adverse event profile than systemic antibiotics, without clinically relevant antibiotic resistance."
- <u>Results of two clinical trials of antibiotics in the treatment of fibromyalgia</u> (*Current Pain and Headache Reports* 2004 Oct[467]): This article discusses the results of two experiments using antibiotics in the treatment of FM: ❶ 96 patients with SIBO diagnosed by lactulose hydrogen breath testing (LHBT) were offered antibiotic treatment for the reduction of gastrointestinal bacteria; 25 of the 96 patients returned for a follow-up LHBT. Neomycin was the most commonly used antibiotic. Eleven of the 25 patients achieved complete transient eradication of SIBO after antibiotic treatment and experienced better improvement in more of their FM symptom scores when compared with the patients who did not achieve complete eradication. This indicates that **a direct relationship exists between the presence of SIBO and intestinal and extraintestinal symptoms in fibromyalgia, and that FM can be alleviated by effective antimicrobial/antibiotic treatment**. ❷ In this double-blind trial of eradication of SIBO in fibromyalgia, 46 patients fulfilling the established criteria for FM were tested for SIBO using LHBT. Forty-two of the 46 patients (91.3%) were positive for SIBO and were randomized to receive placebo or 500 mg of liquid neomycin (a minimally-absorbed gastrointestinal-specific antibiotic drug) twice daily for 10 days. Only six of the 20 patients (30%) in the neomycin group achieved eradication (indicating inefficacy of treatment); thus, no statistically significant difference between groups was available for analysis. Thereafter, 28 patients in the double-blind study testing positive for SIBO went on to receive open-label antibiotic treatment to eradicate SIBO, and this time 17 of the 28 patients (60.7%) achieved eradication of SIBO. When these 23 patients were compared with the 15 patients who failed to eradicate or did not undergo open-label treatment, significant improvement attributable to antibiotic treatment in the FM scores was detected. **Results show that eradication of bacterial overgrowth results in a clinically significant alleviation of FM symptoms.**

Most experienced functional medicine clinicians generally and naturopathic physicians specifically — collectively these are the clinicians with the most experience in the assessment and treatment of various forms of dysbiosis — will agree that ultimate correction of dysbiosis is no easy task. Many aspects must be addressed simultaneously, and these will be detailed in the section on *Therapeutics and Clinical Interventions*. For the here and now, I will simply outline some of the more common and important considerations, in no particular order because order implies importance, which implies effectiveness, which is dependent on the patient's response:

1. <u>Dietary optimization</u>: The diet must be diverse, varied, and plant-based in order to optimize microbial diversity and reduce intake of carbohydrates in general and simple carbohydrates in particular.
2. <u>Probiotic therapy</u>: Administering probiotics to patients who have SIBO can be effective or can cause exacerbations; patients might need to wait until other aspects are optimized before adding "good bacteria."
3. <u>Antimicrobial therapy</u>: Botanicals such as berberine, oregano oil, and others are safe and highly efficacious.
4. <u>Stimulation/normalization of peristalsis</u>: Magnesium and ascorbate can be used to promote laxation; optimization of thyroid function/status is of utmost importance. Exercise, water, relaxation, fiber are important.
5. <u>Systemic health optimization, stress reduction, sleep optimization</u>: Fatty acid intake (e.g., n3 fatty acids) and mitochondrial function need to be optimized in order to optimize gut flora. Stress promotes dysbiosis and microbial virulence; sleep deprivation impairs mitochondria, redox balance, hormones, and immunity.

In this space before diving into *Therapeutics and Clinical Interventions*, I'll briefly mention three concepts that are interrelated. ❶ Microbial identity and location are irrelevant, given that most of the adverse physiologic effects seen can be mediated by a wide range of Gram-negative bacteria; their identities and locations are irrelevant as long as they produce sufficient LPS. The only aspect that is almost entirely dependent on localization to the gut is the production of the mitochondrial neurotoxins H2S and D-lactate. ❷ Any combination of glial activation and mitochondrial dysfunction would be sufficient to induce the fibromyalgia phenotype; the agents most likely to accomplish both are pesticides and corporate toxins such as glyphosate, widely sprayed internationally and pervasive in the American food, air, and water supply. ❸ The variations in microbial combinations and total load, individual patient fragilities and vulnerabilities, nutritional status and xenobiotic permutations give rise to the nuances and variations in clinical presentations and therapeutic responses. However, the model presented

---

[467] Wallace DJ, Hallegua DS. Fibromyalgia: the gastrointestinal link. *Curr Pain Headache Rep.* 2004 Oct;8(5):364-8

throughout this section is an accurate description of the most important and common pathoetiologic considerations and clinical interventions; once these core considerations are addressed, others can be addressed with greater specific and overall efficacy.

Therapeutics and Clinical Interventions

- Overview: Treatments for FM should be ❶ science-based and should ❷ directly address the cause(s) of the disorder; treatments should be ❹ safe (generally) and ❺ effective and ❻ without potential for serious adverse effects. Drug treatment of FM does not meet these criteria; the integrative, nutritional, and functional medicine approaches outlined below can—when properly employed by a skilled clinician—address the cause(s) of FM in a way that is scientific, direct, safe, effective, and well tolerated by essentially all patients. **Treatments for FM must emphasize eradication of SIBO, prevention of SIBO recurrence, the restoration/establishment of optimal nutritional status, and specific support for optimal mitochondrial function; anything less than this will fail to be effective.** Clinical interventions for the treatment of SIBO include dietary carbohydrate restriction, normalization of slow gastrointestinal transit time (e.g., correction of hypothyroidism), selective use of probiotic supplements to normalize intestinal flora, support of mucosal immunity (with nutrients such as vitamin A, zinc, and L-glutamine), and eradication of bacterial overgrowth with drugs (such as ciprofloxacin, rifaximin, amoxicillin/Augmentin[468], metronidazole) and/or natural products (such as berberine[469], *Artemisia annua*, peppermint oil[470], and emulsified time-released oil of oregano[471])—each of these have been reviewed in greater detail elsewhere by this author.[472] Failure of any monotherapeutic approach to immediately resolve the clinical manifestations of FM can explained by the secondary metabolic, immune, and neurophysiological effects that have generally persisted over periods ranging from years to decades for most patients; in other words, the treatment program must be multifaceted in order to address the numerous major problems that cause FM, and the treatment plan must also be sustained long enough to correct the abnormal physiologic patterns that have been established by the body's response/adaptation to the disease process. The treatment program (examples provided) must be complete in order to facilitate correction of systemic oxidative damage (broad-spectrum antioxidant support), resultant nutritional deficiencies (diet optimization, vitamin and mineral supplementation), immune sensitization and induction of proinflammatory cycles (anti-inflammatory nutrition), alterations in neurotransmission and membrane receptor function (amino acid and fatty acid supplementation), and the inflammation-induced disturbances in pain reception and hypothalamic-pituitary-endocrine function (assess/correct hormonal imbalances; supplement with n-3 fatty acids and olive oil to reduce hypothalamic inflammation[473], etc.). Further, patients treated for SIBO who do not positively change their diets and lifestyles (which probably promoted the genesis and perpetuation of the disease-causing SIBO in the first place) are subject to continual recurrence until such changes are implemented and faithfully maintained.

- **FOOD & NUTRITION** As with many rheumatologic/painful/inflammatory conditions with a strong component of gastrointestinal dysbiosis, the single best diet is a vegan or pesco-vegetarian diet free of gluten and most grains (judicious use of brown rice excepted); clearly the diet must emphasize vegetable intake and avoidance of fermentable substrate to induce quantitative reductions and qualitative improvements in microbial populations and metabolic activity. Beyond the gut, the benefits of minimized carbohydrate intake include weight loss (generally beneficial in this patient population), alleviation of physiologic and psychologic dependence on carbohydrate (over)consumption, and increased endogenous production of beta-hydroxy-butyrate which stimulates the terminal complexes of the mitochondrial electron transport chain while also promoting histone acetylation (via inhibition of histone deacetylase) for enhanced DNA transcription resulting in what has been referred to as a rejuvenative phenotype. My foundational diet-nutritional program—the 5-part "supplemented Paleo-Mediterranean Diet"—can easily be modified to pesco-vegetarian or lacto-pesco-vegetarian variants to

---

[468] Malik BA, Xie YY, Wine E, Huynh HQ. Diagnosis and pharmacological management of small intestinal bacterial overgrowth in children with intestinal failure. *Can J Gastroenterol*. 2011 Jan;25(1):41-5. This is a remarkable article and probably one of the most brilliant articles on the treatment of SIBO with drugs.
[469] [No authors listed] Berberine. *Altern Med Rev*. 2000 Apr;5(2):175-7
[470] "A case report of a patient with SIBO who showed marked subjective improvement in IBS-like symptoms and significant reductions in hydrogen production after treatment with ECPO is presented. While further investigation is necessary, results in this case suggest one of mechanisms by which ECPO improves IBS symptoms is antimicrobial activity in small intestine." Logan et al. Treatment of small intestinal bacterial overgrowth with enteric-coated peppermint oil: a case report. *Altern Med Rev*. 2002 Oct;7(5):410-7
[471] Force M, Sparks WS, Ronzio RA. Inhibition of enteric parasites by emulsified oil of oregano in vivo. *Phytother Res*. 2000 May;14(3):213-4
[472] Vasquez A. Nutritional and Botanical Treatments against "Silent Infections" and Gastrointestinal Dysbiosis. *Nutr Perspect* 2006; Jan: 5-21 ichnfm.academia.edu/AlexVasquez
[473] Milanski et al. Saturated fatty acids produce inflammatory response predominantly through activation of TLR4 signaling in hypothalamus. *J Neurosci*. 2009 Jan;29(2):359-70

enhance high-quality protein intake while continuing to emphasize vegetable, nut, and seed intake to maximize fiber and micronutrient intake. In particular, whey protein isolate is attractive in this patient population due to the high content of tryptophan (to correct the common tryptophan deficiency), the glutathione precursors (to alleviate oxidative stress and enhance mitochondrial function), immunoglobulins (to support mucosal defenses against bacterial overgrowth), and growth factors including whey's insulotrophic effect to promote anabolism with resultant improvements in muscle function and gut mucosal integrity. Systemic alkalinization supported by plant-based potassium citrate promotes xenobiotic excretion (reduction in total load of persistent organic pollutants and toxic metals such as lead and mercury) and endogenous production of endorphins (enhanced mood, pain relief). The increased intake of vitamins (including physiologic doses of vitamin D3), minerals, ALA, GLA, EPA, DHA, and probiotics all synergize to alleviate SIBO, oxidative stress, pain and inflammation while also promoting optimal immune and mitochondrial functions. Allergy identification and avoidance is easily achieved via the very practical and no-cost elimination-and-challenge technique. In particular for patients with fibromyalgia, magnesium intake is increased via consumption of dark green vegetables and nutritional supplements while magnesium absorption is promoted by optimized vitamin D status while renal retention of magnesium is promoted with urinary alkalinization. The consistent and pathologic secondary tryptophan deficiency is foremost corrected by eradicating the causative SIBO and the resultant intraintestinal bacterial tryptophanase-catalyzed degradation of ingested tryptophan and via dietary supplementation with L-tryptophan, 5-hydroxytryptophan, and/or whey protein isolate.

> **Dr Vasquez's summary of the medical indication for gluten-containing grains/foods**
> The only medical indication for the consumption of gluten-containing grains is prevention/treatment of acute starvation, assuming no other food is available.

- Diet optimization with the five-part "supplemented Paleo-Mediterranean Diet" (*Nutr Perspect* 2011 Jan[474]): The "supplemented Paleo-Mediterranean Diet" (SPMD)—the 5-part nutritional wellness protocol—as described in most of my textbooks in "chapter 2" and also in my articles available on-line at ichnfm.academia.edu/AlexVasquez should be implemented for most FM patients; exceptions to this general rule might include patients with renal insufficiency due to the risk for potassium excess (hyperkalemia Note 475). Because any patient might have an allergy or intolerance to any food (even a healthy food like citrus fruit, chicken or eggs), patients and doctors must be aware of the potential for food allergies and will therefore have to customize the Paleo-Mediterranean diet *for each individual patient* to exclude foods to which the patient might be allergic or sensitive/intolerant. Otherwise, this 5-part nutrition protocol is based on ❶ vegetables, nuts, seeds, (berries, fruits, and juices generally have to be avoided during treatment for SIBO due to the high content of sugars and—with juice—the rapid passage through the gastrointestinal tract) and lean sources of protein, ❷ high-potency multivitamin and multimineral supplementation, ❸ physiologic doses of vitamin D3 to optimize blood levels of vitamin D3 (measured as 25-OH-vitamin D), ❹ combination fatty acid supplementation (with flax oil [for ALA], fish oil [for EPA and DHA], and borage oil [for GLA] with oleic acid from olive oil incorporated into the diet), and ❺ probiotics—foods or supplements that contain living bacteria with beneficial qualities. The diet should emphasize strict avoidance of grains in general and gluten-containing grains *especially wheat* in particular. This diet is essential for the provision of sufficient protein, fiber, phytonutrients, and alkalinization—potassium citrate is most concentrated in vegetables and helps the body maintain proper acid-alkaline balance.[476] The diet should be low in carbohydrates to reduce fermentable substrate to intestinal bacteria. The most important books for patients to read in support of this diet are *The Paleo Diet* by Dr Loren Cordain and *Breaking the Vicious Cycle* by Elaine Gottschall.

---

[474] Vasquez A. Revisiting the Five-Part Nutritional Wellness Protocol: Supplemented Paleo-Mediterranean Diet. *Nutr Perspect* 2011 Jan ichnfm.academia.edu/AlexVasquez
[475] Because the kidneys are responsible for excreting potassium, reduced kidney function (kidney failure, renal insufficiency) implies that the kidneys may not be able to perform the function of excreting potassium; thus, consumption of a potassium-rich diet could contribute to a dangerous situation of excess potassium in the blood known as hyperkalemia (*hyper*=too much, *kal*=potassium, *emia*=blood disorder). For patients with renal insufficiency, consumption of an otherwise health-promoting diet rich in fruits and vegetables might cause a problem if potassium accumulates in the blood due to impaired excretion. Blood tests can assess renal function as well as the blood potassium level. This is one example of why a clinician/doctor should be employed by patients before implementing diet modification and nutritional supplementation.
[476] "The modern Western-type diet is deficient in fruits and vegetables and contains excessive animal products, generating the accumulation of non-metabolizable anions and a lifespan state of overlooked metabolic acidosis, whose magnitude increases progressively with aging due to the physiological decline in kidney function." Adeva MM, Souto G. Diet-induced metabolic acidosis. *Clin Nutr*. 2011 Aug;30(4):416-21

- Vegetarian diet: Fibromyalgia syndrome improved using a mostly raw vegetarian diet (*BMC Complementary and Alternative Medicine* 2001 Sep[477]): Diets high in fruits, vegetables, nuts, berries, and seeds provide ample fiber to promote laxation and can be useful as adjunctive treatment for gastrointestinal dysbiosis in general and SIBO in particular (i.e., *quantitative* reduction in GI dysbiosis). Perhaps more importantly, plant-based diets result in *qualitative* benefits by changing microbial behavior and reducing production of irritants, toxins, and bacterial metabolites, including the mitochondrial poisons D-lactate and hydrogen sulfide. Fibromyalgia patients who consume a mostly vegetarian diet have experienced significant improvements in function and reductions in FM symptomatology. Poorly designed dietary interventions that allow abundant intake of whole-grain bread, pasta, rice, and fruit juice[478] would be expected to fail because such high-carbohydrate diets feed intestinal bacteria with an abundance of substrate and would therefore be expected to sustain or exacerbate SIBO. Another advantage to a plant-based mostly-raw diet is the avoidance of dietary advanced glycation end-products (AGEs) which are inflammation-promoting chemical combinations of proteins with sugars, which can be consumed in the diet (e.g., baked deserts) or formed endogenously/internally as a result of oxidative stress and elevated blood sugar levels (e.g., diabetes mellitus). As discussed previously, FM patients show higher levels of AGEs in blood cells and muscle tissue; AGEs promote chronic pain and inflammation[479], and therefore dietary and nutritional strategies that reduce AGE intake and/or AGE formation are 1) without risk, and 2) likely to provide manifold health benefits, including but not limited to reductions in pain and inflammation.

Dr Vasquez's Nutrition Protocol: The 5-part "Supplemented Paleo-Mediterranean Diet" (5pSPMD or SMPD)

1. **Diet: Emphasize fruits, vegetables, nuts, seeds, berries, and lean sources of protein** (fish, grass-fed lamb/beef). Minimize fruit intake due to higher sugar content while treating SIBO; the goal is to deprive the bacteria and yeast in the intestines of their preferred food source (carbohydrates, sugars). Make modifications for patient-specific food allergies and sensitivities; this is especially important for patients with known allergy-related conditions such as migraine headaches. Patients with kidney disease should use caution when consuming a potassium-rich diet. Vasquez A. Revisiting the Five-Part Nutritional Wellness Protocol: The Supplemented Paleo-Mediterranean Diet. *Nutritional Perspectives* 2011 Jan
2. **Multivitamin and multimineral supplement**: Nutrient deficiencies are common and are easily treated with nutritional supplementation. Fletcher and Fairfield. Vitamins for chronic disease prevention in adults. *JAMA* 2002 Jun
3. **Vitamin D dosed at 2,000-10,000 IU per day**: The adult requirement for vitamin D3 is approximately 4,000 IU per day; some patients may achieve optimal blood levels with lower doses, but generally daily doses of 4,000-10,000 IU are necessary. Vasquez et al. Clinical Importance of Vitamin D. *Alternative Therapies in Health and Medicine* 2004 Sep
4. **Combination fatty acid supplementation**: A combination of flax oil, borage oil, and fish oil provides the health-promoting fatty acids (ALA, GLA, EPA, DHA). Patients should consume organic virgin olive oil liberally with foods. Vasquez A. New Insights into Fatty Acid Supplementation and Its Effect on Eicosanoid Production and Genetic Expression. *Nutritional Perspectives* 2005; Jan
5. **Probiotics**: Health-promoting bacteria can be consumed in the form of powders, pills, and fermented foods such as yogurt, kefir, pickles and sauerkraut.

For a video review of this foundational diet and introduction to the functional inflammology protocol, see Dr Vasquez "Functional Inflammology Protocol, part 1: Introduction and Foundational Diet" from the 2013 International Conference on Human Nutrition and Functional Medicine: https://vimeo.com/100089988 Password: "DrVprotocol_volume1"

- Vitamin B12 in the form of hydroxo-/methyl-/adenosyl-cobalamin, should be administered to all FM and CFS patients to alleviate pain, glial inflammation, accelerated neurodegeneration; reasonable doses are at least 2,000 mcg per day orally and/or 2,000 mcg per week by subcutaneous/intramuscular injection: Given that sulfur-containing molecules such as sulfite and hydrogen sulfide bind to vitamin B12[480,481], we should reasonably expect that patients with excess exposure to H2S from the gastrointestinal tract would have an increased prevalence of vitamin B12 deficiency, and indeed this has been documented; vitamin B12 deficiency in CFS and FM patients promotes fatigue and brain dysfunction via the effects of vitamin B12 deficiency directly (ie, vitamin B12 deficiency is well known to cause nerve damage and brain damage) and indirectly via impaired

---

[477] Donaldson et al. Fibromyalgia syndrome improved using a mostly raw vegetarian diet. *BMC Complement Altern Med.* 2001;1:7 biomedcentral.com/1472-6882/1/7
[478] Michalsen et al. Mediterranean diet or extended fasting's influence on changing the intestinal microflora, immunoglobulin A secretion and clinical outcome in patients with rheumatoid arthritis and fibromyalgia: an observational study. *BMC Complement Altern Med.* 2005 Dec 22;5:22
[479] "In the interstitial connective tissue of fibromyalgic muscles we found a more intensive staining of the AGE CML, activated NF-kappaB, and also higher CML levels in the serum of these patients compared to the controls. RAGE was only present in FM muscle." Rüster et al. Detection of elevated N epsilon-carboxymethyllysine levels in muscular tissue and in serum of patients with fibromyalgia. *Scand J Rheumatol.* 2005 Nov-Dec;34(6):460-3
[480] Añíbarro et al. Asthma with sulfite intolerance in children: a blocking study with cyanocobalamin. *J Allergy Clin Immunol.* 1992 Jul;90(1):103-9
[481] Fujita et al. A fatal case of acute hydrogen sulfide poisoning caused by hydrogen sulfide: hydroxocobalamin therapy for acute H2S poisoning. *J Anal Toxicol.* 2011 Mar:119-23

Chapter 5.1—Functional Inflammology Protocol for Metabolic Inflammation: Migraine and Fibromyalgia

metabolism of homocysteine, which then triggers pain sensitivity and accelerated neurodegeneration via activation of NMDA receptors in the brain.

- Patients with FM and CFS have decreased vitamin B12 and increased homocysteine in the fluid surrounding the brain while blood test results are generally normal. (*Scand J Rheumatol.* 1997[482]): "Twelve outpatients, all women, who fulfilled the criteria for both fibromyalgia and chronic fatigue syndrome were rated on 15 items of the Comprehensive Psychopathological Rating Scale (CPRS-15). ... Blood laboratory levels were generally normal. The most obvious finding was that, in all the patients, the homocysteine (HCY) levels were increased in the cerebrospinal fluid (CSF). There was a significant positive correlation between CSF-HCY levels and fatigability, and the levels of CSF-B12 correlated significantly with the item of fatigability and with CPRS-15. The correlations between vitamin B12 and clinical variables of the CPRS-scale in this study indicate that low CSF-B12 values are of clinical importance. Vitamin B12 deficiency causes a deficient remethylation of HCY and is therefore probably contributing to the increased homocysteine levels found in our patient group. We conclude that increased homocysteine levels in the central nervous system characterize patients fulfilling the criteria for both fibromyalgia and chronic fatigue syndrome."

• Folic acid in the form of folinic acid or methylfolate ("5-methyltetrahydrofolate" or "5-MTHF"): Most nutrition-knowledgeable doctors do not use "folic acid" in the form of folic acid due to concerns about increased free radical generation and possible increased risk of cellular damage and malignant disease; we still use the term "folic acid" but nowadays this is—in practice—meant to imply the use of either folinic acid or methylfolate, two forms of folic acid that are considered safer, if not also more effective. Strictly speaking, "folic acid" refers to the synthetic form of the vitamin, whereas "folate" refers to derivatives of tetrahydrofolate that are found in food, especially leafy green vegetables, of which most people do not consume a sufficient amount. Some people develop antibodies against the folic acid transporter (cerebral folate receptor autoantibodies) that facilitates entry of folate into the brain, and they must receive either folinic acid or methylfolate to avoid neurologic devastation due to cerebral folate deficiency, wherein blood/serum levels of folate are normal but the brain (on the other side of the "wall" formed by the blood-brain barrier) is starved for this nutrient.[483] Folic acid from diet and/or supplementation serves many roles and thereby provides numerous benefits, largely centered on the provision of single-carbon methyl groups for metabolic processes (e.g., homocysteine metabolism) and DNA methylation, which regulates/suppresses gene transcription and thereby reduces risk of cancer and viral activation (e.g., cervical cancer following exposure to the human papilloma virus [HPV][484]). In this conversation, we are primarily concerned with optimizing folate intake to optimize "neurologic function" (i.e., generally speaking: normalization of homocysteine-mediated NMDAr activation in the brain, spinal cord and periphery) by reducing homocysteine levels because elevated homocysteine levels will cause excessive pain/fatigue/depression due to activation of the NMDA-receptor, mostly in the brain but also in the periphery. The most important nutrients for reducing homocysteine are folate (vitamin B9), pyridoxine (vitamin B6), cobalamin (vitamin B12) and the amino acid N-acetyl-cysteine (NAC); some people have a defect in their ability to convert folate into its active form via the enzyme methylenetetrahydrofolate reductase (MTHFR) and therefore need more nutritional supplementation to push this sluggish pathway to metabolic completion and reduce/normalize homocysteine levels. Increased consumption of folate from diet and/or supplements can alleviate depression, fatigue, and pain and is therefore recommended for all "pain patients", including those with migraine, fibromyalgia, and chronic fatigue syndrome. Adult doses of folate 1-5 mg (1,000-5,000 mcg) per day are reasonable and should be coadministered with a roughly equal amount of vitamin B12. Anti-seizure drugs (especially phenytoin, carbamazepine, barbiturates[485]), some of which are used in the treatment of

---

[482] Regland et al. Increased concentrations of homocysteine in the cerebrospinal fluid in patients with fibromyalgia and chronic fatigue syndrome. *Scand J Rheumatol.* 1997;26(4):301-7
[483] Gordon N. Cerebral folate deficiency. *Dev Med Child Neurol.* 2009 Mar;51(3):180-2
[484] Piyathilake et al. Indian women with higher serum concentrations of folate and vitamin B12 are significantly less likely to be infected with carcinogenic or high-risk (HR) types of human papillomaviruses (HPVs). *Int J Womens Health.* 2010 Aug 9;2:7-12
[485] Morrell MJ. Folic Acid and Epilepsy. *Epilepsy Curr.* 2002 Mar;2(2):31-34

migraine and chronic pain, are notorious for causing folate deficiency and homocysteine elevation[486]; obviously, the drugs would paradoxically promote pain and seizure if folate deficiency develops—coadministration of folate with anti-seizure drugs should be supervised by the prescribing physician. Diagrams illustrating these pathways tend to be repulsively complex, immemorably curvaceous, or incomplete and thereby unusable; the illustration below is perhaps the most simple for efficient understanding of the means by which nutritional supplementation lowers homocysteine levels.

- Positive response to vitamin B12 and folic acid in myalgic encephalomyelitis and fibromyalgia (*PLoS One*. 2015 Apr[487]): "The individual doses of B12 and folic acid, as well as the form of B12 used (i.e. hydroxocobalamin or methylcobalamin), had been due to individual decisions made by the five doctors in interplay with their patients – to a large extent following common sense in a process of trial and error – and limited by the patient's desire and the doctor's permission. It was a general experience that the patients deteriorated when returning to oral treatment, or when the injection interval was prolonged. After such a dose-finding period, they continued injective B12 therapy and learned how to self-administer the injections. In Sweden, the common form of B12 injective substrate has for more than forty years been hydroxocobalamin, provided in 1 mL ampoules with 1 mg/mL. By the end of last century, also methylcobalamin became available, in 2 mL ampoules with 5 mg/mL; i.e. an ampoule of methylcobalamin contains ten times more cobalamin than an ampoule of hydroxocobalamin. Folate in pharmacological doses is available by using tablets of folic acid (1 mg or 5 mg). ... Frequent injections of high-concentrated vitamin B12, combined with an individual daily dose of oral folic acid, may provide blood saturations high enough to be a remedy for good and safe relief in a subgroup of patients with ME/FM. Moreover, we suspect a counteracting interference between B12/folic acid and certain opioid analgesics and other drugs which have to be demethylated as part of their metabolism. Furthermore, it is important to be alert on co-existing thyroid dysfunction."

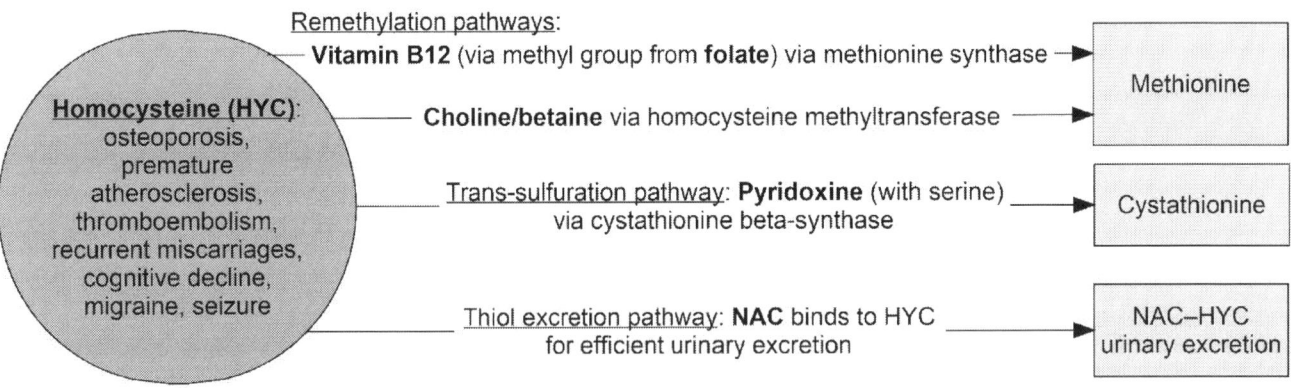

Lowering homocysteine (HYC) via nutritional supplementation: Folate gives methyl group to cobalamin (vitamin B12) to convert HYC via methionine synthase to methionine; choline/betaine can remethylate homocysteine via homocysteine methyltransferase to form methionine. Pyridoxine promotes conversion of HYC via cystathionine beta-synthase to cystathionine. The amino acid N-acetyl-cysteine (NAC) binds to HYC for efficient renal excretion of NAC-HYC.[488]

- **Tryptophan and 5-hydroxytryptophan (5-HTP)**: Tryptophan is an amino acid found in many foods and is essential for human health and survival. Tryptophan is available as a nutritional supplement only by a doctor's prescription; it is available over-the-counter in a nonprescription supplement in the form of 5-hydroxytryptophan (5-HTP), which is commonly sourced from the seeds of *Griffonia simplicifolia*, a woody climbing shrub native to West and Central Africa. Tryptophan is the precursor to the neurotransmitter serotonin, which has antidepressant, anti-anxiety, and analgesic properties. Patients with FM are known to have low blood levels (i.e., functional nutritional insufficiency) of tryptophan, and the severity of the deficiency

---

[486] "Patients who consume antiepileptic drugs are susceptible to high levels of homocysteine and low levels of folate in the blood." Paknahad et al. Effects of Common Anti-epileptic Drugs on the Serum Levels of Homocysteine and Folic Acid. *Int J Prev Med*. 2012 Mar;3(Suppl 1):S186-90
[487] Regland et al. Response to vitamin B12 and folic acid in myalgic encephalomyelitis and fibromyalgia. *PLoS One*. 2015 Apr 22;10(4):e0124648
[488] "NAC intravenous administration induces an efficient and rapid reduction of plasma thiols, particularly of Hcy; our data support the hypothesis that NAC displaces thiols from their binding protein sites and forms, in excess of plasma NAC, mixed disulphides (NAC-Hcy) with an high renal clearance." Ventura et al. N-Acetyl-cysteine reduces homocysteine plasma levels after single intravenous administration by increasing thiols urinary excretion. *Pharmacol Res*. 1999 Oct;40(4):345-50

correlates with the severity of pain.[489,490,491] Blood levels of serotonin are often below normal in FM patients.[492] The accepted *medical-pharmacological* use of selective serotonin reuptake inhibitors (SSRI) drugs to treat the pain, depression, and anxiety associated with FM supports the use of 5-HTP to raise serotonin levels *naturally* by correcting the underlying nutritional insufficiency. As an over-the-counter nutritional supplement, the 5-hydroxylated form of tryptophan (5-HTP) has been used clinically and in numerous research studies. **Supplementation with 5-HTP has been shown to significantly alleviate symptoms of fibromyalgia.**[493] Commonly used doses range from 50 to 300 mg/d, with larger doses divided throughout the day. If tryptophan rather than 5-HTP is used, results are improved when taken on an empty stomach with carbohydrate (such as honey or fruit juice) to induce insulin secretion, which preferentially promotes uptake of tryptophan into the brain. Deficiency of either magnesium or vitamin B6 impairs conversion of 5-HTP into serotonin, and therefore the interventional program must ensure nutritional supra-sufficiency.

- Primary fibromyalgia syndrome and 5-hydroxy-L-tryptophan: a 90-day open study (*Journal of Internal Medicine Research* 1992 Apr[494]): An open 90-day study in 50 fibromyalgia patients showed significant improvement in all measured parameters (number of tender points, anxiety, pain intensity, quality of sleep, fatigue) after treatment with 5-HTP; global clinical improvement assessed by the patient and the investigator indicated a "good" or "fair" response in nearly 50% of the patients during the treatment period.

- Double-blind study of 5-hydroxytryptophan versus placebo in the treatment of primary fibromyalgia syndrome (*Journal of Internal Medicine Research* 1990 May-Jun[495]): A double-blind, placebo-controlled study using 5-HTP in 50 fibromyalgia patients showed significant improvement in all measured parameters, with only mild and transient side effects. *Note by DrV: Again, the common dose range for 5-HTP is 50 to 300 mg per day with doses greater than 100 mg generally best divided throughout the day (e.g., 50 mg thrice per day). I generally recommend starting with 50-100 mg about one hour before bedtime, then adding incremental additions of 50 mg throughout the day for a maximum daily dose of 300 mg. Effectiveness is increased with additional supplementation with magnesium, vitamin B6 (pyridoxine), and the fatty acids found in fish oil (EPA and DHA).*

- **Magnesium**: Magnesium deficiency is epidemic in industrialized societies due to insufficient dietary intake (e.g., from mineral water and leafy green vegetables) and concomitant metabolic-urinary acidosis, which increases urinary magnesium loss.[496,497] Additional causes of magnesium deficiency in fibromyalgia patients include vitamin D deficiency, malabsorption due to SIBO, and the stress of chronic illness. Magnesium deficiency exacerbates the symptoms of fibromyalgia by contributing to impairment of energy/ATP production in skeletal muscle, increased muscle tone and spasms (hypomagnesemic tetany), and anxiety and increased pain sensitivity—hyperalgesia via NMDA receptor overstimulation and neurocortical hyperexcitability. Magnesium deficiency also promotes constipation and intestinal stasis, which exacerbates SIBO. Magnesium supplementation (600 mg or to bowel tolerance to a limit of 1,500 mg in divided doses [bowel tolerance is defined as the dose—commonly of magnesium or vitamin C—that produces slightly loose stools due to the osmotic laxative effect]) should be used routinely in fibromyalgia patients; the primary cautions with magnesium use are renal insufficiency and the use of magnesium-sparing drugs such as the diuretic drug spironolactone. Modest benefits demonstrated in clinical trials with magnesium and malic acid[498] can easily be exceeded with concomitant interventions to address vitamin D deficiency, SIBO, and mitochondrial dysfunction.

- **Vitamin C, ascorbate**: Textbooks[499] and recent metaanalyses[500,501] consistently advocate use of ascorbic acid (ascorbate) 500-1,500 mg/d x50 days for the prevention of CRPS following trauma or surgery, especially to/of

---

[489] Moldofsky H, Warsh JJ. Plasma tryptophan and musculoskeletal pain in non-articular rheumatism ("fibrositis syndrome"). *Pain*. 1978 Jun;5(1):65-71
[490] Yunus MB, Dailey JW, Aldag JC, Masi AT, Jobe PC. Plasma tryptophan and other amino acids in primary fibromyalgia: a controlled study. *J Rheumatol*. 1992 Jan;19(1):90-4
[491] Russell IJ, Michalek JE, Vipraio GA, Fletcher EM, Wall K. Serum amino acids in fibrositis/fibromyalgia syndrome. *J Rheumatol* Suppl. 1989 Nov;19:158-63
[492] Wolfe F, Russell IJ, Vipraio G, Ross K, Anderson J. Serotonin levels, pain threshold, and fibromyalgia symptoms in the general population. *J Rheumatol*. 1997;24(3):555-9
[493] Caruso I, et al. Double-blind study of 5-hydroxytryptophan versus placebo in the treatment of primary fibromyalgia syndrome. *J Int Med Res*. 1990 May-Jun;18(3):201-9
[494] Sarzi Puttini P, Caruso I. Primary fibromyalgia syndrome and 5-hydroxy-L-tryptophan: a 90-day open study. *J Int Med Res*. 1992 Apr;20(2):182-9
[495] Caruso I et al. Double-blind study of 5-hydroxytryptophan versus placebo in the treatment of primary fibromyalgia syndrome. *J Int Med Res*. 1990 May-Jun;18(3):201-9
[496] Cordain L, et al. Origins and evolution of the Western diet: health implications for the 21st century. *Am J Clin Nutr*. 2005 Feb;81(2):341-54
[497] Rylander R, Remer T, Berkemeyer S, Vormann J. Acid-base status affects renal magnesium losses in healthy, elderly persons. *J Nutr*. 2006 Sep;136(9):2374-7
[498] Russell et al. Treatment of fibromyalgia syndrome with Super Malic: randomized, double blind, placebo controlled, crossover pilot study. *J Rheumatol*. 1995 May;22(5):953-8
[499] Papadakis, Maxine; McPhee, Stephen J.; Rabow, Michael W. *Current Medical Diagnosis and Treatment 2014*. McGraw-Hill Education.
[500] Shibuya et al. Efficacy and safety of high-dose vitamin C on complex regional pain syndrome in extremity trauma and surgery. *J Foot Ankle Surg*. 2013 Jan-Feb;52(1):62-6
[501] Meena et al. Role of vitamin C in prevention of complex regional pain syndrome after distal radius fractures. *Eur J Orthop Surg Traumatol*. 2015 May;25(4):637-41

the upper or lower limb. Mechanisms of action include the antioxidant effect[502] (thereby protecting the microvasculature following trauma) and the healing effect (promotion of bone and connective tissue repair, via hydroxyproline). Ascorbate also promotes mitochondrial function at cytochrome c, between complexes 3 and 4. My personal hypothesis is that ascorbate provides analgesic benefits via enhancement of central dopaminergic mechanisms and—via its ability to lower histamine levels (38% reduction following oral administration of ascorbate 2g/d)[503]—alleviation of neurogenic inflammation's vicious cycle, within which histamine plays a major role.[504] Appreciating its safety and efficacy in treating neuropathic, postsurgical, and migraine pain[505,506], I think all patients with pain should receive ascorbate 2-6 grams daily in divided doses. Two potential contraindications to vitamin C supplementation are iron overload and renal insufficiency.

- **S-adenosylmethionine (SAMe)**: Studies using oral or intravenous administration of the nutritional supplement SAMe have reported conflicting results; however, the overall trend seems to indicate that SAMe (800 mg/d orally) is safe and beneficial in the treatment of fibromyalgia.[507] SAMe helps maintain mitochondrial function by preserving glutathione, and its contribution of methyl groups is important for the regulation of gene expression and neurotransmitter synthesis. *Comment: I do not regularly use this supplement, and I would only use it as a last resort if nothing else had worked or if a particular patient had a specific indication for this supplement.*

INFECTIONS & DYSBIOSIS As reviewed in a previous section, fibromyalgia patients have a remarkably high prevalence of occult SIBO, the severity of which directly correlates with the severity of FM symptomatology and the eradication of which directly correlates with alleviation of FM. SIBO is the single primary pathoetiologic mechanism that explains each and every abnormality seen in this condition; the response of fibromyalgia patients to gastrointestinal-specific nonabsorbable antimicrobial treatments such as rifaximin provides diagnostic proof of effective treatment of the causative SIBO. Jejunal aspiration is expensive, inconvenient, cumbersome, invasive, and not completely sensitive, while breath hydrogen and methane testing is also cumbersome and relatively expensive (especially when compared to making a clinical diagnosis and confirmatory treatment) and is likewise not completely reliable. Post-prandial gas and bloating is a reliable clinical indicator of SIBO, and empiric treatment of SIBO with a low-carbohydrate diet (LCD) and antimicrobial agents that results in clinical improvement (which may include alleviation of musculoskeletal pain, improved cognition, alleviation of fatigue, and—especially in elderly patients—alleviation of malabsorption and malnutrition) confirms the diagnosis. Stated more plainly, effective implementation of LCD with antimicrobial treatment is both diagnostic and therapeutic; this allows the diagnosis to be made efficiently and with high specificity, (sensitivity depends upon efficacy) and bypassing expensive/insensitive/nontherapeutic/cumbersome diagnostic methods expedites physicians' efficacy and patients' relief in a manner that is safe and cost-effective, especially when compared to perpetual nontherapeutic symptomatic polypharmacy.

- **Low-carbohydrate diet, specific-carbohydrate diet**: Patients can follow a diet that emphasizes consumption of low-carbohydrate vegetables, nuts, and seeds and excludes grains (especially wheat, which is very highly fermentable), starches from foods such as potatoes, and disaccharides such as lactose and sucrose; most of the characteristics of a competent low-carbohydrate diet can be achieved within a "Paleo diet" such as described and popularized by Cordain. Alternatively or additionally, patients can follow the specific-carbohydrate diet described and popularized by Gottschall.
- **Probiotics**: Probiotics are beneficial bacteria that can be consumed in foods or as nutritional supplements to populate the gut, particularly following antibiotic use or long-term dietary neglect. In addition to their availability in capsules and powders, probiotics are widely consumed in the form of yogurt, kefir, and other cultured foods, and they have an excellent record of safety. Probiotic supplements are available in different strengths (quantity), potencies (viability), and combinations of bacteria (diversity). Some probiotics also contain fermentable carbohydrates (prebiotics) such as fructooligosaccharides (FOS) and inulin, which are substrates to nourish the beneficial bacteria. From a practical clinical perspective, the clinician can choose probiotic foods and supplements and instruct the patient to use these on an ongoing, periodic, or rotational basis. Probiotics

---

[502] Kapoor S. Vitamin C and its emerging role in pain management: beneficial effects in pain conditions besides post herpetic neuralgia. *Korean J Pain*. 2012 Jul;25(3):200-1
[503] Johnston CS, Martin LJ, Cai X. Antihistamine effect of supplemental ascorbic acid and neutrophil chemotaxis. *J Am Coll Nutr*. 1992 Apr;11(2):172-6
[504] Rosa AC, Fantozzi R. The role of histamine in neurogenic inflammation. *Br J Pharmacol*. 2013 Sep;170(1):38-45
[505] Hasanzadeh Kiabi et al. Can vitamin C be used as an adjuvant for managing postoperative pain? A short literature review. *Korean J Pain*. 2013 Apr;26(2):209-10
[506] Mohseni M. Use of vitamin C as placebo in anesthesiology. *Anesth Pain Med*. 2013 Winter;2(3):141
[507] Leventhal LJ. Management of fibromyalgia. *Ann Intern Med*. 1999 Dec 7;131(11):850-8

(i.e., bacteria only) may have a therapeutic advantage over prebiotics or synbiotics (probiotics+prebiotics) when treating SIBO because the fermentable carbohydrate in prebiotics and synbiotics may exacerbate the preexisting bacterial overgrowth by providing already overpopulated bacteria with additional substrate. The benefits of probiotic supplementation have been demonstrated in patients with IBS, rotavirus infection, eczema and increased intestinal permeability, and SIBO associated with renal failure. To date, no studies using probiotics in the treatment of fibromyalgia have been published.

- **Antimicrobial agents**: Antimicrobial agents can be categorized as either natural or pharmaceutical, and as absorbable and systemic or nonabsorbable and gastrointestinal-specific. These can be used empirically, as such treatment for suspected SIBO is well documented in the peer-reviewed clinical medicine literature; however, clinicians must always consider risk-to-benefit ratios especially when using the pharmaceutical antimicrobials which can induce systemic adverse effects (e.g., drug allergy or Stevens-Johnson syndrome or quinolone tendonopathy) or gastrointestinal adverse effects (e.g., nonspecific diarrhea, yeast overgrowth, *Clostridium difficile* diarrhea). Clinicians must always determine the proper choice and dose of therapeutic agents per patient. Two of the best and most important articles on the subject of SIBO are "Lin HC. Small intestinal bacterial overgrowth: a framework for understanding irritable bowel syndrome. *JAMA* 2004 Aug" (concepts and system-wide pathophysiology) and "Malik et al. Diagnosis and pharmacological management of small intestinal bacterial overgrowth in children with intestinal failure. *Can J Gastroenterol* 2011 Jan" (excellent sections on clinical diagnosis and pharmacologic management emphasizing rotational implementation of gut-specific antimicrobials). I prefer to use natural antimicrobial agents continuously for an extended period of time either alone or in conjunction with pharmaceutical antibacterial drugs, which I tend to use on a rotating basis of 7-14 days. Occasionally I will use a short course of an antiparasitic drug such as metronidazole or tinidazole, and I nearly always implement an extended course of antifungal treatment—either oregano oil or nystatin as nonabsorbable agents—punctuated by fluconazole/Diflucan if I suspect treatment resistant gastrointestinal yeast or any dermatologic or sinorespiratory yeast. Clinicians should appreciate that *yeast* colonization of the intestines promotes *bacterial* colonization of the intestines via—for example—elaboration by *Candida albicans* of a sIgA-protease and gliotoxin, an appreciated immunosuppressant. The following list emphasizes the antimicrobial agents I most commonly utilize, always in conjunction with nutritional supplementation (to restore immune function and mucosal defenses) and reduction in dietary carbohydrate intake (to reduce fermentable substrate and thereby "starve the microbes"). Although some dosage and duration suggestions are provided, the clinical reality is that patients need to be treated with *"dose and duration to effect"* or *"titrate to effect"*—meaning that the milligram dose per day and the duration of treatment can and should be customized per the patient's response to treatment. Combination therapy (i.e., more than one treatment at a time), prolonged therapy (treatment of chronic [poly]dysbiosis generally requires longer duration of treatment than does treatment of acute monomicrobial infections), and periodic/punctual treatment (for exacerbations and recurrences).
    - Clinical support favoring "open label" empiric antimicrobial treatment of SIBO in patients with fibromyalgia (*Current Pain and Headache Reports* 2004 Oct[508]): This article discusses the results of two experiments using antibiotics in the treatment of FM: ❶ 96 patients with SIBO diagnosed by lactulose hydrogen breath testing (LHBT) were offered **antibiotic treatment** for the reduction of gastrointestinal bacteria; 25 of the 96 patients returned for a follow-up LHBT. **Neomycin** was the most commonly used antibiotic. Eleven of the 25 patients achieved complete transient eradication of SIBO after antibiotic

---

[508] Wallace DJ, Hallegua DS. Fibromyalgia: the gastrointestinal link. *Curr Pain Headache Rep.* 2004 Oct;8(5):364-8

treatment and experienced better improvement in more of their FM symptom scores when compared with the patients who did not achieve complete eradication. This indicates that a direct relationship exists between the presence of SIBO and intestinal and extraintestinal symptoms in fibromyalgia, and that FM can be alleviated by effective antimicrobial/antibiotic treatment. ❷ In this double-blind trial of eradication of SIBO in fibromyalgia, 46 patients fulfilling the established criteria for FM were tested for SIBO using LHBT. Forty-two of the 46 patients (91.3%) were positive for SIBO and were randomized to receive placebo or 500 mg of **liquid neomycin** (a minimally-absorbed gastrointestinal-specific antibiotic drug) twice daily for 10 days. Only six of the 20 patients (30%) in the neomycin group achieved eradication (indicating inefficacy of treatment); thus, no statistically significant difference between groups was available for analysis. Thereafter, 28 patients in the double-blind study testing positive for SIBO went on to receive **open-label antibiotic treatment** to eradicate SIBO, and this time 17 of the 28 patients (60.7%) achieved eradication of SIBO. When these 23 patients were compared with the 15 patients who failed to eradicate or did not undergo open-label treatment, significant improvement attributable to antibiotic treatment in the FM scores was detected. Results suggest that eradication of bacterial overgrowth results in a statistically and clinically significant alleviation of FM symptoms.

- Nonprescription antimicrobial agents—examples: Most of the agents listed below have their primary or exclusive area of effectiveness within the lumen of the gastrointestinal tract. These agents are generally broad-spectrum and nonspecific, which is perfectly appropriate when treating nonspecific SIBO.
- Prescription-restricted antimicrobial agents—examples: In this section specific for the SIBO of fibromyalgia, nonabsorbable agents (rifaximin, vancomycin, nystatin) are appropriate; agents with systemic absorption (Augmentin and fluconazole) are listed here due to their high efficacy and frequent clinical utilization.
    - Rifaximin/Xifaxan: Rifaximin (gut-specific antibacterial drug) should not be confused with rifampin (systemic antibacterial drug, often used in the treatment of mycobacterium infections such as tuberculosis but also used for other bacterial infections); remember to "get your **facts/fax** right by using ri**fax**imin/Xi**fax**an" while you "use rif**amp**in to **amp**lify the effectiveness of systemic antibiotics in the treatment of chronic infections but it also **amps up** cytochrome p450 and adverse drug effects."
    - Vancomycin orally administered: Clinicians should consider this non/poorly-absorbed antibiotic which is effective against Gram-positive bacteria. Human studies have shown effectiveness in the treatment of IBS, constipation, and primary sclerosing cholangitis. In one particularly remarkable case of a patient with rheumatoid arthritis, I prescribed 125mg/d with great success with the intention to target Gram-positive Th17-inducing segmented filamentous bacteria.
    - Augmentin 1-2g BID: This combination of amoxicillin and clavulanate shows efficacy against more than 90% of gastrointestinal bacteria which contribute to SIBO.

A practical summary of SIBO: small intestine bacterial overgrowth

1. Definition: Generalized nonspecific overpopulation of bacteria (commonly with other microbes such as yeast) in the small intestine (and large intestine, too).
2. Frequency: Very common in clinical practice and the general population.
3. Primary symptoms: Gas and bloating, especially after carbohydrate consumption; may also have constipation and/or diarrhea.
4. Secondary symptoms: Fatigue, muscle aches, difficulty with concentration and cognition ("brain fog"), nutritional deficiencies due to malabsorption, immune activation due to absorption of microbial debris and metabolites, muscle pain due to dysbiotic mitochondriopathy and LPS- and cytokine-induced central sensitization.
5. Diagnosis: ❶ Based on the symptoms above, ❷ jejunal aspiration is the gold standard but is expensive, cumbersome, and potentially hazardous, ❸ measurement of fermentation products (hydrogen and methane) in breath following consumption of a carbohydrate such as glucose, sucrose, or lactulose; the amount of "gas" produced is proportional to the bacterial population, ❹ may find elevated short chain fatty acids (SCFA) in stool or elevated folate in blood, but not all cases of SIBO produce high levels of SCFA or folate, ❺ clinical response to low-carbohydrate diet and/or antibiotic drugs or antimicrobial herbs. The current author (AV) uses #1 in conjunction with #5 most commonly.
6. Treatments: Low-carbohydrate diet with antibiotic drugs (e.g., Xifaxan/Rifaximin (200 or 550 mg each) 400-550 mg tid po [1,200-1,650 mg daily] for 10-30 days) or antimicrobial herbs (e.g., time-released emulsified oregano oil 600 mg daily for 4-6 weeks, and/or berberine 400-1,500 mg daily for 4 weeks). Restoration of normal flora with probiotics, plant-based dietary diversity and authentically fermented foods.

- Nystatin 500,000 units BID-TID PO duration as needed (e.g., 1-6 months empirically or with any use of antibacterial drugs): Nystatin is a safe nonabsorbable gentle and commonly effective antifungal agent originally derived from a natural source of soil microorganisms. Its lack of significant intestinal absorption reduces the incidence of adverse effects while also prohibiting systemic antifungal effectiveness, except for reducing the total microbial load (TML) by reducing gastrointestinal fungal population. Nystatin is safe for long-term use, is inexpensive, and should generally be used anytime that antibiotic/antibacterial drugs are employed.
- Fluconazole/Diflucan 100-150-200 mg every other day for 4-5 doses over 8-10 days: The long half-life of 30 hours allows discontinuous alternate-day dosing without loss of efficacy for most routine outpatient applications. This drug is absorbed systemically with excellent tissue penetration for the delivery of multifocal antifungal effectiveness (e.g., alleviation of sinus and genitourinary fungal infections/colonization)

NUTRITIONAL IMMUNOMODULATION My use of the term and technique "nutritional immunomodulation" refers to a specific protocol designed to induce epigenetic modifications in undifferentiated Th-0 cells for their preferential promotion into the T-regulatory (Treg) FOXp3+ phenotype while shifting immune (im)balance away from the proinflammatory Th-1, Th-2, and Th-17 phenotypes. The primary components of this protocol can be safely implemented in essentially any and all patients without adverse effect; these fundamental components include low-carbohydrate plant-based diet to promote a systemic anti-inflammatory state, vitamin D3, combination fatty acid supplementation for n-3 fatty acids and GLA, probiotics (note that the first four components of the protocol are already represented in the foundational five-part nutritional protocol), vitamin A, lipoic acid, green tea, and a low-sodium diet. More assertive antidysbiotic interventions to promote healthy microbial balance in the gastrointestinal lumen may include botanical/nutritional/pharmacologic antimicrobial interventions, with some preferential utilization of orally administered vancomycin based on research supporting its effectiveness against segmented filamentous bacteria which are specific inducers of the Th-17 phenotype. Since fibromyalgia in its pure form is not directly due to an immune imbalance in the way considered here, this component of the functional inflammology protocol is not specifically relevant; however, for the many patients with concomitant diagnoses of fibromyalgia with another systemic/inflammatory/autoimmune disease (such as diabetes mellitus, rheumatoid arthritis, multiple sclerosis, or psoriasis) then of course this nutritional immunomodulation protocol should be implemented.

DYSFUNCTIONAL MITOCHONDRIA The basic view that mitochondria are the "powerhouses" of the cell responsible for the formation of cellular energy in the form of ATP is what most people learn in high school biology, and little if any additional knowledge is added to medical physicians' appreciation of the diversity of mitchondria's biologic roles in medical school. Lack of appreciation of the importance of the role of mitochondrial in general and mitochondrial dysfunction in particular in health and disease has left a huge blind spot in the therapeutic vision of most clinicians; by failing to appreciate and correct mitochondrial dysfunction, clinicians have missed a valuable component to the treatment plans of many and probably most of their patients. In addition to the well-known role that mitochondria have in the formation of energy/ATP, mitochondria also play major roles in pancreatic insulin secretion, peripheral insulin reception, microbial surveillance, and maintenance of inflammatory balance, insofar as mitochondrial dysfunction clearly contributes to a pro-diabetic and insulin-insulin resistant state, as well as enhanced pro-inflammatory responsiveness to microbial (including viral) stimuli. Fibromyalgia is clearly identified with mitochondrial dysfunction, and while the secondary mitochondrial dysfunction is one of the major causes of fibromyalgic muscle pain and fatigue, the mitochondrial dysfunction does not itself cause fibromyalgia, the primary cause of which is SIBO. Thus, SIBO's generation of LPS, D-lactate, and other mitochondrial toxins is the primary/direct cause of the mitochondrial dysfunction; effective treatment must emphasize SIBO eradication and mitochondrial resuscitation. We can compartmentalize major components of mitochondrial structure and function into these three main components: ❶ citric acid cycle, ❷ electron transport chain, ❸ and the structural integrity of the inner and outer mitochondrial membranes. The main area of clinical importance can be discussed within a conversation of the electron transport chain (ETC) since this is fed by the citric acid cycle and is structurally interwoven into the inner mitochondrial membrane and fully dependent upon the nonpermeability of the outer mitochondrial membrane for the maintenance of the electromechanical proton gradient. Primary treatment must always be directed at the primary cause of any disease—not its secondary complications; in the case of FM, the SIBO must always be treated. Among mitochondria-specific treatments for fibromyalgia, supplementation with CoQ10, melatonin, and acetyl-carnitine (preferably with lipoic acid) are the best studied and most efficacious.

- Coenzyme Q10 (CoQ10): An endogenous antioxidant, vitamin-like substance, and essential component of the mitochondrial electron transport chain, oral supplementation with CoQ10 has been used therapeutically in numerous studies for the successful treatment of migraine, heart failure, hypertension, and renal failure. Additional data have shown immunomodulatory roles for CoQ10, and many clinicians employ it as adjunctive treatment for viral infections, cancer, and allergies.[509,510] The electron transport chain is the terminal step in mitochondrial energy/ATP production; as readers can see in the following diagram, each step or "complex" of the electron transport chain requires nutrients, without which energy/ATP production will be impaired, and provision of which (via supplementation) will generally enhance mitochondrial energy/ATP production. Per Cordero et al[511] in 2012, **CoQ10 levels are 40% lower in blood cells of patients with FM compared with levels in healthy persons**, and reduced levels of CoQ10 correlate with markers associated with expedited destruction of mitochondria (mitophagy).
    - Clinical investigation: Mitochondrial dysfunction and mitophagy activation in blood mononuclear cells of fibromyalgia patients (*Arthritis Research Therapy* 2010 Jan[512]): The authors studied 2 male and 18 female FM patients and 10 healthy controls. They evaluated mitochondrial function in blood mononuclear cells from FM patients measuring CoQ10 levels with high-performance liquid chromatography (HPLC) and measuring mitochondrial membrane potential with flow cytometry. Oxidative stress was determined by measuring mitochondrial superoxide production and lipid peroxidation in blood mononuclear cells and plasma from FM patients. Autophagy activation was evaluated in blood mononuclear cells; mitophagy was confirmed by measuring citrate synthase activity and electron microscopy examination of blood mononuclear cells. The authors **found reduced levels of CoQ10, decreased mitochondrial membrane potential, increased levels of mitochondrial superoxide in blood mononuclear cells (indicating increased oxidative stress and reduced antioxidant defense),** and increased levels of lipid peroxidation in both blood mononuclear cells and plasma from FM patients. Importantly, the authors note that "mitochondrial dysfunction was also associated with increased expression of autophagic genes and the elimination of dysfunctional mitochondria with mitophagy." *What this means in practical terms is that the biochemical aberrations that cause mitochondrial dysfunction lead to destruction of mitochondria via "mitophagy" which literally means "mitochondrial consumption", a process by which dysfunctional mitochondria are eliminated by degradative processes.*
    - Case series of FM patients treated with CoQ10 (*Mitochondrion* 2011 Jul[513]): The authors note that CoQ10 is an essential electron carrier in the mitochondrial respiratory chain and a strong antioxidant and that **low CoQ10 levels have been detected in patients with FM**. The authors found that "**FM patients with CoQ10 deficiency showed a statistically significant reduction in symptoms after CoQ10 treatment during 9 months (300 mg/day). Determination of deficiency and consequent supplementation in FM may result in clinical improvement.**" *This is a small but important study documenting 1) that CoQ10 deficiency is common in FM patients, and 2) that CoQ10 supplementation alleviates the clinical manifestations/symptoms of FM, consistent with the integrated model of FM presented in this book, which includes the components of nutrient deficiency and mitochondrial dysfunction. Although standardized blood testing for CoQ10 levels is widely available, testing for and documentation of CoQ10 deficiency is not necessary before the use of CoQ10 supplementation.*
    - Clinical trial using a combination of Ginkgo biloba and CoQ10 (*Journal of Internal Medicine Research* 2002 Mar[514]): In an open trial of 23 fibromyalgia patients, the combination of 200 mg CoQ10 and 200 mg *Ginkgo biloba* (for a total dose of 48 mg flavone glycosides and 12 mg terpene lactones) daily for 84 days was shown

---

[509] Gaby AR. The role of Coenzyme Q10 in clinical medicine: Part 1. *Altern Med Rev* 1996;1:11-17
[510] Gaby AR. The role of Coenzyme Q10 in clinical medicine: Part 2. *Altern Med Rev* 1996;1:168-175
[511] Cordero MD, De Miguel M, Moreno Fernández AM, et al. Mitochondrial dysfunction and mitophagy activation in blood mononuclear cells of fibromyalgia patients: implications in the pathogenesis of the disease. *Arthritis Res Ther*. 2010;12(1):R17
[512] Cordero et al. Mitochondrial dysfunction and mitophagy activation in blood mononuclear cells of fibromyalgia patients. *Arthritis Res Ther*. 2010;12(1):R17
[513] Cordero MD et al. Coenzyme Q(10): a novel therapeutic approach for Fibromyalgia? case series with 5 patients. *Mitochondrion*. 2011 Jul;11(4):623-5
[514] Lister. Open, pilot study to evaluate potential benefits of coenzyme Q10 combined with Ginkgo biloba extract in fibromyalgia syndrome. *J Int Med Res* 2002 Mar-Apr;30:195-9

Chapter 5.1—Functional Inflammology Protocol for Metabolic Inflammation: Migraine and Fibromyalgia

to provide clinical benefit in 64% of patients. CoQ10 is often deficient in FM patients, and this deficiency both *causes* and *results from* mitochondrial dysfunction; stated differently, CoQ10 depletion and mitochondrial dysfunction form a vicious cycle, a relationship of reciprocal causality. *Ginkgo biloba* extract is an extensively researched botanical medicine with a long history of safe and effective clinical use for various conditions, especially those associated with reduced blood flow and impaired mitochondrial function. *Ginkgo biloba* is a botanical/herbal medicine with a long history of human use; the three most important physiologic effects of *Ginkgo biloba* are ❶ vasodilation—improves blood circulation (which is often compromised in FM patients), ❷ improves mitochondrial function and ATP/energy production, and ❸ antioxidant benefits—quenches/absorbs free radicals, which are oxygen-containing molecules that cause damage to cell structures and body tissues. Given these therapeutic benefits, *Ginkgo* would appear to be a reasonable therapeutic agent to address the secondary pathophysiology in fibromyalgia. *Ginkgo biloba* products are generally standardized for the content of flavone glycosides (approximately 24%) and terpene lactones (approximately 6%) with adult doses ranging from 60-240 mg/d and generally 120 mg/d. *Comment by Dr Vasquez: Ginkgo biloba and CoQ10 are very safe and appropriate for use by nearly all FM patients.*

- Clinical investigation and clinical trial: Oxidative stress, headache symptoms in fibromyalgia and the role of CoQ10 in clinical improvement (*PLoS One* 2012 Apr[515]): The authors introduce this study by noting that FM is a chronic pain syndrome with "unknown etiology" and a wide spectrum of symptoms such as allodynia (perception of pain from stimuli that are not normally painful), debilitating fatigue, joint stiffness, and migraine headaches. The authors note a link between oxidative stress and the clinical symptoms in FM. In this study, the researchers examined oxidative stress and bioenergetic status in blood mononuclear cells (BMCs) and the association with headache symptoms in FM patients. Following this correlative analysis, the authors assessed the effects of oral CoQ10 supplementation on biochemical markers and clinical improvements. In 20 FM patients and 15 healthy controls, a variety of validated clinical and biochemical parameters was assessed; specifically for the biochemical component, measurements were performed for serum CoQ10, catalase, lipid peroxidation (LPO) levels and ATP levels in BMCs. In patients with FM, the authors found lower CoQ10 (CoQ10 deficiency), lower catalase (reduced antioxidant defenses) and lower ATP levels (reduced energy production) in BMCs while FM patients also showed elevated LPO (evidence of free-radical damage) in BMCs. Lower levels of CoQ10 and catalase levels in BMCs correlated with greater severity-frequency of headache. **In this clinical trial using CoQ10 300 mg/d for 3 months, CoQ10 supplementation caused significant reductions in pain and tender points, significant reductions in headache impact, significant elevations in cellular levels of CoQ10, a reduction in malondialdehyde (marker of lipid peroxidation) from 30nmol to 5 nmol (normal 6 nmol), an increase in catalase levels from 35 U/mg to 85 U/mg (normal 96 U/mg), and an increase in BMC production of ATP/energy from 61 nmol/mg to 191 nmol/mg (normal 202 nmol/mg).** Supplementation with CoQ10 300 mg/day divided in three doses for 3 months "restored biochemical parameters and induced a significant improvement in clinical and headache symptoms." *Note by Dr Vasquez: The dose of CoQ10 used clinically is generally approximately 100 mg per day, and occasionally a patient or doctor might decide to use a higher dose, which might be up to 300 mg per day. Higher doses generally provide better results with excellent safety, but CoQ10 tends to be one of the most expensive nutritional supplements and as such the lowest effective dose—again approximately 100 mg/d as the standard—is used. Some patients will not respond to 100 mg/d and will respond well to 300 mg/d; these more challenging patients might also need additional/different treatments, such as supplementation with synergistic nutrients, hormonal correction, xenobiotic depuration, or assertive treatment of dysbiosis/SIBO.*

---

[515] Cordero et al. Oxidative stress correlates with headache symptoms in fibromyalgia: coenzyme Q_{10} effect on clinical improvement. *PLoS One*. 2012;7(4):e35677

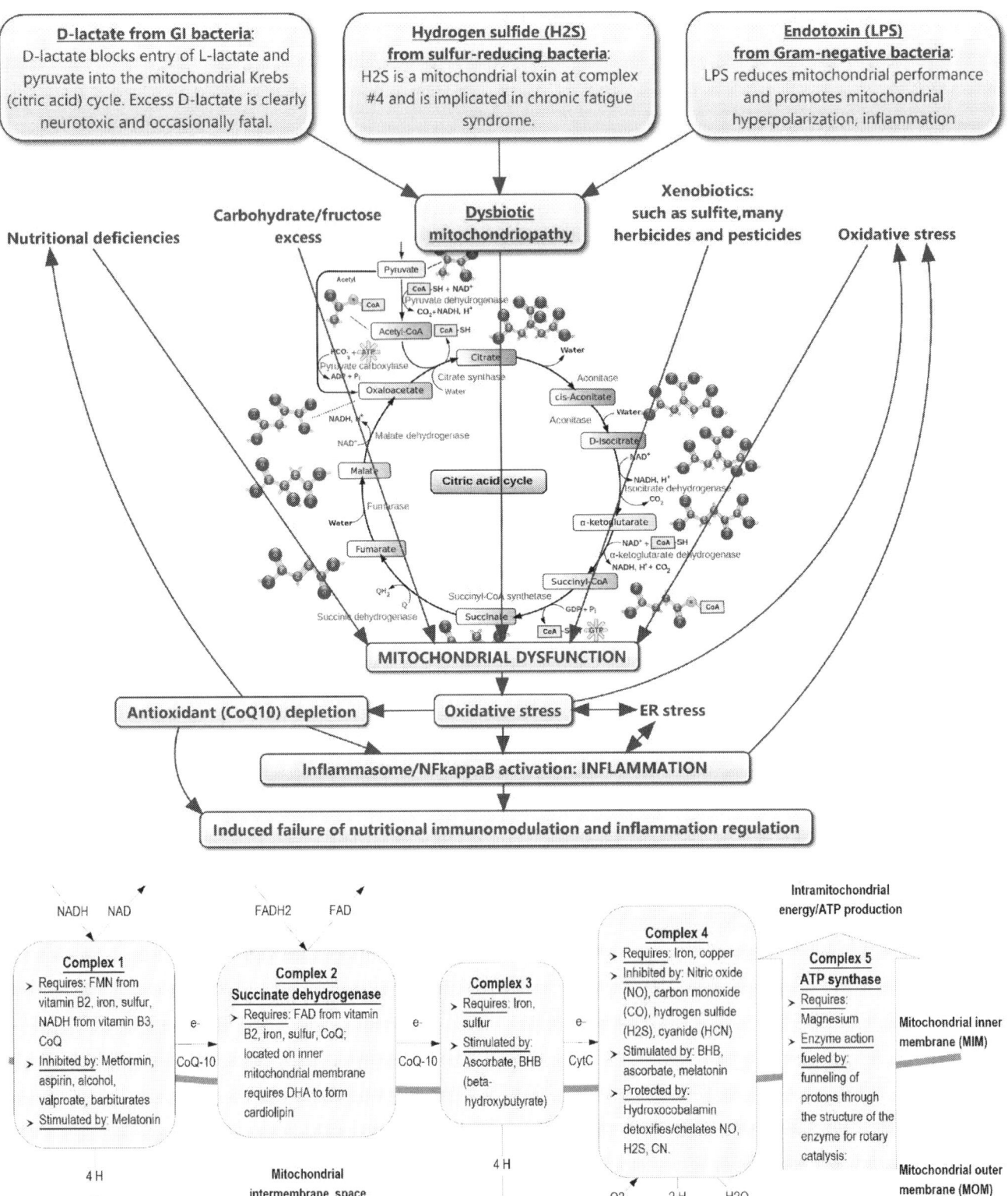

Microbial debris (LPS) and metabolites (D-lactate and H2S) and the subsequent inflammatory response lead to mitochondrial dysfunction: Nutritional deficiencies, genetic faults, and xenobiotic exposure/accumulation exacerbate mitochondrial impairment and also exacerbate the peripheral and central inflammatory responses. Copyright © 2015 by Dr Alex Vasquez. All rights reserved and enforced; this image may not be used, copied, or distributed without written permission. Citric acid cycle from en.wikipedia.org/wiki/Citric_acid_cycle. "Citric acid cycle with aconitate 2" by Narayanese, WikiUserPedia, YassineMrabet, TotoBaggins licensed under Creative Commons Attribution-Share Alike 3.0. See www.inflammationmastery.com/pain

- NLRP3 inflammasome is activated in fibromyalgia and reduced by coenzyme Q10 (*Antioxid Redox Signal* 2014 Mar[516]): "Mitochondrial dysfunction was accompanied by increased protein expression of interleukin (IL)-1, NLRP3 (NOD-like receptor family, pyrin domain containing 3) and caspase-1 activation, and an increase of serum levels of proinflammatory cytokines (IL-1 and IL-18). CoQ10 deficiency induced by p-aminobenzoate treatment in blood mononuclear cells and mice showed NLRP3 inflammasome activation with marked algesia. A placebo-controlled trial of CoQ10 in FM patients has shown a reduced NLRP3 inflammasome activation and IL-1 and IL-18 serum levels. ...CONCLUSION: These findings provide new insights into the pathogenesis of FM and suggest that NLRP3 inflammasome inhibition represents a new therapeutic intervention for the disease. ... After CoQ10 supplementation [300 mg/day CoQ10 divided into three doses], NLRP3 and IL-1 gene were downregulated. ... IL-1 and IL-18 serum levels were significantly reduced with respect to placebo." The most complete/consistent model of FMS is that it is caused by SIBO, leading to mitochondrial dysfunction, central sensitization, tryptophan/serotonin/melatonin deficiencies; these interconnections are illustrated later in this chapter and detailed elsewhere.[517]
  - Mitochondrial dysfunction in CRPS: Patients with CRPS show evidence of impaired oxygen diffusion and mitochondrial impairment[518], and Tan et al[519] specifically noted that mitochondrial ETC "complex II activity in the CRPS I patients was significantly lower." From a "mitochondrial micromanagement" perspective, one might consider the use of riboflavin 400 mg/d and CoQ10 100-300 mg/d to support ETC complex #2, but obviously the entire mitochondria—indeed the entire body—works together as a unit.
- Melatonin: Melatonin is a hormone produced in the pineal gland of the brain; melatonin is synthesized from the neurotransmitter serotonin, and production of both serotonin and melatonin are dependent on the nutritional availability of tryptophan and/or 5-HTP as discussed above. Patients with FM show decreased nocturnal secretion of melatonin.[520] Melatonin benefits FM patients through a wide range of mechanisms, including promotion of restful sleep and reduction in LPS-induced mitochondrial impairment. As a powerful antioxidant, melatonin scavenges oxygen and nitrogen-based reactants generated in mitochondria and thereby limits the loss of intramitochondrial glutathione, the most important component of antioxidant defense; this prevents damage to mitochondrial protein and DNA. **Melatonin increases the activity of Complexes 1 and 4 of the mitochondrial electron transport chain, improving mitochondrial respiration and increasing ATP synthesis** under various physiological and experimental conditions.[521] Successful treatment with melatonin or its precursor tryptophan/5-HTP should not deter the clinician from addressing other contributing or causative problems such as vitamin D deficiency, gastrointestinal dysbiosis including SIBO, magnesium deficiency, and chronic psychoemotional stress. The adult physiologic dose which mimics natural internal (endogenous) production is approximately 200-500 mcg [micrograms] nightly. In adults, supplementation with melatonin has a wide therapeutic index and is used safely and effectively in doses up to 20 to 40 mg [milligrams] nightly.
  - Case series (n=4): Melatonin therapy in fibromyalgia (*Journal of Pineal Research* 2006 Jan[522]): Melatonin (3–6 mg per night, administered orally 1 hour before bedtime) has been reported to normalize sleep, alleviate pain and fatigue, and resolve many other clinical manifestations of FM. The authors report, "After 15 days of treatment with melatonin, all patients developed a sleep/wake cycle that was considered normal. They also mentioned a significant reduction of pain. At this time, the patients were taken off hypnotics. Thirty days after the initiation of melatonin, other medications were withdrawn and thereafter they only took melatonin." *Comment by Dr Vasquez: These results are impressive, but—again—the other components of FM such*

---

[516] Cordero et al. NLRP3 inflammasome is activated in fibromyalgia: the effect of coenzyme Q10. *Antioxid Redox Signal.* 2014 Mar 10;20(8):1169-80
[517] Vasquez A. *Naturopathic Rheumatology and Integrative Inflammology v3.5*, 2014 and *Fibromyalgia in a Nutshell*, 2012
[518] "The mean venous oxygen saturation (S(v)O(2)) value (94.3% ± 4.0%) of the affected limb was significantly higher than S(v)O(2) values found in healthy subjects (77.5% ± 9.8%) pointing to a severely decreased oxygen diffusion or utilization within the affected limb. ... Ultrastructural investigations of soleus skeletal muscle capillaries revealed thickened endothelial cells and thickened basement membranes. Muscle capillary densities were decreased in comparison with literature data. High venous oxygen saturation levels were partially explained by impaired diffusion of oxygen due to thickened basement membrane and decreased capillary density. ...The abnormal skeletal muscle findings points to severe disuse but only partially explain the impaired diffusion of oxygen; mitochondrial dysfunction seems a likely explanation in addition." Tan et al. Impaired oxygen utilization in skeletal muscle of CRPS I patients. *J Surg Res.* 2012 Mar;173(1):145-52
[519] "We observed that mitochondria obtained from CRPS I muscle tissue displayed reduced mitochondrial ATP production and substrate oxidation rates in comparison to control muscle tissue. Moreover, we observed reactive oxygen species evoked damage to mitochondrial proteins and reduced MnSOD levels." Tan et al. Mitochondrial dysfunction in muscle tissue of complex regional pain syndrome type I patients. *Eur J Pain.* 2011 Aug;15(7):708-15
[520] Wikner J, et al. Fibromyalgia—a syndrome associated with decreased nocturnal melatonin secretion. *Clin Endocrinol* (Oxf). 1998 Aug;49(2):179-83
[521] León J, Acuña-Castroviejo D, Escames G, Tan DX, Reiter RJ. Melatonin mitigates mitochondrial malfunction. *J Pineal Res.* 2005 Jan;38(1):1-9
[522] Acuna-Castroviejo D, Escames G, Reiter RJ. Melatonin therapy in fibromyalgia. *J Pineal Res.* 2006 Jan;40(1):98-9

as SIBO and CoQ10 deficiency should also be treated assertively to reduce the risk of relapse and to treat the underlying problems; good healthcare and good self-care should extend beyond mere symptom alleviation.
- Clinical trial (n=101): Adjuvant use of melatonin for treatment of fibromyalgia (*Journal of Pineal Research.* 2011 Apr[523]): group A (24 patients) treated with 20 mg/day fluoxetine alone; group B (27 patients) treated with melatonin 5 mg alone; group C (27 patients) treated with 20 mg fluoxetine plus 3 mg melatonin; group D (23 patients) treated with 20 mg fluoxetine plus 5 mg melatonin for 8 weeks. "Using melatonin (3 mg or 5 mg/day) in combination with 20 mg/day fluoxetine resulted in significant reduction in both total and different components of Fibromyalgia Impact Questionnaire score compared to the pretreatment values. In conclusion, **administration of melatonin, alone or in a combination with fluoxetine, was effective in the treatment of patients with FMS.**"

- Acetyl-L-carnitine (ALC): Acetyl-L-carnitine is a form of the amino acid L-carnitine, most notable for its critical role in supporting mitochondrial energy/ATP production by supporting the metabolism (beta oxidation) of fatty acids in the mitochondria. A large study with 102 patients showed that ALC (administered by oral and parenteral routes, 1500 mg/d) was beneficial in patients with fibromyalgia.[524] Given the role of ALC in supporting and improving mitochondrial function, this supplement probably benefits fibromyalgia patients by compensating for LPS-induced skeletal muscle dysfunction.

- D-ribose: D-ribose is a naturally occurring pentose carbohydrate available as a dietary supplement. When administered orally (5 g thrice daily), it safely provides numerous benefits to fibromyalgia patients, according to a recent pilot study with 41 patients.[525] Improvements are seen in energy, sleep, mental clarity, pain intensity, and well-being, as well as global assessment. Among its beneficial mechanisms of action is enhancement of mitochondrial ATP production. Thus, the benefits of D-ribose supplementation may be mediated by restoration or preservation of mitochondrial impairment caused by LPS in fibromyalgia patients.

- Creatine monohydrate: Skeletal muscle levels of phosphocreatine and ATP are reduced in patients with fibromyalgia compared with normal controls; thus, oral supplementation with creatine would appear to be an obvious intervention to restore these depressed levels to normal. Artimal et al[526] reported that a patient with severe refractory fibromyalgia attained sustained alleviation of depression and pain, as well as improvements in sleep and quality of life, following oral administration of creatine monohydrate for 4 weeks (3 grams daily in the first week, then 5 grams daily). Creatine supplementation has been shown to improve ATP production and oxygen utilization in brain and skeletal muscle in humans.[527]

- Oxygen: One-hundred percent oxygen delivered by facial mask at 8 L/min for 10 minutes can help abort an attack of cluster headache. Oxygen is the required electron and proton acceptor of the mitochondrial electron transport chain (ETC) for ATP production; thus, supraphysiologic oxygen, like supraphysiologic doses of mitochondria-specific nutrients, generally improves mitochondrial energy (ATP) production. The model of pain sensitization in so-called "chronic pain syndromes"—specifically migraine, cluster headache, fibromyalgia, myofascial pain syndrome and complex regional pain syndrome (CRPS)—that I have developed includes a tripartite vicious cycle of mitochondrial dysfunction, glial activation, and neuronal hyperexcitation, all of which promote brain inflammation, central sensitization, and perpetuation and amplification of pain. Strong support for this model comes from both its biochemical and physiologic rationale as well as the efficacy of corresponding treatments that address each of the main components: mitochondrial dysfunction, glial activation, neuronal hyperexcitation, brain inflammation.

  Appreciation of the efficacy of oxygen therapy—whether as normobaric (more available and affordable) or hyperbaric (more effective and more expensive; hyperbaric oxygen therapy [HBOT])—in various pain states invites us to revisit the naturopathic profession's hierarchy of therapeutics. Symptomatic therapy clearly has a role in patient care but, in order to avoid repeated use of urgent/emergency care and the creation of unnecessary medical dependency, repeated acute care or even "maintenance therapy" should never replace treatment of the underlying cause(s). Patients need to receive treatment aimed at the underlying pathophysiology so that health is optimized and patients are moved toward better general well-being and disease-specific health. Oxygen

---

[523] Hussain SA, Al-Khalifa II, Jasim NA, Gorial FI. Adjuvant use of melatonin for treatment of fibromyalgia. *J Pineal Res.* 2011 Apr;50(3):267-71
[524] Rossini M, et al. Double-blind, multicenter trial comparing acetyl l-carnitine with placebo in treatment of fibromyalgia patients. *Clin Exp Rheumatol.* 2007 Mar-Apr;25:182-8
[525] Teitelbaum JE, Johnson C, St Cyr J. The use of D-ribose in chronic fatigue syndrome and fibromyalgia: a pilot study. *J Altern Complement Med.* 2006 Nov;12:857-62
[526] Amital D, Vishne T, Rubinow A, Levine J. Observed effects of creatine monohydrate in a patient with depression and fibromyalgia. *Am J Psychiatry.* 2006 Oct;163(10):1840-1
[527] Watanabe A, Kato N, Kato T. Effects of creatine on mental fatigue and cerebral hemoglobin oxygenation. *Neurosci Res.* 2002 Apr;42(4):279-85

therapy is abortive of pain and shows some contribution to breaking the vicious cycles of mitochondrial impairment (immediate treatment of headaches, current-prospective treatment of fibromyalgia and CRPS) but this therapy does not address the other aspects of mitochondrial dysfunction (e.g., CoQ10 deficiency) and does not address the cause (e.g., small intestine bacterial overgrowth [SIBO] in fibromyalgia) and therefore should remain as supplemental, abortive/acute, and adjunctive therapy not as they foundation of therapy. Obviously, hyperbaric therapy makes more money for doctors/clinics and—(except/including) when patients buy their own home hyperbaric units—will therefore receive more press and more endorsement than therapies that are curative and empowering (ie, autonomous, without medical dependence). Given that most patients with persistent inflammation and pain are antioxidant deficient and therefore at increased risk for oxygen toxicity (including subclinical damage), antioxidant repletion should occur prior to oxygen therapy; the idea of administering supraphysiologic doses of oxygen to pull more protons and electrons through an *already damaged* and *pro-inflammatory* and *ROS-generating* electron transport chain is not meritorious, although some will defend it—weakly—on the theoretic basis of hormesis.

- High-flow oxygen therapy for **all types of headache** (*Am J Emerg Med* 2012 Nov[528]): "We performed a prospective, randomized, double-blinded, placebo-controlled trial of patients presenting to the ED with a chief complaint of headache. The patients were randomized to receive either 100% oxygen via nonrebreather mask at 15 L/min or the placebo treatment of room air via nonrebreather mask for 15 minutes in total. ... A total of 204 patients agreed to participate in the study and were randomized to the oxygen (102 patients) and placebo (102 patients) groups. Patient headache types included tension (47%), migraine (27%), undifferentiated (25%), and cluster (1%). Patients who received oxygen therapy reported significant improvement in visual analog scale scores at all points when compared with placebo: 22 mm vs 11 mm at 15 minutes, 29 mm vs 13 mm at 30 minutes, and 55 mm vs 45 mm at 60 minutes. ... In addition to its role in the treatment of cluster headache, high-flow oxygen therapy may provide an effective treatment of all types of headaches in the ED setting.

- Hyperbaric oxygen therapy for **fibromyalgia**; randomized n=60, crossover n=24 (*PLoS One.* 2015 May[529]): "The HBOT protocol comprised 40 sessions, 5 days/week, 90 minutes, 100% oxygen at 2ATA. ... HBOT in both groups led to significant amelioration of all FMS symptoms, with significant improvement in life quality. Analysis of SPECT imaging revealed rectification of the abnormal brain activity: decrease of the hyperactivity mainly in the posterior region and elevation of the reduced activity mainly in frontal areas. No improvement in any of the parameters was observed following the control period. CONCLUSIONS: The study provides evidence that HBOT can improve the symptoms and life quality of FMS patients. Moreover, it shows that HBOT can induce neuroplasticity and significantly rectify abnormal brain activity in pain related areas of FMS patients." Why would (hyperbaric) oxygen provide more benefit for fibromyalgia than for migraine, given the both are largely due to mitochondrial dysfunction?—Because in migraine, the mitochondrial dysfunction is generally low, and then acute with exacerbations, and it is most notable only in the brain; in contrast, in fibromyalgia, the mitochondrial dysfunction is more moderate-severe and therefore more amenable to treatment during the course of the disease (ie, one does not have to wait for an exacerbation or attack). Also in fibromyalgia, the mitochondrial dysfunction and the pain are both central in the brain as well as peripheral in the muscles; both locations contribute partly to the pain sensations, and oxygen therapy addresses both components, therefore leading to more opportunity for symptomatic improvement.

- Hyperbaric oxygen therapy for **fibromyalgia**; randomized n=50 (*J Int Med Res.* 2004 May[530]): "We conducted a randomized controlled study to evaluate the effect of hyperbaric oxygen (HBO) therapy in FMS (HBO group: n = 26; control group: n = 24). Tender points and pain threshold were assessed before, and after the first and fifteenth sessions of therapy. Pain was also scored on a visual analogue scale (VAS). There was a significant reduction in tender points and VAS scores and a significant increase in pain threshold of the

---

[528] Ozkurt et al. Efficacy of high-flow oxygen therapy in all types of headache: a prospective, randomized, placebo-controlled trial. *Am J Emerg Med.* 2012 Nov;30(9):1760-4
[529] Efrati et al. Hyperbaric oxygen therapy can diminish fibromyalgia syndrome—prospective clinical trial. *PLoS One.* 2015 May 26;10(5):e0127012
[530] Yildiz et al. A new treatment modality for fibromyalgia syndrome: hyperbaric oxygen therapy. *J Int Med Res.* 2004 May-Jun;32(3):263-7

HBO group after the first and fifteenth therapy sessions. There was also a significant difference between the HBO and control groups for all parameters except the VAS scores after the first session. We conclude that HBO therapy has an important role in managing FMS."
- Hyperbaric oxygen therapy for complex regional pain syndrome (*J Int Med Res*. 2004 May[531]): "In this double-blind, randomized, placebo-controlled study we aimed to assess the effectiveness of hyperbaric oxygen (HBO) therapy for treating patients with complex regional pain syndrome (CRPS). Of the 71 patients, 37 were allocated to the HBO group and 34 to the control (normal air) group. Both groups received 15 therapy sessions in a hyperbaric chamber. Pain, edema and range of motion (ROM) of the wrist were evaluated before treatment, after the 15th treatment session and on day 45. In the HBO group there was a significant decrease in pain and edema and a significant increase in the ROM of the wrist. When we compared the two groups, the HBO group had significantly better results with the exception of wrist extension. In conclusion, HBO is an effective and well-tolerated method for decreasing pain and edema and increasing the ROM in patients with CRPS."
- Hyperbaric oxygen therapy for myofascial pain syndrome (*J Natl Med Assoc* 2009 Jan[532]): "Thirty patients with the diagnosis of MPS were divided into HBO (n=20) and control groups (n=10). Patients in the HBO group received a total of 10 HBO treatments in 2 weeks. Patients in the control group received placebo treatment in a hyperbaric chamber. Pain threshold and visual analogue scale (VAS) measurements were performed immediately before and after HBO therapy and 3 months thereafter. Additionally, Pain Disability Index (PDI) and Short Form 12 Health Survey (SF-12) evaluations were done before HBO and after 3 months. HBO therapy was well tolerated with no complications. In the HBO group, pain threshold significantly increased and VAS scores significantly decreased immediately after and 3 months after HBO therapy. PDI, Mental and Physical Health SF-12 scores improved significantly with HBO therapy after 3 months compared with pretreatment values. In the control group, pain thresholds, VAS score, and Mental Health SF-12 scores did not change with placebo treatment; however, significant improvement was observed in the Physical Health SF-12 test. We concluded that HBO therapy may be a valuable alternative to other methods in the management of MPS."

SOCIOLOGY, SLEEP, STRESS, SOMATIC TREATMENTS, SWEAT/EXERCISE, SPECIAL SUPPLEMENTATION Common clinical and lifestyle considerations are listed in the following sections.

- Sociology/psychology, and stress management/reduction: Everyone—patients as well as clinicians—can benefit from developing self-awareness, emotional intelligence, and other core life skill and insights; since much of our perception of stress has a psycho-epistemological basis, enhanced self-awareness in this key area can help to deconstruct the phenomenon of stress and its secondary consequences. Because this consideration is self-evident in terms of safety, efficacy, broad applicability, and life-enhancement, specific literature will not be reviewed here.
- Sleep: Sleep deprivation induces immune suppression, enhanced sensitivity to pain, and an objectively documentable proinflammatory state evidenced by increases in serum hsCRP. Patients should be encouraged to optimize sleep by avoiding late-in-the-day exercise, overstimulation, caffeine (which generally has a half-life of six hours), and overuse of bright lights following nightfall; items that are conducive to sleep are having a dark and quiet room, relaxing music or reading, and using melatonin. Enhancement of sleep quality and duration have been shown to alleviate systemic inflammation, tendency toward insulin resistance, and pain perception/sensitivity.
- Sweating and exercise: Obesity/overweight and physical inactivity are consistently associated with elevated risk for and experience of depression, low self-esteem, social isolation, systemic inflammation, cardiometabolic disease and diabetes mellitus type-2, cancers of various types, and inflammatory disorders such as asthma and psoriasis. Weight optimization and physical activity promote enhanced self-confidence, self-efficacy, skill-building, social interaction, and reductions in cause-specific and all-cause mortality. Mechanistically, exercise—defined here as physical activity of sustained duration and intensity to promote diaphoresis/sweating—promotes lipolysis (for mobilization of adipose-stored toxins, weight reduction, and enhanced BHB production

---

[531] Kiralp et al. Effectiveness of hyperbaric oxygen therapy in the treatment of complex regional pain syndrome. *J Int Med Res*. 2004 May-Jun;32(3):258-62
[532] Kiralp et al A novel treatment modality for myofascial pain syndrome: hyperbaric oxygen therapy. *J Natl Med Assoc*. 2009 Jan;101(1):77-80

for induction of histone acetylation and ECT stimulation), promotes glycolysis (to promote induction of enhanced mitochondrial function and insulin sensitivity), and hyperventilation which results in respiratory alkalosis and secondary urinary alkalinization (which promotes mineral retention, xenobiotic excretion, endorphin elevation, and cortisol reduction). Commonly accepted international guidelines as well as common sense advocate 30-60 minutes of daily exercise that should globally include components such as aerobic training, resistance training, skill-building, balance, and flexibility; intensity, duration, and variety are tailored to patient needs and preferences. Because this consideration is self-evident in terms of safety, efficacy, broad applicability, and life-enhancement, specific literature will not be reviewed here.

- Somatic treatments (chiropractic, acupuncture, osteopathic manipulation, qigong, balneotherapy): In a randomized, controlled clinical trial among 24 female fibromyalgia patients, balneotherapy (warm bath) in daily 20-minute sessions 5 days per week for 3 weeks (total of 15 sessions; water temperature: 96.8°F = 36°C), resulted in statistically significant reductions in measured inflammatory mediators (PGE2, interleukin-1, LTB4) and amelioration of clinical symptoms among treated FM patients.[533] The symptomatic benefits of balneotherapy for FM patients have been corroborated in other trials.[534,535,536] Chiropractic treatment (including spinal manipulation, stretching, soft tissue treatments, and therapeutic ultrasound) has shown modest symptomatic benefit in several fibromyalgia case series and clinical trials.[537,538] A short-term trial showed that osteopathic manipulative therapy with standard medical care was superior to medical care alone for FM patients.[539] Acupuncture (including traditional, nontraditional, and electrical stimulation) also has been found beneficial for fibromyalgia patients.[540,541,542] Acupuncture may relieve fibromyalgia pain by improving regional blood flow, in addition to other mechanisms.[543,544] Because specific needle placement does not appear to be important[545], the conclusion that true acupuncture is ineffective because it may not differ markedly from the results obtained by sham acupuncture[546] may not be logical. A similar conundrum is seen in other clinical trials involving physical interventions such as manual osseous manipulation, wherein authentic treatments and sham treatments may both be effective by virtue of common physiological responses.[547] Qigong was found helpful for 10 fibromyalgia patients, and benefits were still apparent at three months' follow-up.[548]

- Special treatment, somatic treatment—Intramuscular needling and anesthesia: Myofascial pain is commonly received by needing (inserting a sterile needle into the muscle), whether or not the location is specific (e.g., acupuncture) and whether or not anesthetic agents, saline, or nothing (ie, dry needling) accompany the needle; having said that, more accurate localization (e.g., available trigger points or tender points) and the use of anesthetic agents tends to yield better results.

    - Analgesic and anti-hyperalgesic effects of muscle injections with lidocaine or saline in fibromyalgia syndrome. (*Eur J Pain*. 2014 Jul[549]): "We enrolled 62 female patients with FM into a double-blind controlled study of three groups who received 100 or 200 mg of lidocaine or saline injections into both trapezius and gluteal muscles. ...[Each subject received 2 muscle injections into the center of each trapezius muscle and 2 injections into the upper medial quadrants of both gluteus maximus muscles. ...Each syringe used for muscle injections contained either 5 ml of 1% lidocaine (50 mg) or 5 ml of normal saline.] RESULTS: Primary

---

[533] Ardiç F, Ozgen M, Aybek H, et al. Effects of balneotherapy on serum IL-1, PGE2 and LTB4 levels in fibromyalgia patients. *Rheumatol Int*. 2007 Mar;27(5):441-6
[534] Evcik D, Kizilay B, Gökçen E. The effects of balneotherapy on fibromyalgia patients. *Rheumatol Int*. 2002 Jun;22(2):56-9
[535] Fioravanti A, Perpignano G, Tirri G, et al. Effects of mud-bath treatment on fibromyalgia patients: a randomized clinical trial. *Rheumatol Int*. 2007 Oct;27(12):1157-61
[536] Dönmez A, Karagülle MZ, Tercan N, et al. SPA therapy in fibromyalgia: a randomised controlled clinic study. *Rheumatol Int*. 2005 Dec;26(2):168-72
[537] Citak-Karakaya I, et al. Short and long-term results of connective tissue manipulation and combined ultrasound therapy in patients with fibromyalgia. *J Manipulative Physiol Ther*. 2006 Sep;29(7):524-8
[538] Blunt KL, et al. The effectiveness of chiropractic management of fibromyalgia patients: a pilot study. *J Manipulative Physiol Ther*. 1997 Jul-Aug;20(6):389-99
[539] Gamber et al. Osteopathic manipulative treatment in conjunction with medication relieves pain associated with fibromyalgia syndrome. *J Am Osteopath Assoc*. 2002 Jun:321-5
[540] Martin DP, et al. Improvement in fibromyalgia symptoms with acupuncture: results of a randomized controlled trial. *Mayo Clin Proc*. 2006 Jun;81(6):749-57
[541] Singh BB, et al. Effectiveness of acupuncture in the treatment of fibromyalgia. *Altern Ther Health Med*. 2006 Mar-Apr;12(2):34-41
[542] Deluze C, Bosia L, Zirbs A, Chantraine A, Vischer TL. Electroacupuncture in fibromyalgia: results of a controlled trial. *BMJ*. 1992 Nov 21;305(6864):1249-52
[543] Sandberg M, Larsson B, Lindberg LG, Gerdle B. Different patterns of blood flow response in the trapezius muscle following needle stimulation (acupuncture) between healthy subjects and patients with fibromyalgia and work-related trapezius myalgia. *Eur J Pain*. 2005 Oct;9(5):497-510
[544] Sandberg M, Lindberg LG, Gerdle B. Peripheral effects of needle stimulation (acupuncture) on skin and muscle blood flow in fibromyalgia. *Eur J Pain*. 2004 Apr;8(2):163-71
[545] Harris RE, Tian X, Williams DA, Tian TX, Cupps TR, Petzke F, Groner KH, Biswas P, Gracely RH, Clauw DJ. Treatment of fibromyalgia with formula acupuncture: investigation of needle placement, needle stimulation, and treatment frequency. *J Altern Complement Med*. 2005 Aug;11(4):663-71
[546] Assefi NP, et al. A randomized clinical trial of acupuncture compared with sham acupuncture in fibromyalgia. *Ann Intern Med*. 2005 Jul 5;143(1):10-9
[547] Mein EA, et al. Manual medicine diversity: research pitfalls and the emerging medical paradigm. *J Am Osteopath Assoc*. 2001 Aug;101(8):441-4
[548] Chen KW, Hassett AL, Hou F, et al. A pilot study of external qigong therapy for patients with fibromyalgia. *J Altern Complement Med*. 2006 Nov;12(9):851-6
[549] Staud et al. Analgesic and anti-hyperalgesic effects of muscle injections with lidocaine or saline in patients with fibromyalgia syndrome. *Eur J Pain*. 2014 Jul;18(6):803-12

mechanical hyperalgesia at the shoulders and buttocks decreased significantly more after lidocaine than saline injections (p = 0.004). Similar results were obtained for secondary heat hyperalgesia at the arms (p = 0.04). After muscle injections, clinical FM pain significantly declined by 38% but was not statistically different between lidocaine and saline conditions. Placebo-related analgesic factors (e.g., patients' expectations of pain relief) accounted for 19.9% of the variance of clinical pain after the injections. ... CONCLUSION: These results suggest that muscle injections can reliably reduce clinical FM pain, and that peripheral impulse input is required for the maintenance of mechanical and heat hyperalgesia of patients with FM. Whereas the effects of muscle injections on hyperalgesia were greater for lidocaine than saline, the effects on clinical pain were similar for both injectates."

- Special supplementation in the treatment of FM—targeting (micro)glia activation and glutaminergic/NMDAr-mediated neuroexcitation: In my previous publications (prior to 2015) and consistent with the bulk of the basic science and clinical research, the emphasis of my fibromyalgia protocol has been on the treatment of SIBO and mitochondrial dysfunction, and the ever-necessary fine-tuning of the treatment protocol per patient. Progressively throughout 2015 as I further developed my understanding of the nuances of glial activation and the increasingly popular "gut-brain" concept (reviewed in printed monograph[550] and CE/CME videos[551]), I have become convinced that we should be—and already have been—addressing the glial activation directly. *How can we have already been doing this if we did not know that we were doing it?*—Simply by using nutrients such as anti-inflammatory fatty acids (e.g., EPA and DHA), nutrient-dense diets with minimal/moderate carbohydrate intake, vitamin and mineral supplementation (especially pyridoxine, magnesium, zinc, and vitamin D), phytonutrients, probiotics, CoQ10 and melatonin). All of these nutrients and substances have safety and efficacy for patients generally and FM patients particularly. What we know now is that these and other nutrients lessen the severity and duration of glial activation—brain inflammation—as well, and they therefore can be used to this effect. As such, I will summarize here that central sensitization is easily understood as a combination of **microglial inflammation**, which results in a) formation of the NMDAr agonist **QUIN**, b) formation of NO- which increases glutamate release while also causing mitochondrial dysfunction, and c) astrocyte activation leading to increased glutaminergic neurotransmission. As such, addressing the microglial activation *directly* while also addressing the dysbiotic and mitochondrial components is expected to enhance efficacy of the overall protocol; further, this discussion enhances our understanding of the mechanisms of action of and the clinical rationale for these interventions.

    - Dousing glial inflammation with vitamin D, fatty acids EPA and DHA, melatonin, phytonutrients: Various specific nutritional supplements and "over the counter remedies" have evidence—per in vitro, experimental, or human studies—to reduce glial inflammation; from these can be selected interventions which safely reduce glial activation in patients. All of these have proven safety for human use; the utility in reducing clinically relevant glial activation is established by the combination of available research plus the response of individual patients. As expected given its numerous anti-inflammatory properties, **DHA** reduces (micro)glia-induced inflammation, and the effect is enhanced with aspirin; the combination of **DHA with (low-dose) aspirin** is increasingly appreciated as synergistically anti-inflammatory and proresolutory, specifically but not exclusively via enhanced production of neuroprotective and anti-inflammatory resolvins.[552] Two paradoxes are worth noting: 1) Although immunostimulatory in the periphery[553], **melatonin reduces glial activation** in experimental models of brain injury.[554] 2) **Vitamin D**

---

[550] Vasquez A. *Human Microbiome and Dysbiosis in Clinical Disease*. ICHNFM, 2015.
[551] Vasquez A. "Microbiome and Dysbiosis in Clinical Disease" available CE/CME at NutritionAndFunctionalMedicine.org and pay-per-view at vimeo.com/ichnfm/vod_pages
[552] "Docosahexaenoic Acid increased total Glutathione levels in microglia cells and enhanced their anti-oxidative capacity. It reduced production of the pro-inflammatory cytokines TNF-α and IL-6 induced through TLR-3 and TLR-4 activation. Furthermore, it reduced production of Nitric Oxide. Aspirin showed similar anti-inflammatory effects with respect to TNF-α during TLR-3 and TLR-7 stimulation. ... Combination of Aspirin and Docosahexaenoic Acid showed augmentation in total Glutathione production during TLR-7 stimulation as well as a reduction in IL-6, TNF-α and Nitric Oxide. CONCLUSIONS: Collectively, these findings highlight the combination of Docosahexaenoic Acid and Aspirin as a possible measure against inflammation of the nervous system, thus leading to protection against neurodegenerative diseases with an inflammatory etiology." Pettit LK, Varsanyi C, Tadros J, Vassiliou E. Modulating the inflammatory properties of activated microglia with Docosahexaenoic acid and Aspirin. *Lipids Health Dis*. 2013 Feb 11;12:16
[553] This is a clinical trial showing anti-infective efficacy of melatonin, while other articles have specifically documented increases in inflammatory cytokines following melatonin administration. "Administration of melatonin as an adjuvant therapy in the treatment of neonatal sepsis is associated with improvement of clinical and laboratory outcomes." Gitto et al. Effects of melatonin treatment in septic newborns. *Pediatr Res*. 2001 Dec;50(6):756-60
[554] "Melatonin administration was associated with markedly restrained microglial activation, decreased release of proinflammatory cytokines and increased the number of surviving neurons at the site of peri-contusion. Meanwhile, melatonin administration resulted in dephosphorylated mTOR pathway." Ding et al. Melatonin reduced microglial activation and alleviated neuroinflammation induced neuron degeneration in experimental traumatic brain injury. *Neurochem Int*. 2014 Oct;76:23-31

reduces glial activation[555], and this is highly consistent with the clinical benefits seen of vitamin D against depression and other neuropsychiatric conditions, clearly including chronic pain; yet, antimicrobial peptide LL-37, production of which is at least partly dependent on vitamin D adequacy, induces glial-mediated neuroinflammation[556], perhaps thereby explaining the rare and possibly transient/inconsequential exacerbation of "sickness behavior" in some patients upon commencement of vitamin D supplementation. Many **phytonutrients—especially curcumin, quercetin, green tea catechins, baicalein, and luteolin**—show anti-inflammatory and neuroprotective benefits, some of which are mediated via reducing microglial activation/inflammation; we can endlessly debate the bioavailability of agents such as curcuminoids and quercetin, or we can accept them as low-cost high-safety nutrients that merit clinical utilization based on mechanistic studies and successful multicomponent clinical trials.[557,558]

- Alleviating **NO-induced glutaminergic neurotransmission and mitochondrial dysfunction** with **vitamin B12, especially in the form of hydroxocobalamin**: Vitamin B12 in general and hydroxocobalamin in particular bind with nitric oxide (NO-); supplemental (hydroxo)cobalamin has a pharmaconutritional effect of alleviating migraine[559] and low-back pain[560,561], two conditions known to have a component of neuroinflammatory central sensitization. Likely, the clinically observed analgesic effect of (hydroxo)cobalamin is mediated partly if not largely via its "chelation" or "detoxification" of NO-, thereby reducing the mitochondrial dysfunction and NMDAr activation that would have otherwise been triggered by NO-.

- Alleviating astrocyte-induced and QUIN-triggered glutamate/NMDA receptor activation with pyridoxine, magnesium, zinc: As previously reviewed, microglial activation promotes NMDAr activation via QUIN and glutamate. Regarding glutamate's activation of the NMDAr, sufficient biochemical, experimental, and clinical data allows us to conclude that we can reduce glutamate's excitatory effect by reducing glutamate itself via supplemental pyridoxine (vitamin B6), either in its active phosphorylated form of P5P (pyridoxal 5'-phosphate) or by supporting its magnesium-dependent requirement for conversion to the active P5P form when pyridoxine itself is used. In addition to promoting conversion of pyridoxine to P5P, magnesium also partly blocks calcium passage through the NMDAr-associated calcium channel (as does zinc) and also offsets the effects of increased intracellular calcium, in addition to supporting mitochondrial function, which is easily compromised by both inflammation and increased intracellular calcium. The clinical benefit of pyridoxine supplementation in migraine headache[562], seizures/epilepsy[563], neuropsychiatric symptoms of premenstrual syndrome (depression, irritability and tiredness)[564] is likely mediated via several different mechanisms, primary among which is the enhanced conversion of glutamate to gamma-amino-butyric acid (GABA), thereby synergistically reducing neuroexcitation and enhancing neuroregulation. A generalized

---

[555] "According to the results of the present study, activated microglia might increase the expression of 1-α-hydroxylase and VDR. 25(OH)D3 is converted into 1,25(OH)2D3 by 1-α-hydroxylase, which then stimulates VDR signaling and inhibits the phosphorylation of p38 in activated microglia. This cascade might inhibit the inflammatory reaction of activated microglia. In conclusion, the present study suggests that vitamin D3 might have an important role in the negative regulation of microglial activation." Hur et al. Regulatory Effect of 25-hydroxyvitamin D3 on Nitric Oxide Production in Activated Microglia. *Korean J Physiol Pharmacol.* 2014 Oct;18(5):397-402

[556] "We blocked the inflammatory stimulant action of LL-37 by removing it with an anti-LL-37 antibody. The inflammatory effect was also prevented by treatment with inhibitors of PKC, PI3K and MEK-1/2 as well as with the intracellular Ca(2+)-chelator, BAPTA-AM. This indicates involvement of these intracellular pathways. Our data suggest that LL-37, in addition to its established roles, may play a role in the chronic neuroinflammation which is observed in neurodegenerative diseases such as Alzheimer's and Parkinson's disease." Lee et al. Human antimicrobial peptide LL-37 induces glial-mediated neuroinflammation. *Biochem Pharmacol.* 2015 Mar 15;94(2):130-41

[557] Blaylock RL, Maroon J. Natural plant products and extracts that reduce immunoexcitotoxicity-associated neurodegeneration and promote repair within the central nervous system. *Surg Neurol Int* 2012;3:19

[558] Bredesen DE. Reversal of cognitive decline: a novel therapeutic program. *Aging* (Albany NY). 2014 Sep;6(9):707-17

[559] "Drugs which directly counteract nitric oxide (NO), such as endothelial receptor blockers, NO-synthase inhibitors, and NO-scavengers, may be effective in the acute treatment of migraine, but are also likely to be effective in migraine prophylaxis. In the underlying pilot study the prophylactic effect of the NO scavenger hydroxocobalamin after intranasal administration in migraine was evaluated. ... 1 mg intranasal hydroxocobalamin daily. ... A reduction in migraine attack frequency of >/ or = 50% was seen in 10 of 19 patients... A reduction of > or = 30% was noted in 63% of the patients. The mean attack frequency in the total study population showed a reduction from 4.7 +/- 1.7 attacks per month to 2.7 +/- 1.6. ." van der Kuy et al. Hydroxocobalamin, a nitric oxide scavenger, in the prophylaxis of migraine: an open, pilot study. *Cephalalgia.* 2002 Sep;22(7):513-9

[560] "The efficacy and safety of parenteral Vitamin B12 in alleviating low back pain and related disability and in decreasing the consumption of paracetamol was confirmed in patients with no signs of nutritional deficiency." Mauro et al. Vitamin B12 in low back pain. *Eur Rev Med Pharmacol Sci.* 2000 May-Jun;(4)3:53-8

[561] "Intramuscular methylcobalamin is both an effective and safe method of treatment for patients with nonspecific low back pain, both singly or in combination with other forms of treatment." Chiu et al. The efficacy and safety of intramuscular injections of methylcobalamin in patients with chronic nonspecific low back pain: a randomised controlled trial. *Singapore Med J.* 2011 Dec;52(12):868-73

[562] Sadeghi et al. Effects of pyridoxine supplementation on severity, frequency and duration of migraine attacks in migraine patients with aura. *Iran J Neurol.* 2015 Apr 4:74-80

[563] "An 8-year-old girl treated at our facility for superrefractory status epilepticus was found to have a low pyridoxine level at 5μg/L. After starting pyridoxine supplementation, improvement in the EEG for a 24-hour period was seen. ... A selective pyridoxine deficiency was seen in 94% of patients with status epilepticus (compared to 39.4% in the outpatients) which leads us to believe that there is a relationship between status epilepticus and pyridoxine levels." Dave et al. Pyridoxine deficiency in adult patients with status epilepticus. *Epilepsy Behav.* 2015 Nov;52(Pt A):154-8

[564] Doll H, Brown S, Thurston A, Vessey M. Pyridoxine (vitamin B6) and the premenstrual syndrome: a randomized crossover trial. *J R Coll Gen Pract.* 1989 Sep;39(326):364-8

schematic—mostly direct but very clearly clinically accurate—is provided; the illustration connects NMDAr activation with neuropsychiatric complications while providing insight into clinical remediation.

- <u>In the treatment of pain—including headaches and fibromyalgia—reducing the effects of glutamate-mediated neurotransmission and cellular effects is of very high importance</u>: Excess glutaminergic neurotransmission very clearly promotes anxiety, depression, fibromyalgia pain, myofascial pain and myofascial trigger points[565], migraine and headaches, seizures/epilepsy, and neurodegeneration. Our therapeutic goals are to 1) reduce glutamate levels with vitamin B6 and by avoiding/treating microglial activation, 2) reduce glutamate-triggered influx of calcium with zinc and magnesium, also vitamin D, alkalinization (increased consumption of base-forming foods, such as fruits and vegetables which contain citrate which is converted to bicarbonate to promote alkalinization, one effect of which is to promote magnesium retention, thereby alleviating pain[566]), omega-3 fatty acids such as from fish oil, 3) reduce the effects of glutamate/NMDA receptor activation by counterbalancing with benzodiazepine/GABA receptor activation by promoting conversion of glutamate to GABA and perhaps also by using niacinamide and the botanicals that act as ligands for the GABA receptor.

- <u>Reduce homocysteine levels, and recall that homocysteine may be elevated in the central nervous system (cerebrospinal fluid) of patients with chronic fatigue syndrome and fibromyalgia even when levels in blood/serum/plasma are normal[567]; safety and benefit of folate and vitamin B12 administration have been documented</u>[568]: Homocysteine contributes to NMDA receptor activation resulting in—identically as with glutamate—increased intracellular calcium and neurotoxicity.[569]

    - <u>Folinic acid or methylfolate 2-5 mg/d</u>: Use in combination with other vitamins, especially vitamin B12, in the form of hydroxocobalamin, adenosylcobalamin, or methylcobalamin—cyanocobalamin is obviously to be avoided because of its clinically relevant content of cyanide.
    - <u>Vitamin B12 >2,000 mcg per day orally, or 1-2 mg per week by injection</u>: Use in the form of hydroxocobalamin, adenosylcobalamin, or methylcobalamin—cyanocobalamin is obviously to be avoided because of its clinically relevant content of cyanide.
    - <u>Vitamin B6, pyridoxine 50-250 mg/d</u>: The phosphorylated form (P5P) can also be used; when the HCL form is used, additional attention must be given to magnesium status/supplementation and urinary alkalinization. As a rule, B6 supplementation should always be used with magnesium supplementation.
    - <u>Riboflavin 20-400 mg/d</u>: Small doses of 2 mg/d have been shown to significantly reduce homocysteine levels, and doses of 400 mg/d are common and well-tolerated in the treatment of migraine.
    - <u>Thyroid optimization</u>: Hypothyroidism causes elevated homocysteine and promotes insulin resistance[570] and should be treated appropriately per Chapter 1.
    - <u>NAC 600 mg per day and upward to 500-1,500 mg thrice daily</u>: Doses of NAC 4,800 mg/d have been used with success and safety in the treatment of SLE.
    - <u>Avoidance of homocysteine-elevating factors</u>: High coffee intake (>5 cups per day), ethanol, tobacco smoking, and medications/treatments (such as methotrexate, metformin, niacin and fibrate drugs); fish oil can raise homocysteine levels in some patients. Metformin is well-known to cause malabsorption of vitamin B12 and to thereby exacerbate "diabetic neuropathy" and promote depression and dementia/psychosis.
    - <u>Choline, phosphatidylcholine, lecithin (approximately 2.6 g choline/d)</u>: Each TBS (tablespoon, approximately 15 mL) of lecithin contains 275 mg of choline; thus, if the goal is to get to 2.6 g choline, one would need to use 10 TBS (150 mL) per day of granulated lecithin.
    - <u>Betaine, trimethylglycine 6–12 g/day</u>: Effects are weak/modest; likely more relevant for patients taking drugs such as metformin and fibrates that promote loss of betaine in urine.

---

[565] Wang et al. Spatial pain propagation over time following painful glutamate activation of latent myofascial trigger points in humans. *J Pain*. 2012 Jun;13(6):537-45
[566] Vormann et al. Supplementation with alkaline minerals reduces symptoms in patients with chronic low back pain. *J Trace Elem Med Biol*. 2001;15(2-3):179-83
[567] Regland et al. Increased concentrations of homocysteine in the cerebrospinal fluid in patients with fibromyalgia and chronic fatigue syndrome. *Scand J Rheumatol*. 1997;26(4):301-7
[568] Regland et al. Response to vitamin B12 and folic acid in myalgic encephalomyelitis and fibromyalgia. *PLoS One*. 2015 Apr 22;10(4):e0124648
[569] Abushik et al. The role of NMDA and mGluR5 receptors in calcium mobilization and neurotoxicity of homocysteine in trigeminal and cortical neurons and glial cells. *J Neurochem*. 2014 Apr;129(2):264-74
[570] Yang N et al. Novel Clinical Evidence of an Association between Homocysteine and Insulin Resistance in Patients with Hypothyroidism or Subclinical Hypothyroidism. *PLoS One*. 2015 May 4;10(5):e0125922

Chapter 5.1—Functional Inflammology Protocol for Metabolic Inflammation: Migraine and Fibromyalgia

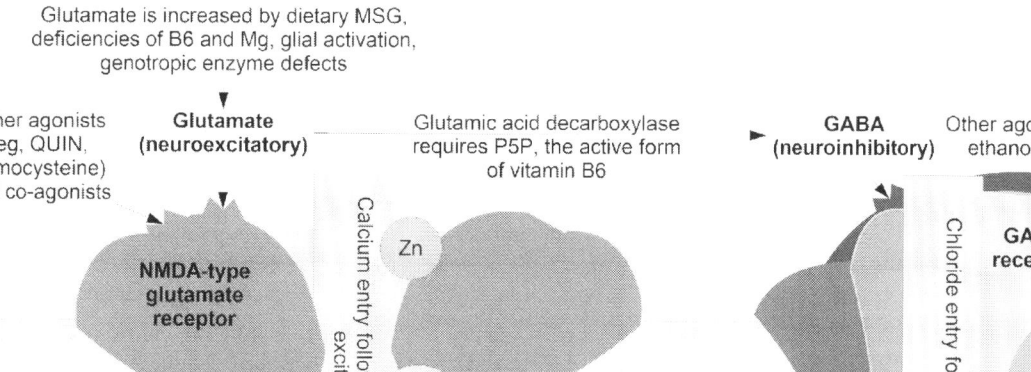

Illustration of the NMDA-type glutamate receptor, its activation, effects, and nutritional modulation: The NMDA receptor is activated by glutamate, QUIN, and other substances which act as agonists (e.g., homocysteine) or co-agonists (e.g., glycine). Different forms of the NMDA receptor exist; thus, the image presented here is a generalized version that is conceptually accurate (rather than all-inclusive; for more details see reviews[571]) and clinically relevant. Neuroexcitatory glutamate is converted to neuroinhibitory GABA by the enzyme glutamic acid decarboxylase, which shows vitamin B6 dose-responsiveness in its reduction of glutamate levels. Magnesium and zinc (and perhaps copper) retard the passage of calcium through this channel, thereby mitigating some of the effects of NMDAr activation. Quenching nitric oxide (for example with hydroxocobalamin), which would otherwise trigger glutamate release, and dousing glial activation are important considerations not included in this illustration. For updates and additional information and explanations, see videos and articles at www.inflammationmastery.com/pain

---

[571] Vyklicky et al. Structure, function, and pharmacology of NMDA receptor channels. *Physiol Res*. 2014;63 Suppl 1:S191-203

ENDOCRINE IMBALANCE & OPTIMIZATION Peptide-based and steroid-based hormones have wide-ranging effects beyond those with which they are classically and thus simplistically associated. The "main" hormones that we consider in most chronic inflammatory disorders are the three pro-inflammatory hormones (prolactin, estradiol, and insulin) and the three anti-inflammatory hormones (DHEA, cortisol, and testosterone); each of these hormones can be objectively assessed with serologic testing and modulated with therapeutic intervention. A full thyroid evaluation—including history, physical examination (with particular scrutiny for cold extremities [DDX: hypothyroidism, hypogonadism, vasoconstriction/vaso-obstruction, Raynaud's disorder, peripheral vascular disease, H2S-producing GI dysbiosis], relative bradycardia [DDX, hypothyroidism, heart block, beta-blocker medications], and delayed Achilles reflex return [considered diagnostic of hypothyroidism]), and laboratory evaluation (including TSH, free T4, total or free T3, rT3, and antithyroid antibodies) is warranted in any patient whose concerns include fatigue, depression, systemic inflammation and chronic pain; musculoskeletal manifestations of hypothyroidism include muscle pain, weakness, myopathy, and adhesive capsulitis. The pineal hormone melatonin was discussed in a previous section.

- Testosterone to treat central sensitization of chronic pain in fibromyalgia patients (*Int Immunopharmacol* 2015 Aug[572]): "Considering these mechanisms together, abnormally low testosterone levels are likely to result in amplified ascending/descending facilitation of nociception and reduced descending inhibitory control, resulting in a widening pain field and neuronal plasticity which can turn into the entrenched chronic pain states found in fibromyalgia patients." In this article, the author's review the role of testosterone as an antiinflammatory hormone that reduces glial activation and thereby mitigates the neuro-inflammatory basis (detailed previously) of central sensitization and the resulting pain. Testosterone is well-known to have anti-inflammatory and immunomodulatory properties with clinical utility in rheumatoid arthritis and cluster headache.
- Treatment of pain in fibromyalgia patients with testosterone gel (*Int Immunopharmacol.* 2015 Aug[573]): "Assessment of the typical symptoms of fibromyalgia by patient questionnaire and tender point exam demonstrated significant change in: decreased muscle pain, stiffness, and fatigue, and increased libido during study treatment. These results are consistent with the hypothesized ability of testosterone to relieve the symptoms of fibromyalgia. Symptoms not tightly related to fibromyalgia were not improved."

XENOBIOTIC ACCUMULATION & DETOXIFICATION The term "xenobiotics" is generally used to refer to carbon-based foreign chemicals such as persistent organic pollutants (POPs) including herbicides, pesticides, phthalates, parabens, dioxin-related chemicals, and many others; used more casually, the term may also be used to include noncarbon-based foreign substances such as toxic metals like lead, mercury, cadmium, and arsenic. Thus, "xenobiotics" has become somewhat synonymous with "toxins" in both professional-level and vernacular conversations. Laboratory assessments for chemical and metal toxins are commercially available through specialized medical laboratories and are based on analysis of blood and urine. The many biochemical and physiologic components of detoxification/depuration have been reviewed in chapter 4 of this book. Essentially everyone—all humans on the planet worldwide—have biochemical evidence of xenobiotic chemical/metal accumulation, generally with numerous xenobiotics, which have additive and synergistic adverse effects on physiology and health. Thus, scientifically, since xenobiotic accumulation is pandemic, consideration of and treatment for xenobiotic accumulation via therapeutic detoxification programs and lifestyle interventions should be routine. Easy and effective means for promoting detoxification of chemicals and metals include plant-based diet to promote bowel and renal excretion of toxins (via reduced enterohepatic recycling [better microflora, more fiber for adsorption, more frequent fecal excretion] and reduced renal resorption [urinary alkalinization], respectively), NAC for arsenic chelation and GSH production, sweating/exercise (lipolysis promotes mobilization of lipophilic toxins from adipose tissue, diaphoresis promotes direct toxin excretion), sufficient micronutrient and protein intake supports phase 1 and phase 2 of the oxidation and conjugation processes in the liver. Chemical xenobiotics can be bound in the gut during the normal process of enterohepatic recycling/recirculation with periodic or rotational use of activated charcoal, cholestyramine, and chlorella; anecdotal reports from clinical practices support the use of phytochelatin (metal-binding peptides from plants, used by plants for protection from metal toxicity) in the

---

[572] White HD, Robinson TD. A novel use for testosterone to treat central sensitization of chronic pain in fibromyalgia patients. *Int Immunopharmacol.* 2015 Aug;27(2):244-8
[573] White et al. Treatment of pain in fibromyalgia patients with testosterone gel: Pharmacokinetics and clinical response. *Int Immunopharmacol.* 2015 Aug;27(2):249-56

prevention/treatment of metal toxicity in humans but no formal clinical studies have been performed to document the effectiveness of this approach although its safety is clinically appreciated.

- Pilot study: *Chlorella pyrenoidosa* for patients with fibromyalgia syndrome (*Phytother Res* 2000 May[574]): *Chlorella pyrenoidosa* is a unicellular green alga that grows in fresh water. It is a dense source of nutrients, particularly vitamin D (500 IU vitamin D per 1.35 g *Chlorella*). *Chlorella* may have value in treating some fibromyalgia patients, but overall the efficacy is low. Thus, *Chlorella* should not be used as monotherapy for fibromyalgia, although it may be a useful adjunct either as a source of vitamin D, as a means to help modify gut flora, or as an aid in the detoxification of xenobiotics due to its ability to bind ingested and bile-excreted toxins and prevent their absorption and reabsorption in a manner similar to that of cholestyramine, a drug used to bind cholesterol in the gut, promote its excretion, and thereby lower blood cholesterol levels.[575,576,577] This "detoxifying" effect of *Chlorella* in humans is supported by 2 clinical trials showing that nursing mothers who supplement with *Chlorella* during lactation transfer less dioxin in their breast milk compared to nursing mothers who do not consume *Chlorella*.[578,579]

- Clinical investigation: Reduced exposure to xenobiotics (cosmetics) alleviates fibromyalgia (*Journal of Women's Health* 2004 Mar[580]): Women use more cosmetic products than do men, and fibromyalgia is more common in women. Cosmetic products generally contain skin-absorbable xenobiotics with potentially adverse effects; therefore this study was conducted to determine if avoidance of cosmetics would alleviate symptoms of FM. The author of this report describes a prospective, randomized, controlled trial of 48 women with FM (some of whom had a rheumatic condition) who were regular users of cosmetics was carried out to investigate if a reduced use of cosmetics would reduce the symptoms. The patients were told to avoid or completely abstain from using all ointments, creams, skin lotions, pain-relieving liniments, cleaning lotions, oil treatments, hair-coloring chemicals, and tanning lotion; they were also advised to reduce their use of soap and shampoo, both of which—like skin creams—are generally formulated with perfumes and other chemicals and applied to large regions of the body. This research showed that, after 2 years, FM patients who reduced their exposure to chemicals/xenobiotics/cosmetics experienced significant reductions in pain, sleep disturbances, and musculoskeletal stiffness ($p < 0.02$), together with better physical function and improved sense of well-being as measured by the Fibromyalgia Impact Questionnaire (FIQ). Thus, avoiding chemical exposure appears to provide no-cost no-risk therapeutic benefit to FM patients by alleviating pain and improving several indicators of overall health.

Conclusion: In sum, current research indicates that fibromyalgia results from impairment of cellular energy/ATP production (mitochondrial dysfunction) and induction of pain hypersensitivity (peripheral and central sensitization) due to absorbed metabolic toxins from bacterial/microbial overgrowth of the gastrointestinal tract; this is complicated by induction of tryptophan deficiency which is most likely caused by tryptophan degradation by bacterial tryptophanase activity and which leads to serotonin and melatonin insufficiencies, which lead to associated biochemical and clinical consequences, discussed previously. Available studies have shown that SIBO is ubiquitous among fibromyalgia patients and that antimicrobial interventions—whether pharmaceutical or nutritional—are efficacious. Secondary physiological effects such as mitochondrial impairment, pain sensitization, nutritional deficiencies, oxidative stress, and reduced tissue perfusion are treated with combined use of select therapeutics as reviewed previously. Patients presenting with widespread pain should be screened for causative underlying disease; if no other explanation can be found, then the diagnosis of fibromyalgia should be made, and the condition should be treated with the nondrug therapeutics discussed above. The first visit can include history, physical examination, and laboratory tests; initial laboratory assessment should include complete blood count

---

[574] Merchant RE, et al. Nutritional supplementation with Chlorella pyrenoidosa for patients with fibromyalgia syndrome: a pilot study. *Phytother Res* 2000 May;14:167-73
[575] Pore RS. Detoxification of chlordecone poisoned rats with chlorella and chlorella derived sporopollenin. *Drug Chem Toxicol*. 1984;7(1):57-71
[576] Morita K, Ogata M, Hasegawa T. Chlorophyll derived from Chlorella inhibits dioxin absorption from the gastrointestinal tract and accelerates dioxin excretion in rats. *Environ Health Perspect*. 2001 Mar;109(3):289-94
[577] Morita K, Matsueda T, Iida T, Hasegawa T. Chlorella accelerates dioxin excretion in rats. *J Nutr*. 1999 Sep;129(9):1731-6
[578] Nakano S, Noguchi T, Takekoshi H, Suzuki G, Nakano M. Maternal-fetal distribution and transfer of dioxins in pregnant women in Japan, and attempts to reduce maternal transfer with Chlorella (Chlorella pyrenoidosa) supplements. *Chemosphere*. 2005 Dec;61(9):1244-55
[579] Nakano et al. Chlorella (pyrenoidosa) supplementation decreases dioxin and increases immunoglobulin a concentrations in breast milk. *J Med Food*. 2007 Mar:134-42
[580] Sverdrup B. Use less cosmetics—suffer less from fibromyalgia? *J Womens Health* (Larchmt). 2004 Mar;13(2):187-94

(CBC), metabolic/chemistry panel, serum 25-hydroxyvitamin D, C-reactive protein (CRP), anti-nuclear antibodies (ANA), antibodies against cyclic citrullinated proteins (anti-CCP antibodies), ferritin, muscle enzymes aldolase and creatine kinase, and a complete thyroid assessment including TSH, free T4, free T3, total T3, reverse T3 (rT3), and antithyroid peroxidase and antithyroglobulin antibodies. First-day interventions can include dietary optimization, multivitamin-multimineral supplementation (including vitamin D3 and magnesium), tryptophan/5-HTP, CoQ10, mixed tocopherols, and combination fatty acids including gamma-linolenic acid (GLA), eicosapentaenoic acid (EPA) and docosahexaenoic acid (DHA). SIBO should be treated empirically; otherwise, it can be objectively assessed with breath hydrogen and methane testing, stool analysis, culture, microscopy, and parasitology. At follow-up visits, additional assessments and interventions (such as for toxic metals and chronic occult infections) can be used to fine-tune the diagnosis and further discover and define its contributors in order to maximize the patient's response to treatment and promote optimal recovery and health.

MIGRAINE HEADACHES, HYPOTHYROIDISM, AND FIBROMYALGIA:

ASSESSMENTS AND THERAPEUTIC APPROACHES USING INTEGRATIVE CHIROPRACTIC, NATUROPATHIC, OSTEOPATHIC, AND FUNCTIONAL MEDICINE

Objective *real* non-neurologic non-psychiatric abnormalities in fibromyalgia: the case against primary central sensitization.

DR. ALEX VASQUEZ
FUNCTIONALINFLAMMOLOGY.COM
INFLAMMATIONMASTERY.COM

Video access: www.inflammationmastery.com/pain
Archive: https://vimeo.com/56334919

MIGRAINE HEADACHES, HYPOTHYROIDISM, AND FIBROMYALGIA:

ASSESSMENTS AND THERAPEUTIC APPROACHES USING INTEGRATIVE CHIROPRACTIC, NATUROPATHIC, OSTEOPATHIC, AND FUNCTIONAL MEDICINE

Is "central sensitization" in fibromyalgia a bunch of c.r.a.p. (Commercial Rationalization Advocating Pharmaceuticals)?

DR. ALEX VASQUEZ
FUNCTIONALINFLAMMOLOGY.COM
INFLAMMATIONMASTERY.COM

Video access: www.inflammationmastery.com/pain
Archive: https://vimeo.com/56437367

How does small intestinal bacterial/microbial overgrowth (SIBO) cause fibromyalgia?

1. **Bacterial endotoxin/lipopolysaccharide/LPS**: causes inflammation, mitochondrial/muscle impairment, and increased sensitization to pain.
2. **Bacteria-produced D-lactic acid**: neurotoxin and metabolic poison; causes fatigue, muscle pain, dyscognition.
3. **Bacteria-produced hydrogen sulfide (H2S)**: neurotoxin and metabolic poison; causes fatigue, muscle pain, dyscognition.
4. **Bacteria-produced tryptophanase**: leads to tryptophan deficiency (documented in FM patients) and "serotonin deficiency" (pain, fatigue, carbohydrate cravings, depression) and "melatonin deficiency" (sleep disturbance, mitochondrial impairment, oxidative stress, muscle fatigue).

Video access: www.inflammationmastery.com/pain
Archive: https://vimeo.com/56334918

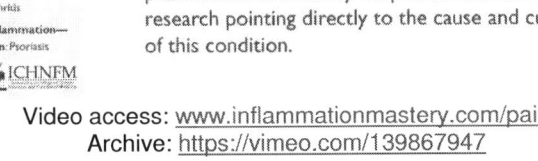

Fibromyalgia
▸ One of my favorite clinical disorders because it blends 3 of my favorite topics:

1. **Dysbiosis**—these patients have a high prevalence of SIBO, which completely explains every aspect of the condition
2. **Mitochondrial dysfunction**—these patients have mitochondrial dysfunction caused by SIBO
3. **Social (in)justice**—perfectly exemplifies the extent to which the medical profession and medical practice can be easily coopted—hijacked—by the commercial interests of the pharmaceutical industry despite a mountain of research pointing directly to the cause and cure of this condition.

Video access: www.inflammationmastery.com/pain
Archive: https://vimeo.com/139867947

Intracellular Hypercalcinosis: A Functional Nutritional Disorder with Implications Ranging from Myofascial Trigger Points to Affective Disorders, Hypertension, and Cancer

This article was originally published in Naturopathy Digest in 2006
naturopathydigest.com/archives/2006/sep/vasquez.php

Introduction: Let us explore the possibility that elevated levels of calcium *within the cell* (intracellular hypercalcinosis) might predispose toward a wide range of clinical problems including migraine, hypertension, myofascial trigger points, inflammation, and cancer. Further, let's review the data showing that several commonly employed nutritional interventions can be used synergistically to counteract and correct this problem. By the time readers complete this article, they will have 1) an understanding of this problem, 2) a protocol for how to correct this problem, and 3) be able to explain the biochemical rationale for using these nutritional protocols in patients who might otherwise be treated with drugs in general and calcium-channel-blocking drugs in particular.

Although prescription drugs are often used by medical doctors in a "willy-nilly manner" (according to Harvard Medical School Professor Dr. Jerry Avorn[581]), let's assume for a moment that legitimate reasons exist for the widespread use of drugs that block calcium channels in cell membranes—the "calcium-channel-blocking drugs." Although it is counterintuitive to promote health by interfering with the body's natural function, calcium-channel-blocking drugs are routinely used in pharmaceutical medicine for a broad range of problems including hypertension, heart rhythm disturbances, bipolar disorder, and anxiety/panic disorders. Widespread medical use of calcium-channel-blocking drugs appears to validate the supposition that excess intracellular calcium is an important contributor to these and perhaps other problems. Therefore, if intracellular hypercalcinosis is the problem, then any safe and cost-effective treatment that can correct this problem should be met with the same widespread acceptance given to calcium-channel-blocking drugs, which are universally accepted and utilized in the allopathic "conventional medicine" society. At the very least, we can generally state that all phenomena that contribute to calcium deficiency result in an increase in intracellular calcium levels (the "calcium paradox") due to the effect of parathyroid hormone, which specifically promotes calcium uptake in cells while mobilizing calcium from bone. Additionally, a few other nutritional influences (such as fatty acid imbalances) modulate cellular calcium balance, and these will be discussed in the section on clinical interventions.

The Problem of Excess Intracellular Calcium: Although the current author is the first to coin the phrase "intracellular hypercalcinosis", several other authors have pointed to the problem of the "calcium paradox" and the means by which *body-wide calcium deficiency* can result in *intracellular calcium overload*, which triggers a cascade of events leading to adverse health effects. Most notably, the work of Takuo Fujita[582,583] stands out in its clarity and specificity in linking intracellular hypercalcinosis with disorders such as hypertension, arteriosclerosis, diabetes mellitus, neurodegenerative diseases, malignancy, and degenerative joint disease.

Mechanisms by which intracellular hypercalcinosis contributes to disease have been defined, at least partially. However, we must remember that nutritional disorders never occur in isolation, and that the effects of intracellular hypercalcinosis observed clinically are overlaid with manifestations of the primary nutritional/metabolic disorder. Stated differently, contrary to what the pharmaceutical paradigm's monotherapeutic use of calcium-channel-blocking drugs would imply, intracellular hypercalcinosis never occurs by itself. For example, if intracellular hypercalcinosis is contributed to by vitamin D3 deficiency, then some of the observed clinical complications of that condition are due to and yet independent from the excess intracellular calcium since the primary problem (vitamin D3 deficiency) causes adverse effects and deficiency symptoms that are independent of its effect on intracellular calcium levels. To better understand the specific effects of excess intracellular calcium, a brief review of a few specific biochemical/physiologic mechanisms by which intracellular hypercalcinosis can contribute to disease is warranted. We must start by realizing that calcium is much more than a "bone nutrient" and that it functions as an electrolyte, intracellular messenger, and regulator of cell replication and metabolism. Let's talk about five pathways by which increased intracellular calcium promotes disease:

---

[581] America the Medicated. cbsnews.com/stories/2005/04/21/health/main689997.shtml
[582] Fujita T. Calcium paradox: consequences of calcium deficiency manifested by a wide variety of diseases. *J Bone Miner Metab*. 2000;18(4):234-6
[583] Fujita et al. Calcium paradox disease: calcium deficiency prompting secondary hyperparathyroidism and cellular calcium overload. *J Bone Miner Metab*. 2000;18(3):109-25

1. <u>Adverse effects on membrane receptors and intracellular transduction</u>: The concentration of extracellular calcium exceeds the concentration of intracellular calcium by a ratio of 10,000 to one. When intracellular calcium levels rise even slightly, receptors and messaging systems in the cell membrane fail to function optimally. Thereby, increased intracellular calcium can predispose to insulin resistance (via interference with insulin receptors) and can promote neurodegeneration by amplifying the intracellular cascade of effects that follows activation of the brain's NMDA-receptors (excitoneurotoxicity). More specifically, we must note that the recently discovered "calcium-sensing receptor" (CaR, a G protein-coupled plasma membrane receptor) senses minute alterations in serum calcium levels and then ultimately translates these variations into changes in cellular function, notably alterations in cell replication (think cancer) and eicosanoid production (think inflammation).[584,585] Given that CaR are found in a wide range of cell types, including those found in bone, the kidneys, and immune system, we can see a pathway by which alterations in calcium balance could be implicated in a wide range of diseases. CaR-mediated alterations in cell function are likely to be complicated by disorders of vitamin D3 nutrition and metabolism (that commonly complicate disorders of calcium homeostasis), which affect an even wider range of cell types including those of the breast, prostate, ovary, lung, skin, lymph nodes, colon, pancreas, adrenal medulla, brain (pituitary, cerebellum, and cerebral cortex), aortic endothelium, and immune system, including monocytes, transformed B-cells, and activated T-cells. This is an example of the complexity involved in understanding nutrition in general and the effects of nutritional deficiency (always multifaceted) in particular.

2. <u>Mitochondrial failure and cell death</u>: According to the most recent edition of the classic text *Robbins Pathologic Basis of Disease* (pages 15-16), increased intracellular calcium is a major cause of cell death. When calcium levels are increased within the cell, one adverse effect is the inhibition of mitochondrial function. Since calcium is pumped out of the cell in an energy-dependent process, and because dysfunctional mitochondria pour calcium into the intracellular space, calcium-induced mitochondrial failure results in an additional increase in intracellular calcium. Further complicating this problem is the fact that the cell membrane becomes increasingly permeable to calcium as calcium levels increase. Elevated intracellular calcium levels activate enzymes such as ATPase, phospholipase, proteases, and endonucleases that promote cell death.

3. <u>Pro-inflammatory effects of intracellular calcium</u>: The recent finding that intracellular calcium activates NF-kappaB[586] has obvious implications given the pivotal role of NF-kappaB in the promotion of systemic inflammation and diseases such as rheumatoid arthritis.[587] Thus, increased intracellular calcium appears to promote inflammation. This may explain in part how vitamin D3 supplementation (which lowers intracellular calcium levels) exerts its clinically impressive anti-inflammatory and immunomodulatory benefits.[588]

4. <u>Enhanced production of lipid peroxides</u>: Fujita notes that lipid peroxides lead to an increase in cell membrane permeability to calcium, which results in increased intracellular calcium; this activates metabolic pathways that increase oxidative stress, thus leading to a vicious cycle stimulated by the production of additional lipid peroxides. Thus, intracellular hypercalcinosis promotes oxidative stress, which becomes self-perpetuating by this and other mechanisms. Of course, we all know by now that increased production of free radicals contributes to the development of many health problems, such as cancer, cardiovascular disease, arthritis, autoimmunity, diabetes, and other forms of rapid biological aging.

5. <u>Myofascial trigger points, chronic muscle spasm, and increased vascular tone (hypertension)</u>: The release of calcium from the sarcoplasmic reticulum triggers muscle contraction and plays a role in hypertension (hence the use of calcium-channel-blocking drugs in the treatment of hypertension), chronic muscle spasm (especially when complicated by magnesium deficiency), and the perpetuation of myofascial trigger points.[589] Reducing the levels of cytosolic and sarcoplasmic calcium promotes muscle relaxation.

---

[584] Peterlik M, Cross HS. Vitamin D and calcium deficits predispose for multiple chronic diseases. *Eur J Clin Invest*. 2005 May;35(5):290-304
[585] Heaney RP. Long-latency deficiency disease: insights from calcium and vitamin D. *Am J Clin Nutr*. 2003 Nov;78(5):912-9
[586] "Furthermore, a calcium chelator, BAPTA-AM, attenuated the NF-kappaB activation... CONCLUSIONS: Induction of NF-kappaB within 30 min by TNF-alpha- and IL-1beta was mediated through intracellular calcium but not ROS." Chang JW, Kim CS, Kim SB, Park SK, Park JS, Lee SK. Proinflammatory cytokine-induced NF-kappaB activation in human mesangial cells is mediated through intracellular calcium but not ROS: effects of silymarin. *Nephron Exp Nephrol*. 2006;103:e156-65
[587] Tak PP, Firestein GS. NF-kappaB: a key role in inflammatory diseases. *J Clin Invest*. 2001 Jan;107(1):7-11
[588] Timms et al. Circulating MMP9, vitamin D and variation in the TIMP-1 response with VDR genotype: mechanisms for inflammatory damage in chronic disorders? *QJM*. 2002 Dec;95(12):787-96. See also: Vasquez A, Manso G, Cannell J. The clinical importance of vitamin D. *Altern Ther Health Med*. 2004 Sep-Oct;10(5):28-36
[589] Simons DG. Cardiology and myofascial trigger points: Janet G. Travell's contribution. *Tex Heart Inst J*. 2003;30(1):3-7

Nutritional Interventions to Ameliorate Intracellular Hypercalcinosis: Now that we've reviewed the data implicating intracellular hypercalcinosis as a legitimate contributor to a wide range of clinical disorders and diseases, let's explore some nutritional solutions.

1. Correction of vitamin D deficiency: Vitamin D deficiency causes calcium deficiency which increases parathyroid hormone production resulting in increased intracellular calcium levels. Vitamin D deficiency is common (40-80% of most populations) and can be established via history and more objectively by measurement of serum 25-hydroxyl-vitamin D. Replacement doses are in the range of 1,000 IU per day for infants, 2,000 IU per day for children, and 4,000 IU per day for adults.[590] Vitamin D2 (ergocalciferol) should be avoided, and vitamin D3 (cholecalciferol) should be used, preferably in emulsified form to facilitate absorption, especially in older patients and those with impaired digestion and absorption.[591]
2. Reduction in dietary arachidonic acid intake: Arachidonic acid promotes intracellular calcium uptake, as demonstrated in a recent study using human erythrocytes.[592] Rich sources of arachidonic acid include beef, liver, pork, lamb, and cow's milk.
3. Increase intake of eicosapentaenoic acid (EPA): EPA reduces intracellular calcium levels in experimental models[593] and anticancer, antihypertensive, and anti-inflammatory effects of EPA are seen clinically. One to three grams per day is reasonable for adults.
4. Urinary alkalinization: Diet-induced chronic metabolic acidosis[594] promotes loss of calcium in urine[595] and thus indirectly contributes to calcium deficiency and the resultant rise in parathyroid hormone and intracellular calcium levels. An alkalinizing plant-based Paleo-Mediterranean diet should be the foundational treatment for numerous reasons[596]; however some patients may need to supplement with vegetable culture, potassium citrate, potassium bicarbonate, and/or sodium bicarbonate either regularly or "as needed"/PRN.
5. Ensuring adequate intake of calcium: A healthy diet can supply upwards toward 1,000 mg of calcium per day, and some people may choose to supplement with an additional 500 to 1,500 mg daily. Calcium supplementation should be used with magnesium, vitamin D and other components of the supplemented Paleo-Mediterranean diet.
6. Avoiding other dietary and lifestyle factors that promote calcium loss in urine: Caffeine, sugar, alcohol/ethanol, and psychoemotional stress all increase calcium loss in urine and thus contribute to secondary hyperparathyroidism and intracellular hypercalcinosis.

Conclusions: In this brief article, I have introduced and reviewed important concepts related to diet-induced alterations in cellular calcium balance. Notice that this discussion of calcium has transcended the usual conversation of simple "deficiency" and "excess." What I've done here is review data showing that we can indirectly modulate certain aspects of intracellular nutrition to promote optimal biochemical balance within the cell in order to optimize health and prevent and correct disease and dysfunction. Next time someone tells you that there is no scientific basis for interventional nutrition, sit them down and give them a lecture on causes and treatments for intracellular hypercalcinosis. Tell them it is only the tip of the iceberg, and that they'd be wise to take interventional nutrition seriously. Just because we buy groceries and nutritional supplements without a prescription (for now), this does not mean that these choices are not powerful or lacking in scientific merit. Amazing results can be achieved with diet modification and nutritional/botanical supplementation.

---

[590] Vasquez A, Manso G, Cannell J. The clinical importance of vitamin D (cholecalciferol). *Altern Ther Health Med*. 2004 Sep-Oct;10(5):28-36
[591] Vasquez A. Subphysiologic Doses of Vitamin D are Subtherapeutic: Comment on the Study by The Record Trial Group. *The Lancet* 2005 Published on-line May 6
[592] "The Ca(2+) influx rate varied from 0.5 to 3 nM Ca(2+)/s in the presence of AA and from 0.9 to 1.7 nM Ca(2+)/s with EPA." Soldati L, Lombardi C, Adamo D, Terranegra A, Bianchin C, Bianchi G, Vezzoli G. Arachidonic acid increases intracellular calcium in erythrocytes. *Biochem Biophys Res Commun*. 2002 May 10;293(3):974-8
[593] "This is a consequence of the ability of EPA to release Ca2+ from intracellular stores while inhibiting their refilling via capacitative Ca2+ influx that results in partial emptying of intracellular Ca2+ stores and thereby activation of protein kinase R." Palakurthi SS, Fluckiger R, Aktas H, Changolkar AK, Shahsafaei A, Harneit S, Kilic E, Halperin JA. Inhibition of translation initiation mediates the anticancer effect of the n-3 polyunsaturated fatty acid eicosapentaenoic acid. *Cancer Res*. 2000 Jun 1;60(11):2919-25
[594] Maurer M, Riesen W, Muser J, Hulter HN, Krapf R. Neutralization of Western diet inhibits bone resorption independently of K intake and reduces cortisol secretion in humans. *Am J Physiol Renal Physiol*. 2003 Jan;284(1):F32-40
[595] Sellmeyer et al. Potassium citrate prevents increased urine calcium excretion and bone resorption induced by a high sodium chloride diet. *J Clin Endocrinol Metab*. 2002 May;87(5):2008-12
[596] Vasquez A. A Five-Part Nutritional Protocol that Produces Consistently Positive Results. *Nutritional Wellness* 2005 September inflammationmastery.com/reprints

Index:

2010, ACR guidelines for the diagnosis and assessment of FM, 910
3-3-hydroxyphenyl-3-hydroxypropionic acid, 935
5-hydroxytryptophan, 960
Acetaminophen, 913
Acetyl-L-carnitine, 970
Acupuncture, 973
Alcohol, 929
Algal chlorovirus ATCV-1, 935
Amitriptyline, 912
Aneurysm (intracranial), 875
Antimetabolites, 928
Antimicrobial treatment of SIBO, 963
Augmentin, 964
Carboxy-methyl-lysine, 904
Chiropractic, 973
Chlamydophila pneumoniae, 918
Chlorella, 979
Chlorovirus ATCV-1, 935
Citrate synthase, 890
Clinical criteria—description and contrast of the 1990 criteria and the 2010 criteria, 909
Cluster headache, 874
Cluster headache - differential diagnosis of head pain, 875
Cluster headaches, treatment with melatonin, 899
Coenzyme Q10, 966
Conclusions and Therapeutic Approach, 979
CoQ10, 969, CoQ-10 treatment for migraine, 890
Cortical spreading depression, 866
Cosmetics, 979
Cough headache - differential diagnosis of head pain, 875
Cranial nerve V, 863
Cranial nerve V, pain sensation in migraine, 863
Creatine monohydrate, 970
Cyclobenzaprine, 913
Cymbalta, 912
Cytochrome-c-oxidase, 890
DA-type glutamate receptor (NMDAr), 871
Detoxification, 921
Diet optimization, 957
Diflucan, 965
Dimercaptosuccinic acid, 920
D-lactic acid, 924, D-lactic acidosis, 929
D-lactic acid intestinal bacteria in chronic fatigue syndrome, 929
DMSA, 920
D-ribose, 970
Duloxetine, 912
Enhanced processing of autoantigens, 934
Feverfew, 897
Fibromyalgia, 901
Fibromyalgia disease, 902
Fluconazole, 965
Functional Medicine (FxMed) perspectives, 916
Glial activation, 935
Glutamate and the NMDA receptor in headache, 870
Glutamate promotes pain and inflammation, 872
Glutamate/NMDA receptor, 935
Glycolytic pathways, 894
Hemochromatosis, 919
HPHPA, 935
Hydrogen sulfide, 924, 929
Hydroxocobalamin, 883
Hypothyroidism, 917
Indole, 929
Intracranial aneurysm, 875
Intracranial mass lesion, 876

Iron overload, 919, 876, iron overload as a cause of headaches, 877
Lead, 920
L-tryptophan, 949
Lyrica, 912
Magnesium, 961
Melatonin, 969
Melzack and Wall, 908
Meningitis, 876
Mercurial myopathy, 920
Mercury, 920
Microglial activation, 935
Migraine - differential diagnosis of head pain, 876
Migraine with aura, 874
Milnacipran, 913
Mitochondrial dysfunction, 904, 905
Mitochondrial dysfunction promotes central sensitization, 928
Mitochondrial impairment in migraine and cluster headache, 864
Mitochondrial myopathy, 905
Mitophagy, 906, 925, 949, 951, 965, 966
Mycoplasma species, 918
NADH-cytochrome-c-reductase, 890
NADH-dehydrogenase, 890
Neomycin, 964
Neurologic deficit in the evaluation of head pain - clinical management, 877
NLRP3 inflammasome is activated in fibromyalgia, 969
NMDA receptor, 935
Occult infections, 918
Osteopathic manipulation, 973
Oxygen, for cluster headaches, 892, 970, 973
Patient (mis)education in standard medicine, 914
Pentosidine, 904
Pregabalin, 912
Probiotics, 962
Putrescine, 929
Pyridoxine, 885
Pyridoxine lowers serum/blood glutamate levels, 886
Pyruvate dehydrogenase complex, 894
Qigong, 973
Restless leg syndrome, 949
Rifaximin, 954, 964
Rifaximin as treatment for SIBO and IBS, 954
S-adenosylmethionine, 962
Savella, 913
SIBO, 922
Skatole, 929
Small intestine bacterial overgrowth, 922
Small intestine bacterial overgrowth in fibromyalgia, 923
Standard Medical Treatment for Fibromyalgia, 912
Tanacetum parthenium, 897
Tartaric acid, 929
Therapeutic Interventions, 956
Tramadol, 913
Trigeminal nerve, pain sensation in migraine, 863
Tryptamine, 929
Tryptophan, 960
Tyramine, 929
Vagal stimulation, 923
Vancomycin orally administered, 964
Vegetarian diet for fibromyalgia, 958
Vitamin D, 982, 983
Vitamin D deficiency, 916
Xenobiotics, 920
Xifaxan, 964

Printed in Great Britain
by Amazon